T0178401

History of Computing

The *History of Computing* series publishes high-quality books which address the history of computing, with an emphasis on the 'externalist' view of this history, more accessible to a wider audience. The series examines content and history from four main quadrants: the history of relevant technologies, the history of the core science, the history of relevant business and economic developments, and the history of computing as it pertains to social history and societal developments.

Titles can span a variety of product types, including but not exclusively, themed volumes, biographies, 'profile' books (with brief biographies of a number of key people), expansions of workshop proceedings, general readers, scholarly expositions, titles used as ancillary textbooks, revivals and new editions of previous worthy titles.

These books will appeal, varyingly, to academics and students in computer science, history, mathematics, business and technology studies. Some titles will also directly appeal to professionals and practitioners of different backgrounds.

More information about this series at http://www.springer.com/series/8442

Simon Lavington

Early Computing in Britain

Ferranti Ltd. and Government Funding, 1948–1958

Simon Lavington
School of Computer Science
and Electronic Engineering
University of Essex
Colchester, UK

ISSN 2190-6831 ISSN 2190-684X (electronic)
History of Computing
ISBN 978-3-030-15105-8 ISBN 978-3-030-15103-4 (eBook)
https://doi.org/10.1007/978-3-030-15103-4

Library of Congress Control Number: 2019936292

This Springer imprint is published by the registered company Springer Nature Switzerland AG
The registered company address is: Gewerbestrasse 11, 6330 Cham, Switzerland

Preface and Introduction

There is a tendency to view computer history either from a top-down or a bottom-up perspective. Enthroned at the top of the historical landscape are the policymakers: government mandarins and board-level industrialists who view each new generation of computers through their shifting requirements of economics, politics and national defence. Far below these policymakers, and secure at the foundational level, are the technology researchers and developers: the creators of theories and laboratory prototypes, whose motivation is the satisfaction of responding to challenges issued by their fellow academics. In between these two extremes is a largely unsung and heterogeneous collection of engineers, programmers and marketing staff who turn ideas into products, features into benefits and problems into solutions. Theirs is the song that you'll hear throughout this book.

The focus of the story is particularly on those who worked for one manufacturing company, Ferranti Ltd., and for the nine end-user organisations who purchased the first Ferranti computers in the period 1951–1957. It is a story grounded in technology but brought alive by personal experiences and practical compromise. Here you'll read of short-term social impacts and longer-term evolutionary adaptations, as new and untried equipment began to have an impact.

Why is this story of relevance to the emergence of modern computers? The 10-year period 1948–1958 was of great significance. The first Ferranti computer, which was prosaically called the Mark I, was the first production machine to have been delivered anywhere. A copy of this and improved versions called the Mark I* (*Mark One Star*) were the first substantial computers to have been delivered in Canada, Holland and Italy. And whilst American companies such as UNIVAC and IBM were selling tens of computers to their home market during the early 1950s, Ferranti seemingly had the rest of the world to itself.

Nevertheless, marketing was a struggle. In 1950, Ferranti employed Dr. B. V. (Vivian) Bowden, a Cambridge graduate with good connections in government and industry, to promote these new computers. Bowden remembers[1] that "I had to determine if more than one machine of this type would ever be required, or if, as it was at that time widely believed, a single machine would be able to do all the computation which would ever be needed in this country. We have now [May 1953] sold half a dozen of these great machines and it has been my responsibility to discuss their possible application to a very wide range of subjects…. I have interviewed a large number of people whose interests are in the fields of science, engineering and commerce, from the chief designers of aircraft firms and the presidents of insurance companies to the organisation and methods branch of the Bank of England and the Treasury. I have discussed their problems with scores of accountants, actuaries, engineers and scientists of all kinds …"

The scientific and engineering sectors were the more receptive to begin with. But Bowden found "that they all faced the same problems. If the calculations were short, they were done on desk calculating machines operated by clerks. If the calculations were long and complicated, there were three overwhelming objections to the use of a computer. The machine usually broke down before it finished the calculation, its memory was too small to hold all the data it needed, and, worst of all, it took so long for the mathematicians to get the program right that it was usually quicker (and cheaper) to get the calculation done by clerks … We were more worried by the shortage of programmers than by anything else ….".

Users did eventually take the plunge but the uptake was slow. Travelling hopefully was the name of the game in the early days of the digital revolution. As remarked by Dr. G. W. (Gerry) Morgan of the Government Communications Headquarters (GCHQ) in 1951: "It is very clearly recognised that quite 90% of the value of a computer does not become apparent until it is in use".[2]

Stepping back from the efforts of Vivian Bowden in 1950, it is now clear that defence was the market driver in both the UK and the USA at the time. Britain felt especially hard-pressed with the passing of the McMahon Act 1946, which cut off American cooperation in the development of nuclear weapons. The Berlin Airlift and the radical changing of Soviet encryption methods marked 1948 as the year when the Cold War really became a threat. Then 1949 saw the Yangtze Incident and the emergence of the Peoples Republic of China. 1950 marked the start of the

[1] B. V. Bowden (1910–1989) obtained his Ph.D. from the Cavendish Lab., University of Cambridge. His wartime work on radar included a spell with a joint UK–US research team at the Naval Research Lab., Washington, followed by post-war work at MIT and the UK's Atomic Energy Authority. He joined Ferranti in 1950. These quotations come from two sources: (a) Four-page letter written by Bowden on 8 May 1953, when applying for the post of Principal of Manchester College of Science and Technology (later UMIST). Bowden's application was successful. He remained at UMIST until retiring in 1976. In 1963, he was created Lord Bowden of Chesterfield and held the appointment of Minister of State for Education and Science until 1965. He died in 1989. (b) Bowden, B. V. 1970. The language of computers. *American Scientist* 58 (1): 43–53. This paper was originally delivered at the Brighton College of Technology as the first Richard Goodman lecture, 2 May 1969.

[2] *GCHQ & Ferranti.* G. W. Morgan. GCHQ Minute X/565/1802, 7th November 1951. GCHQ ordered a Ferranti Mark I* computer towards the end of 1951 and took delivery of the machine in the autumn of 1953.

Korean War. With this backdrop, it was two government agencies, the Ministry of Supply and the National Research Development Corporation, that underwrote the development of all the early Ferranti computers.

A Reader's Guide

The first Ferranti computer did not spring fully-formed from the ground. It incorporated ideas that had first emerged from several research laboratories in the period 1946–1949 and on particular techniques developed at the University of Manchester. These came together in a 1948 prototype called the Small-Scale Experimental Machine, or simply *The Baby*. With government encouragement, this Baby was picked up and nurtured by Ferranti Ltd., a well-established Manchester electrical engineering firm. Chap. 1 provides the historical background to these events, with supporting technological details being conveniently confined to Chaps. 14, 15 and 16. It is no surprise that world-renowned luminaries such as John von Neumann and Alan Turing appear in these pages.

The transfer of technologies from academia to industry and the establishment of Ferranti's computer production resources are covered in Chaps. 2 and 4. A giant machine was being created that, to quote Ferranti's Computing Department Manager, contained "about 4,000 valves [thermionic tubes], 2,500 capacitors, 15,000 resistors, 100,000 soldered joints, six miles of wire and consumed 27 kW of power .... In point of fact, no one knew what the problems of maintenance would be, and the solution of these problems, together with that of improving the general efficiency, gave rise to a considerable amount of difficulty". In Chap. 4 we see that one set of solutions to the Mark I's software inefficiencies was incorporated into an upgraded version called the Mark I* (*Mark One star*).

They were exacting times for those whose lives were touched by the project. As one of the Ferranti programmers remembered, "It was a very good team—very exciting! And we were a good group, too, socially". The programs being written were largely applied to science and engineering applications, though the human side occasionally shone through: computer music, computer games and even the generation of love letters.

The Ferranti Mark I and Mark I* computers were proving difficult to sell to business and commerce. They were extremely expensive and their input/output facilities were at first ill-suited to accounting and management sectors dominated by punched-card equipment. Chapter 5 describes Ferranti's efforts to adapt and to introduce new products. Here and in Chap. 18 are presented the performance, cost and sales statistics of rival machines emanating from both UK and American computer manufacturers. 1955 was the critical year in which competitor computers began to appear.

The sales of early Ferranti computers were largely to government establishments, with defence applications predominating as the clouds of the Cold War loomed overhead. The UK's Ministry of Supply was the procurement entity responsible for the installation of Ferranti computers at the Government Communications Headquarters (GCHQ) at Cheltenham, the Armaments R & D Establishment at Fort Halstead in Kent and at the Atomic Weapons Research Establishment at Aldermaston. The decisions, installations, applications, personalities and final shut-down of Ferranti Mark I* computers at these three locations are fully covered in Chaps. 10–12. The stories cover relevant requirements over the period 1945–1967.

Government funding secured the delivery of Ferranti Mark I computers at two Universities: Manchester and Toronto. The decisions, installations, applications, personalities and final shut-down of machines at these two sites are covered in Chaps. 2, 3 and 13. Another site with university and national Research Council connections was the Instituto Nazionale per le Applicazioni Calcolo (INAC) in Rome. The deployment and use of the Ferranti Mark I* at INAC is described in Chap. 9.

The Ferranti Mark I* ordered by the Amsterdam Laboratories of the Shell company in December 1952 is believed to have been the first computer in the world to be bought as a commercial venture rather than with direct or indirect government funding. The choice of this machine, its installation and its impact are covered in Chap. 8.

Both in the UK and in America, the aerospace industry was the first market sector to adopt the new type of digital computer with enthusiasm. Ferranti Mark I* machines were installed at the Chadderton factory of A. V. Roe (of Avro Lancaster and Avro Vulcan fame) and at the Ansty factory of Armstrong Siddeley Motors in, respectively, early 1956 and October 1957. The lives and times of these computers are covered in Chaps. 6 and 7. Since these two companies were largely working on government defence contracts, the Ferranti Mark I* installations can be said to have been partly financed by indirect government funds.

As for Ferranti, where did it all end? How did the company progress from being the first and largest British computer manufacturer in the early 1950s to losing its mainframe computer department in 1963 and finally becoming extinct in 1993? The saga, which covers successors to the Mark I series such as Pegasus, Mercury, Orion and Atlas, is revealed progressively in Chaps. 4, 5 and 13. Finally, the names and personal anecdotes of many of Ferranti's unsung employees from the Mark I days are rounded up in Chap. 17.

In summary, enjoy the photographs and enjoy the emerging hardware and software technologies and their applications but, above all, enjoy the personal reminiscences of those who made many small but useful contributions to the promotion of the modern computer in the decade from 1948.

Ipswich, UK Simon Lavington

Acknowledgements

The author is especially fortunate to have established contact with a heroic band of former Ferranti employees, now in their nineties, who have come up with the personal anecdotes, original documents and photos which have breathed life into this book. Less heroically perhaps, but equally useful, has been the information given to the author by relatives of former Ferranti employees, two well-known computer pioneers and a couple of fellow historians. The following alphabetical list gives the names of all those to whom the author owes an especial thank you.

Judie Adnett, Peter Barnes, Reg Boor, Robin Bowden, Tony Brooker, Martin Campbell-Kelly, Andrea Celli, Olaf Chedzoy, Tony Comer, Sheila Cooper, Geraldine Cutler, Dai Edwards, Rosalie van Egmond, Allan Ellson, Brian Jeffrey, Len Hewett, Glynn Libberton, David Link, Margaret Marrs, Graham McLean, John McNamara, Jim Miles, Elisabetta Mori, Johnny Mudge, Ginny Murray, Gordon Pattison, James Peters, Tacye Phillipson, Alan Sercombe, Jan Shearsmith, Jonathan Swinton, Joan Travis, Len Whalley, Mike Williams, Onno Zweers.

Credits for image permissions are given towards the end of the book.

Contents

Chapter 1
The Small Seeds of Innovation

1.1 Post-war Britain

This is the story of emerging technology and, in due course, the creation of a family of novel computers. It is a story with a quiet start, away from public attention but of keen interest to a handful of government scientific advisers who foresaw a need. The need, initially at any rate, was to support the science underpinning the nation's defence. Meanwhile, the UK's industry and commerce had other preoccupations, with post-war reconstruction predominating. War-time rationing of commodities such as food, clothes and petrol had persisted and Cold War tensions, initiated in June 1948 with the Berlin Air Lift, had intensified. We'll see that June 1948 also saw the appearance of an innovative project that blossomed, three years later, into the family of computers that lie at the heart of our story.

Why is it worthwhile recalling the life and times of a family of inanimate machines? The tale is fascinating not so much because of the new technologies these computers contained but because of their impact on the strategic and socio-economic fabric of post-war Britain. This was the dawn of the *Digital Age* and the people involved with these early computers found themselves projected to the forefront of a quiet and creeping revolution. Many of the seeds of this revolution had been sown in America. We shall concentrate on some particularly British seeds which also bore fruit in Canada, Holland and Italy.

To understand why the word *revolution* seems appropriate, it is pertinent to ask what went before? How did enterprises manage before the arrival of the modern digital computer? How did people compute before the computer? Who, if anyone apart from the government's defence advisers, saw a need for improvement?

And who conceived the particular family of computers which lie at the heart of our story? Did these machines spring fully-formed from the production line of one far-sighted company? And why concentrate on this particular family? Was it indeed world-class?

S. Lavington, *Early Computing in Britain*, History of Computing,
https://doi.org/10.1007/978-3-030-15103-4_1

This chapter sets out to answer these questions in the context of 1950s Britain. We start with images of old-fashioned computing methods, to establish the mood. We then set the industrial scene by introducing the electrical engineering company Ferranti Ltd., which was the producer and nurturer of our family. Finally, we introduce the small research group in Manchester University where the family of machines was first conceived.

1.2 Old-Style Computing

The Second World War had produced a huge legacy of technological innovations but had also left huge economic and social problems. As the Cold War loomed, there were strategic imperatives to apply innovative technologies to a range of computational problems concerned with defence—the development of nuclear weapons being an obvious example. The pursuit of computational advances in the defence arena went largely unnoticed by industry and commerce, where pre-war computational equipment generally continued in the old ways. Times were tough in Civvy Street.

On the bright side, the UK government had organised the Festival of Britain in the summer of 1951 as an intimation of post-war recovery. On show were many examples of British design and innovation though there were no general-purpose digital computers amongst the many exhibits. Journalists had for a couple of years been feeding the public headlines about *Giant Brains* and *Memory Machines* but the laboratory prototypes of modern computers that inspired these headlines had not yet been turned into robust commercial products.

At the start of the 1950s most scientific and engineering organisations carried out their day-to-day numerical computations manually, with the aid of electro-mechanical desk calculators or floor-mounted punched card tabulators and sorters. Commercial and administrative organisations often favoured the punched-card equipment because information was held semi-permanently on the punched cards. For certain special problems such as aircraft simulations, analogue computers were employed. The Differential Analyser is one type of analogue computer. Here, war-time innovations had begun to have an impact. Whereas pre-war Differential Analysers had been mechanical giants equipped with conventional electrical motors, the post-war equivalent was more compact and used electronics and servomechanism technologies originally developed for the control of guns.[1] Figure 1.1 shows such a Differential Analyser on display in the Festival of Britain.

[1](a) The Elliott Differential Analyser of Fig. 1.1 is described in: Ashdown, G.L., and K.L. Selig. 1951. A General Purpose Differential Analyser: Part I, Description of Machine. *The Elliott Journal* 1 (2): 44–48; (b) The background to this, and similar post-war DAs, is discussed in Chap. 4 of: Lavington, Simon. 2011. *Moving Targets: Elliott-Automation and the Dawn of the Computer Age in Britain, 1947–67*. Berlin: Springer. ISBN: 978-1-84882-932-9.

Fig. 1.1 An exhibit in the Dome of Discovery at the 1951 Festival of Britain. The equipment in the background is a Differential Analyser, a type of analogue computer. Although the basic design-principles of Differential Analysers date from the 1930s, they were still proving very effective for certain types of computation until the end of the 1950s

Analogue computers aside, those who needed to compute would normally perform step-by-step calculations manually, using a variety of electro-mechanical equipment. Figures 1.2 and 1.3 show some examples of desk calculators of the 1950s; Fig. 1.4 shows some typical punched card equipment.

The replacement of manual equipment by automatic digital computers—the modern *stored-program,* or *universal* computers—happened gradually over a ten-year period. First to adopt the new equipment were scientific and engineering establishments. For a number of reasons, business and commerce took longer to be convinced of the advantages in changing from old to new. One reason was simply that early digital computers did not have the capacity to meet the data-processing requirements of large organisations. An example is the British Army's payroll task —see also Fig. 1.3. By 1958, the Royal Army Pay Corps' computerisation was still the most ambitious of the UK government's planned clerical computer projects. Setting up and running a computer centre to provide a reliable payroll system for an army of several hundred thousand soldiers was a major challenge which broke new ground. The RAPC issued requirements in 1956 to ten British and four American computer manufacturers. From these 14 suppliers only two proposed computer systems, the Elliott 405 and the IBM 705, went through to the detailed evaluation

Fig. 1.2 A scientist at Shell's Amsterdam Laboratory explains the mathematical background to a research problem requiring a numerical solution. The two female computing clerks known as *computers*, each equipped with an electro-mechanical Marchant desk calculator, would carry out step-wise manual computations to solve the problem. This photo was taken in about 1952. By 1954 the Shell Laboratory had acquired a Ferranti computer and the human *computers* in the photo had been re-trained as programmers. Lidy Zween-de Ronde (on the left) led the way and went on to run programming courses for other Laboratory staff on the Ferranti machine. The full Shell story is told in Chap. 8

stage.[2] IBM won the contract, finally installing their IBM 705 system at the end of 1960. This was almost ten years after the first modern production computer, the Ferranti Mark I, was installed at Manchester University.

Let us now introduce Ferranti Ltd., the company who built the computers whose stories lie at the heart of this book.

1.3 Ferranti Ltd. in the UK's Industrial Context

The name *Ferranti* has now vanished from the list of British manufacturers so some historical background is in order. Ferranti Ltd. was a family-run electrical engineering company. It had been founded in 1882 by Sebastian Ziani de Ferranti, an

[2](a) Royal Army Pay Corps: Electronic Computer Investigating Committee: *Report of Working Party. Application of automatic data-processing to soldiers' pay accounting*. The War Office, August 1957. Also: Royal Army Pay Corps: *Report of the Electronic Computer Investigation Committee*. The War Office (F9), December 1957; (b) The wider background to this RAPC procurement is discussed in Chap. 9 of: *Moving Targets* (see b in Footnote 1).

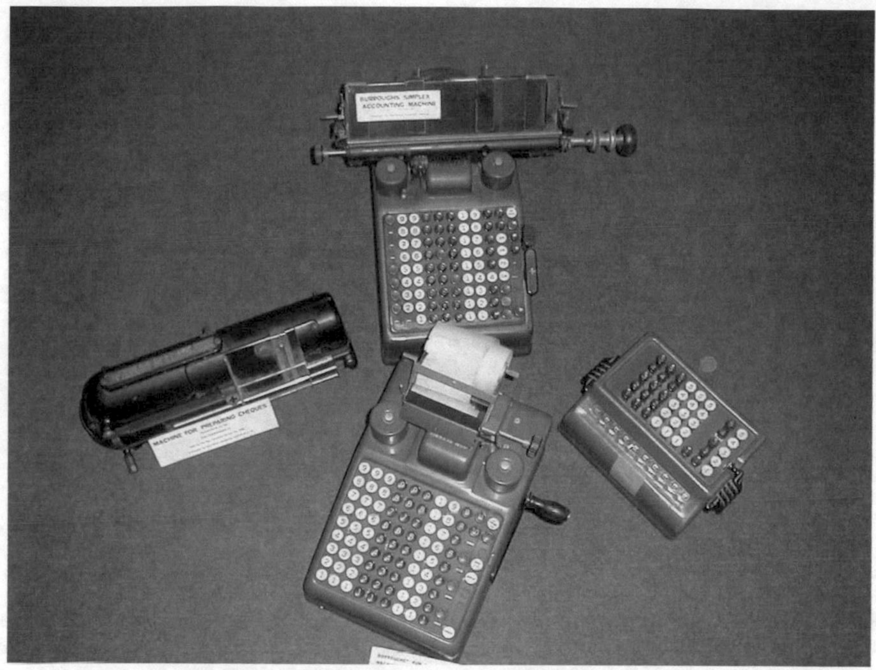

Fig. 1.3 Four types of desk-top equipment used by the Royal Army Pay Corps for calculating soldiers' pay in the 1950s. The device on the left is a machine for preparing cheques. The unit at the top is a Burroughs Simplex Accounting machine. It was not until 1961 that the army replaced manual payroll methods with an IBM digital computer

inventive engineer born in Liverpool in 1864 (Fig. 1.5). Ferranti Ltd. was one of several large industrial enterprises in Britain. Many of these were long-established firms such as the General Electric Company Ltd. (GEC) founded in 1889, Associated Electrical Industries Ltd. (AEI) first registered in 1899, and English Electric Co. Ltd. which was formed from four smaller firms in 1918.

By 1950 any one of the companies such as GEC, AEI, English Electric, etc. could in principle have started to manufacture the new type of modern digital computer—at least, that is what the UK's National Research Development Corporation thought. The 1948 Development of Inventions Act had set up the National Research Development Corporation (NRDC), to encourage and underwrite the exploitation of British ideas. In Chap. 5 we will look more closely into NRDC's attempts to promote a UK computer industry, attempts that seemed at times to be akin to "Trying to drive mules uphill".[3] For now we follow the fortunes

[3]Lord Halsbury, who was the Director of the National Research Development Corporation from 1949 to 1959, made this oft-quoted remark in a lecture given in London sometime after his retirement. The exact date has yet to be determined.

Fig. 1.4 IBM punched card equipment installed in the University of Toronto Computation Centre towards the end of 1948. In May 1952 a Ferranti computer was delivered to Toronto University, as described in Chap. 3, and the punched card equipment was made redundant

Fig. 1.5 Sebastian Ziani de Ferranti (1864–1930). He began constructing electrical apparatus aged 13 and had amassed almost 200 patents by the time he died. He founded the Ferranti company in 1882 and was the London Electric Supply Co's Chief Engineer from 1886 to 1905. He was elected a Fellow of the Royal Society in 1927

of just one company, Ferranti Ltd., which did manage to achieve some early market success.

Ferranti Ltd. delivered its first production computer, called the Ferranti Mark I, to Manchester University on 12th February 1951. This is believed to have been a world first—a few weeks ahead of the delivery of the first UNIVAC I computer in

America, though a few weeks behind the special ERA ATLAS computer delivered to the US Navy's secret CSAW establishment in Washington—(see Chap. 10). Why was Ferranti in the vanguard? Basically because it just happened to be in the right place at the right time, as we will soon reveal.

1.4 Ferranti: A Powerhouse of the Northwest

The company's founder set out to promote the 'all electric' age and his company soon achieved success in the field of electrical generation and supply. In 1887 Sebastian de Ferranti was the design engineer who built the Deptford Generating Station in London. Completed in 1890, it was the first truly modern power station. It supplied high-voltage AC power which was transformed to a lower voltage for consumer use where required. Ferranti's company started producing electrical equipment (especially transformers) for sale. Needing more space, the company moved to Hollinwood in Oldham, just north of Manchester, in 1896 because land prices in the northwest were lower than those in the London area.

The Ferranti company gradually diversified into *light-current* electrical engineering, typified by electronic circuits based on thermionic valves (or tubes). In 1935 Ferranti purchased a disused wire drawing mill at Moston, a couple of miles closer to Manchester than Hollinwood (Fig. 1.6). At Moston Ferranti started manufacturing electronic products such as radios, televisions and scientific instruments. By the start of the Second World War, Ferranti was well-placed to benefit from government defence contracts in areas such as radar, instrumentation and control.

The number of Ferranti employees expanded from 5800 in 1939 to over 11,000 in 1945, dropping sharply after the end of hostilities but then recovering to 7000 by 1950. Profits also took a dip after the end of hostilities but had recovered to £349,000 by 1947 and had reached £551,000 by 1950. Ferranti's gradual move from *heavy-current* to *light-current* electrical engineering activity is illustrated in Table 1.1, which is based on the company's Patent Applications.[4]

In 1948 Ferranti's growing electronic expertise and its location on the northern perimeter of Manchester made the company a credible contender when, as we shall see in Sect. 1.5, the government wished to place a contract to build a large computer based on a research prototype being developed at Manchester University.

---

[4]Patent figures have been extracted from relevant volumes of *Index to names of applicants in connection with published complete specifications*. This Index is published periodically by the UK Patent Office. This Ferranti analysis is discussed in more detail in Chap. 6 of *Moving Targets* (see b in Footnote 1).

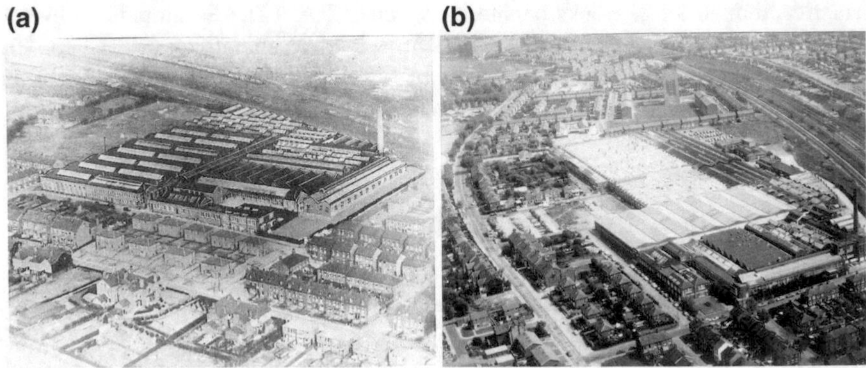

Fig. 1.6 **a** The former Wire Works at Moston in 1934, looking east. In 1935 Ferranti Ltd. took over this factory and used it for the manufacture of electronic equipment and instrumentation. **b** The same site looking north-east in 1984, after being significantly extended to the north. Ferranti's large factory, which no longer exists, lay to the east of St. Mary's Road, Moston. The area is now devoted to housing

Although initially lacking any relevant in-house R&D, Ferranti had certainly begun to think about the possibilities of electronic computers. The mathematician D. G. Prinz had joined Ferranti's Instrument Department at Moston in 1947 (Fig. 1.7). In September 1948 Prinz was sent on a fact-finding mission to America to acquaint Ferranti with modern computer developments.[5] Upon his return Prinz had urged Ferranti to become involved in the area. Prinz's initial emphasis seems to have been the use of computers for control, a topic in which Ferranti had become involved during the war. Prinz also reported that the Americans had made him aware that Professor F. C. (Freddie) Williams was working at Manchester University on a novel system for computer storage.[6] Ferranti was urged to make contact with Professor Williams.

Until mid-1948 it seems that Ferranti were largely unaware of the research into computer memory technologies going on just a few miles away at Manchester University, though the leader of the University's team, Professor Williams, was certainly known to the company as a consultant on control matters. Liaison between the company and the University was accelerated in October 1948, when Ferranti

[5] *Report of a visit made by D. G. Prinz to USA in September 1948, on behalf of Ferranti*. National Archive for the History of Computing (NAHC) document NAHC/PRI/C1a.

[6] There are many papers describing the Williams/Kilburn electrostatic storage scheme, for which the first patent was filed by Williams on 11th December 1946. Here are two of the early reports: (a) Kilburn, T. 1948. *A Storage System for use with Binary Digital Computing Machines*. Ph.D. thesis, University of Manchester. A first version of this thesis with the same title was written as an internal report for TRE. This report was then circulated by the Dept. of Electrotechnics, University of Manchester, dated 1st December 1947. Several copies are known to have reached the USA; (b) Williams, F.C., and T. Kilburn. 1949. A storage system for use with binary digital computing machines. *Proceedings of IEE* 98 (Part 2, 30): 183.

Table 1.1 Analysis of Ferranti Patent applications before and after the Second World War, showing the move away from *heavy current* products and towards *light current* products

Area of electrical engineering products	No. of new patents, 1937–38	No. of new patents, 1944–46
Electrical instrumentation, measuring and indicating apparatus	13	11
Electronics (mainly radio and radar) and light-current apparatus	24	38
Electrical machines (generators, motors, transformers); heavy-current and high-voltage devices	7	3
Other (eg domestic appliances; non-electrical devices)	10	8

received a letter from the government's Chief Scientist stating that the company would be given a substantial amount of money "*to construct an electronic calculating machine to the instructions of Professor F. C. Williams*".[7] Just how this letter came to be written is described in Sects. 1.5 and 1.6.

The early Ferranti computers therefore did not spring magically out of the ground. Their production was preceded by about five years' worth of experimentation and university-led research. Lots of historians have written—sometimes carefully, sometimes carelessly—about the origins of the modern computer and the part played by Manchester in the period 1945–48. Indeed so much has already been written that some readers may wish to skim the next preamble about academic projects and go rapidly to Chap. 2 where we describe the installation and use of Ferranti's first fully-engineered production computer, the Ferranti Mark I.

1.5 A Baby Is Born

The prototype for the first Ferranti computer was developed at Manchester University in three phases, over the period 1947–1949. The pre-history is fascinating. Here we give an overview based on Lavington.[8] A more detailed historical account, which includes an analysis of the probable influence of the American mathematician John von Neumann on the project, is presented in Chap. 14.

In the autumn of 1946 Williams, who was always called FC by colleagues and Freddie by outsiders, was still employed in his war-time post as a leading

[7]Copy of letter from Sir Ben Lockspeiser to Eric Grundy, 26 October 1948, giving Ferranti authority "to construct an electronic calculating machine to the instructions of Professor F. C. Williams". National Archive for the History of Computing, document NAHC/FER/B3.

[8]Lavington, Simon. 1975. *A History of Manchester Computers*. 1st ed. Second edition published by the British Computer Society, 1998. ISBN: 0-902505-01-8. See also the history site of the School of Computer Science, University of Manchester: www.computer50.org/.

Fig. 1.7 Dr. Dietrich G. Prinz (1903–1989) joined Ferranti Ltd. in 1947. He was an advocate of the new type of *stored-program* computers and became Ferranti's leading programmer for scientific applications. In 1952 as a hobby Prinz developed a program to solve simple chess problems. He is seen in the photo demonstrating the program on the Ferranti Mark I computer at Manchester University in about 1955. The box of switches in the right foreground is an extra facility that has no particular connection with the chess program

electronics engineer in the government's Telecommunications Research Establishment (TRE). Whilst at TRE Williams had an idea for a novel form of digital storage. Although in 1946 TRE was not interested in building their own computer they were certainly keen to support Williams' ideas. Williams moved to Manchester at the end of December 1946 to take up the Chair of Electro-Technics at the University. (The Department's name was changed to *Electrical Engineering* 18 months later). Williams arranged for his TRE colleague, Tom Kilburn, to be seconded to the University and to draw on TRE supplies so that the digital storage work could continue. Another TRE engineer, Geoff Tootill, was to follow.[9]

[9]Letter from R. A. Smith (TRE's Superintendent Physics) to F. C. Williams, dated 25th November 1946. Smith proposes that 'T. Kilburn and one other' be seconded to Manchester University to work with Williams. The first 'one other' was Arthur Marsh but he soon asked to be returned to TRE because he apparently saw no future in computers. He was replaced by Geoff Tootill, who came to Manchester in September 1947. See also the interview of G. C. Tootill by Thomas Lean. British Library *National Life Stories,* reference C1379/02, 3rd Dec 2009–9th April 2010.

Fig. 1.8 An advert for Ferranti computers in the Dutch trade journal *Algemeen Handelsblad* on 3rd February 1955. Ferranti had delivered a Mark I* computer to Shell's Amsterdam Laboratory in June 1954—see Chap. 8—which probably accounts for the emphasis on *Electronische rekenmachine (electronic calculator)* in the advertisement. By February 1955 Ferranti had delivered a total of seven production computers—more than all other British computer manufacturers combined. By February 1955 Elliott Brothers Ltd. and English Electric Ltd. had each delivered just one computer

F. C. Williams was an inventor, deeply proud of being an engineer. He has been called a genius. Here's what Sir Bernard Lovell, of Jodrell Bank radio-telescope fame, wrote about FC: "During the war we worked together on the 10 centimetre air interception automatic lock-follow blind firing system. We had many uncomfortable flights together in a Blenheim during the development of the system and I soon grasped that he [Williams] was a genius. He had the idea of the automatic strobe which searched the CRT time base for an echo, then locked on and gave us the information to drive the paraboloid [aerial] using a MetVic metadyne system".[10] The main point of this quotation is to demonstrate that, by the time he turned his hand to digital storage, Williams had become an expert in electronic pulse techniques, especially those involving Cathode Ray Tube (CRT) waveforms and control systems. All of these skills were used in the development of the memory systems for early Manchester computers.

A technical explanation of Williams' novel storage system, based on a Cathode Ray Tube is given in Chap. 16. It is sufficient here to state that the system used readily-available components and crafty circuitry to provide cost-effective RAM (random access memory). Figure 1.9 shows the basic components of what became known as the Williams/Kilburn storage system.

The first task for Williams and Kilburn in 1947 was to verify that a reasonable number of binary digits, or bits, could be continuously stored for a long time without error. Providing cost-effective, rapid-access, memory was the most difficult problem facing all designers of early computing equipment in the UK and USA at that time. The second task for Williams and Kilburn was to show that their memory device could perform satisfactorily when storing the code and data for a realistic computation. To do this, they built a Small-scale Experimental Machine, SSEM or the *Baby*, and verified that Baby could execute a convoluted test program which ran for nearly an hour during which several million instructions were correctly obeyed. The first successful demonstration of this *stored-program* machine took place on the morning of Monday 21st June 1948 (Fig. 1.10).

The general ideas for Baby's instruction set and other architectural details came from several sources, not least of which was the Institute of Advanced Study at Princeton University in America—(see Chap. 14). Baby's June 1948 design was certainly general-purpose but its storage capacity, instruction repertoire and input/output facilities were primitive. They quickly grew. Williams' and Kilburn's unique contribution was to make the facilities work at electronic speeds and, within months, to add some novel features such as index registers and a two-level memory system that combined fast RAM with slower magnetic drum storage (similar to a present-day hard disc).

The famous mathematician Alan Turing moved to Manchester University in October 1948 at the invitation of Professor Max Newman of the Mathematics

[10]Letter dated 3rd March 1999 from Sir Bernard Lovell to Dr. John Ponsonby (Jodrell Bank). Ponsonby had come across a rare copy of the Blue Book, a lengthy technical report entitled *A steerable radio telescope* sent to DSIR on 20th March 1951 in support of a grant for what was to become the Jodrell Bank Radio Telescope. This particular copy of the Blue Book had been sent on 31st March 1951 by Patrick Blackett to F. C. Williams for comment. Lovell responded to Ponsonby's find with some background history, including an assessment of Williams' ability as a suitable referee of DSIR research applications.

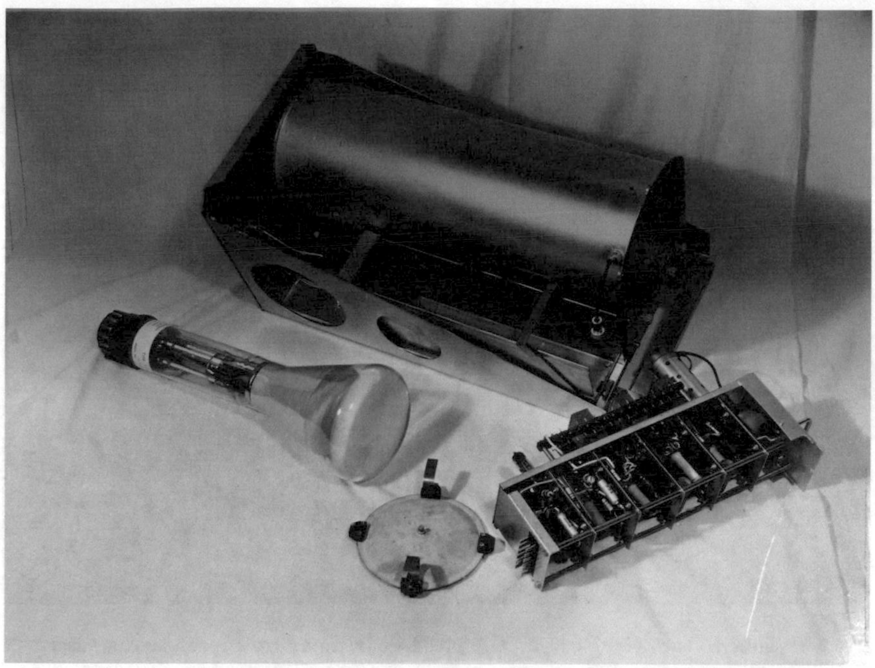

Fig. 1.9 The basic components of a Williams/Kilburn memory tube, as manufactured by Ferranti for their Ferranti Mark I computer in 1951. A technical explanation of how this RAM system works is given in Chap. 16. Briefly, electrostatic charges representing binary 1s and 0s are created on the CRT screen, causing voltage signals to be detected by the circular pick-up plate and fed via the pre-amplifier to the computer

Department. Turing immediately interested himself in the Baby computer. He suggested that 5-track paper tape equipment should be connected to the Manchester computer for input/output, thus replacing Baby's laborious keyboard-and-screen arrangement of June 1948. Turing arranged for 5-track paper tape equipment to be obtained, most probably via his former Bletchley Park contacts, but this was not incorporated into the enhanced Baby until the autumn of 1949. Meanwhile, Baby's instruction repertoire and storage capacity were being enhanced by the engineers.

We shall meet Alan Turing and Max Newman again in the next chapter, when we describe the applications for which the University's 1949 computer—and Ferranti's 1951 follow-on computer—were put to use. For now, we need to go back to the autumn of 1948 and recount how Ferranti's Moston factory became involved with Manchester University.

Fig. 1.10 Tom Kilburn *(left)* and F. C. (Freddie) Williams *(right)* at the controls of the Manchester University prototype computer in 1949. Colleagues remarked that this photo seems to show a ceremony in which "Tom was married to the computer by F C". The marriage lasted a lifetime

1.6 The Amazingly Brief MOS Contract

Sir Henry Tizard, Chief Scientific Advisor to the Ministry of Supply, saw the Small-scale Experimental Machine (aka the Baby computer) in operation at Manchester University in July 1948 and "considered it of national importance that the development should go on as speedily as possible, so as to maintain the lead that this country has thus acquired in the field of big computing machines, in spite of the large amount of effort and material that has been put into similar projects in America".[11]

In October Sir Ben Lockspeiser, Chief Scientist at the Ministry of Supply (MOS), paid the Baby a visit on the suggestion of Professor Patrick Blackett, a Nobel Prize-winning Physicist at the University who was well-known at the MOS. Lockspeiser was sufficiently impressed with the Baby to write on 26th October to

[11]Newman, M.H.A. 1948. *Status Report on the Royal Society Computing Machine Laboratory.* Report prepared for a Senate Sub-committee of the University of Manchester, 15th Oct 1948. A re-typed copy is available as NAHC document NAHC/MUC/2/C/2.

Fig. 1.11 The Manchester University Mark I digital computer in June 1949. This image comes from the Illustrated London News of June 1949, with the caption *A marvel of our time: the 'memory machine' which can solve the most complex mathematical problems.* The article described each sub-unit of the computer, of which two have been outlined in red boxes. The left-hand red box contains the operator's screen and keyboard for manual input/output. The right-hand red box is the random-access memory (RAM). A magnetic drum memory (equivalent to a modern hard drive) is behind the racks to the left. The room containing the whole computer measured about 20 ft × 20 ft (about 6 m × 6 m)

Eric Grundy, Manager of Ferranti's Instrument Department at Moston, as follows[12]:

> Dear Mr Grundy,
>
> I saw Mr Barton [MOS] yesterday morning and told him of the arrangements I made with you at Manchester University. I have instructed him to get in touch with your firm and draft and issue a suitable contract to cover these arrangements.
>
> You may take this letter as authority to proceed on the lines we discussed, namely, to construct an electronic calculating machine to the instructions of Professor F C Williams.
>
> I am glad we were able to meet with Professor Williams as I believe that the making of electronic calculating machines will become a matter of great value and importance.
>
> Please let me know if you meet with any difficulties.

Grundy (Fig. 1.12) wrote back enthusiastically the next day saying, amongst other things, "We [Ferranti's Instrument Department] have been organising our team and expect to start work in earnest on 1st November".[13] The contract had certainly landed on fertile soil though the actual design and construction of the

[12] Copy of letter from Sir Ben Lockspeiser to Eric Grundy, 26 October 1948, giving Ferranti authority "to construct an electronic calculating machine to the instructions of Professor F. C. Williams". National Archive for the History of Computing, document NAHC/FER/B3.

[13] Letter accompanying National Archive for the History of Computing document NAHC/FER/B3.

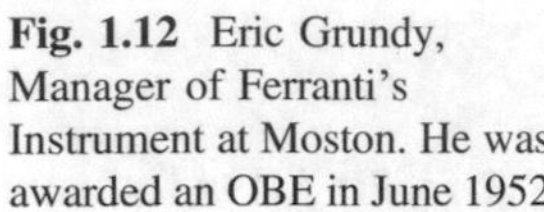

Fig. 1.12 Eric Grundy, Manager of Ferranti's Instrument at Moston. He was awarded an OBE in June 1952

Ferranti computer was soon switched to the Radio Department at Moston, not the Instrument Department.

The formal paperwork was drawn up and signed on 19th February 1949 for an agreed amount of £113,783 2s 7d. The contract-number was 6/WT/4890/CB14(d). As well as providing for the construction, installation and maintenance of a fully-engineered computer for the University of Manchester, the contract required Ferranti to help the University with extending the capacity of its existing prototype computer and with the development of electronic computer techniques.

Ferranti's re-engineered production version of the Manchester University Mark I computer was delivered to the University's new Computing Machine Laboratory on 12th February 1951. The machine was at first simply called the *Manchester Electronic Computer*. The formal title, the *Ferranti Mark I*, was in use from 1952 onwards. Copies of this computer, and of an enhanced version called the Mark I Star (always written *Mark I**) were soon coming out of the Moston factory.

The speed with which the MOS contractual arrangements were set in motion is interesting. In 1948 the Ministry of Supply (MOS) was the UK's single government entity responsible for developing advanced military materials including atomic weapons, and for the running of defence-related research establishments such as TRE, the Armaments Research and Development Establishment (ARDE) and (for some administrative purposes) the Government Communications Headquarters (GCHQ). The MOS interest in the Manchester activities was a sign of the strategic importance the government attached to the new form of general-purpose computer. Indeed, in 1947 MOS had initiated a classified computer project of its own called MOSAIC for the analysis of missile-tracking data.[14] MOSAIC first ran a program in 1953. In October 1946 the Admiralty had placed a contract with Elliott Brother Ltd. for MRS5, a moving target anti-aircraft gunnery control project. This involved a special-purpose digital computer called the Elliott 152, which carried out its first simple computations in mid-1950.[15] The Elliott 152 can lay claim to being the first machine to attempt real-time, on-line, digital process control.

A simple ballistics example[16] will serve to show why the scientific and defence sectors were interested in the potentials of high-speed digital computers. Calculating the flight path of a typical artillery shell (projectile) in 1950 was estimated to take the following times:

24 h, when done manually on an electro-mechanical desk calculator;
20 min, when done on an analogue computer (a Differential Analyser);
10 s if done on the Ferranti Mark I digital computer.

The actual shell in question took 20 s to travel from gun to target, suggesting that real-time digital fire control might one day become a possibility.[17]

[14]Coombs, A.W.M. July & Oct 1955, Jan 1956, March 1953. An Electronic Digital Computer, Parts 1–4. *Post Office E E Journal* 48, 114, 137 & 212: 38–42; 49, 18, 126 (April & July 1956). Summary information also given in: Coombs, A.W.M. 1954. The Ministry of Supply Automatic Computer. In *Automatic Digital Computation, the Proceedings of a Symposium* held at NPL, Published by HMSO.

[15]A secret Admiralty contract for a real-time fire-control system, called MRS5, had been given to Elliott's Borehamwood Laboratory in the autumn of 1946. This contract resulted in the Elliott 152 digital computer, which ran a program under laboratory conditions in September 1950. Classified technical reports of this project exist in the Bodleian Library, University of Oxford. A description, based on these reports, is given in Chaps. 1 and 2 of *Moving Targets* (see b in Footnote 1). The Royal Navy did not, in the end, adopt MRS5 and instead chose an analogue *Fly Plane* system implemented by Ferranti. Ship-borne digital fire control did not go to sea until the end of the 1950s.

[16]Pollard, B.W. 1950. Application of Digital Computing Machines. *Ferranti Journal* 8 (4): 94.

[17]A secret Admiralty contract for a real-time fire-control system, called MRS5, had been given to Elliott's Borehamwood Laboratory in the autumn of 1946. This contract resulted in the Elliott 152 digital computer, which ran a program under laboratory conditions in September 1950. Classified technical reports of this project exist in the Bodleian Library, University of Oxford. A description, based on these reports, is given in Chaps. 1 and 2 of *Moving Targets* (see b in Footnote 1). The Royal Navy did not, in the end, adopt MRS5 and instead chose an analogue *Fly Plane* system implemented by Ferranti. Ship-borne digital fire control did not go to sea until the end of the 1950s.

1.7 The Wider Context

To summarise the Manchester events so far: the basic operation of a new kind of computer had first been demonstrated in June 1948. By the end of that year, enhancements to the design were in hand and government funds had been made available for a production version of the prototype to be built by Ferranti Ltd. Ferranti took some months to assemble a design team at Moston and to set up a robust production and testing framework. The first fruits of the Moston endeavours appeared in February 1951.

To put this into perspective, Manchester's computer activity was only one of many other projects in Britain and America, and one in Australia, addressing the same basic problems. In England two projects in particular were coming to fruition: the EDSAC computer at the University of Cambridge (first working in May 1949) and the PILOT ACE computer at the National Physical Laboratory (first working in May 1950). We shall meet these, and other, early British machines in Chap. 5 when we discuss the market opportunities for computers and the competitors to Ferranti Ltd.

At the heart of all these projects was a machine which is technically, though laboriously, described as the *stored-program universal computer*. To this formal description was often added appropriate adjectives such as *high speed, general-purpose, automatic* and *electronic*, to distinguish the new breed of machines from their electro-mechanical predecessors which were often *sequence-controlled* rather than *stored-program* and often *special-purpose* rather than *universal*. The cognoscenti find that all this terminology is necessary when attempting to pin down the various interpretations of the basic word *computer*—which, to add to the confusion, was also used as a job-description prior to 1950. For our present story all that we need to grasp is that the Manchester University Baby, though very small and fragile, is considered to be the first practical example of the modern computer. In June 1948 Baby ran the first piece of software, making tangible the concept of programming, or coding, as we now understand it. From this small start, Ferranti produced their first commercially-available computer in 1951.

References

Ashdown, G.L., and K.L. Selig. 1951. A General Purpose Differential Analyser: Part I, Description of Machine. *The Elliott Journal* 1 (2): 44–48.

Coombs, A.W.M. 1954. The Ministry of Supply Automatic Computer. In *Automatic Digital Computation, the Proceedings of a Symposium* held at NPL, Published by HMSO.

Coombs, A.W.M. (July & Oct 1955, Jan 1956, March 1953). An Electronic Digital Computer, Parts 1–4. *Post Office E E Journal* 48, 114, 137 & 212: 38–42; and 49, 18, 126 (April & July 1956).

Lavington, Simon. 1975. *A History of Manchester Computers*. 1st ed., NCC.

Lavington, Simon. 2011. *Moving Targets: Elliott-Automation and the Dawn of the Computer Age in Britain, 1947–67*. Berlin: Springer. ISBN 978-1-84882-932-9.

Kilburn, T. 1948. *A Storage System for use with Binary Digital Computing Machines.* Ph.D. thesis, University of Manchester.

Newman, M.H.A. 1948. *Status Report on the Royal Society Computing Machine Laboratory.* Report Prepared for a Senate Sub-committee of the University of Manchester, 15th Oct 1948.

Pollard, B.W. 1950. Application of Digital Computing Machines. *Ferranti Journal* 8 (4): 94.

Williams, F.C., and T. Kilburn. 1949. A Storage System for Use with Binary Digital Computing Machines. *Proceedings of IEE* 98 (Part 2, 30): 183.

Chapter 2
Academic/Industrial Collaboration: From Chorlton-on-Medlock to Moston, and Back

We will leave Ferranti's team at Moston to set up their production facilities and first turn to Manchester University. What was it like to be part of an academic research group in 1949 and 1950, as the prototype computer was refined and put to use? How were the ideas then transferred to Ferranti? How did the University prepare for the arrival of the re-engineered machine? Meanwhile, what was life like at Moston? How long did it take to perfect a robust product? Finally, once the new computer was delivered, were there any teething troubles and what were the applications to which it was put? This chapter covers the critical period 1949–1954, when both academia and industry were learning how to manage the hardware and software of a radically new scientific instrument.

2.1 The University Waits: People and Places

By November 1949 Professor Williams' team at Manchester had completed the final version of the enhanced Baby computer. Alan Turing and a few others were gaining some experience of running programs on this upgraded machine, which became known as the *Manchester University Mark I* or sometimes *MADM*. It was now time to transfer the detailed logical design to Ferranti's Computer Department in the Moston factory. Plans were also made to erect a special building to house the fully-engineered Ferranti computer as soon as it could be delivered to the University. The expected delivery-date was, rather optimistically as it turned out, set for June 1950.

Manchester University's main buildings lay about a mile south of the city centre, in the romantically-named but distinctly un-bucolic district of Chorlton-on-Medlock—see Fig. 2.1. Ferranti's Instrument Department was a bus ride away in Moston about three miles north-east from the city centre, as pictured in Chap. 1. From November 1948 onwards there were regular comings and goings between

S. Lavington, *Early Computing in Britain*, History of Computing,
https://doi.org/10.1007/978-3-030-15103-4_2

Professor Williams' group at the University and Eric Grundy's group at the Moston factory. People were busy: this was the industrial north-west in innovative mode.

On the hardware side, Professor F. C. Williams remained the undisputed figurehead of the academic/industrial computer collaborations. He was elected a Fellow of the Royal Society in 1950. Williams' wider university and national responsibilities meant that day-to-day decisions were gradually being delegated to Tom Kilburn. Kilburn became a University Lecturer towards the end of 1948. Under Kilburn was a small team consisting of his full-time assistant Geoff Tootill (still on secondment from TRE since September 1947), a lecturer (J. C. (Cliff) West, who had graduated in 1943 and joined the staff in 1946), three young research students (see Chap. 17) and a technician. Altogether a modest-sized group, but full of ideas. John

Fig. 2.1 The main buildings of the University of Manchester in about 1957, looking east. Oxford Road runs north/south across the middle of the photograph and most University buildings in 1957 lay to the west of this. A clearer view is provided in Fig. 2.2. The Magnetism Lab., where the Small-scale Experimental Machine (*Baby*) was born in 1948, is indicated by 'A' in the image above. B indicates the Computing Machine Laboratory, where the first Ferranti Mark I was located in 1951. C indicates the new Department of Electrical Engineering opened in 1954, to where the Ferranti Mark I was moved in 1954

Fig. 2.2 Part of the University of Manchester in 1974 looking east, after the stonework of the main building had been cleaned. Oxford Road runs north/south, from upper left to lower right across the photograph. Point A indicates the old Computing Machine Laboratory, opened in 1951; point B indicates the Department of Electrical Engineering in Dover Street, opened in 1954; in the photo it is hidden behind a taller more recent building. Point C indicates the new Computer Building (now called the *Kilburn Building*), opened in 1972

Crawley, the Ministry of Supply's and later the National Research Development Corporation's Patent Agent, was a regular visitor during 1948 and 1949.[1]

By the autumn of 1950 the hardware group had started to think how they might build a faster version of the Mark I computer. The existing prototype machine

[1]*The development of the Manchester University Mark I computer*: an interview with Emeritus Professor D. B. G. (Dai) Edwards, 6th May 2015. Transcript available from the interviewer, Simon Lavington. The interview focuses on the period September 1948–September 1950. Dai Edwards joined the SSEM (Baby) computer team in mid-September 1948 and remained at the University all his working life.

began to be cannibalised as basic units such as power supplies were re-deployed in a project that later emerged as *MEG* ('megacycle machine'). MEG first ran a program in May 1954, as described in Chap. 13.

On the mathematical or software side at the University there was of course the father-figure of Professor Max Newman and, from October 1948, Alan Turing. Max Newman had a long-standing interest in automatic computing machines, as described more fully in Chap. 14. He was elected a Fellow of the Royal Society in 1939. In 1948 the full scale of Turing's genius was only known to a few fellow mathematicians. He certainly made a lasting impact on computing at Manchester but not quite in the manner that the engineers might have expected. Andrew Hodges, in his masterful biography[2] has said: "Turing at Manchester could perhaps have led the world in software development. His partly explored ideas included the use of mathematical logic for program checking, implementing Church's logical calculus on the machine, and other ideas which, combined with his massive knowledge of combinatorial and statistical methods, could have set the agenda in computer science for years ahead. This, however, he failed to do …".

In a way, Turing's genius transcended mere electronic machines. From October 1948 until the delivery of the Ferranti Mark I in February 1951 Turing's correspondence[3] shows him to have been busy writing and reviewing mathematical papers, pursuing his interest in the possibilities of machine intelligence[4] and, from February 1950, starting his unique personal research journey into the shape and form of living things, *Morphogenesis*.[5] In a letter dated 28th November 1952 Turing described his Morphogenesis work as investigating "the manner in which the development of pattern in an embryo may be determined by a combination of chemical reactions and diffusion within the embryo. The partial differential

[2]Hodges, Andrew. 1983. *Alan Turing: The Enigma.* Published by Burnett Books. ISBN 0-09-152130-0. Andrew Hodges' classic 600-page biography is the starting-point for anyone wishing to learn more about Alan Turing.

[3]A collection of Alan Turing's correspondence between April 1949 and June 1954 is held at the University of Manchester Library, Reference GB 133 TUR/Add. The listing can be seen at: https://archiveshub.jisc.ac.uk/search/archives/4f6c3f0c-9a70-33c5-bd03-df331fb06146?terms=%22BBC%22.; Of particular relevance to Chap. 2 are: (a) Items 9 and 31: correspondence with S. P. (Stan) Frankel, Los Angeles, 16th May 1950. Item 31 (14th Sept 1951) says: "Dr. Frankel from Cal tech has been working with us here for a couple of months".; (b) Item 11: Letter dated 2nd Dec 1950 from Sheila Macintyre, Dept. of Mathematics, University of Aberdeen. On 11th Dec 1950 AMT replies: "I am not able to hold out much encouragement".; (c) Item 18. Letter dated 20th April 1951, from E. G. Cox, Chemistry Dept. University of Leeds, mentioning D. W. J. Cruickshank.; (d) Item 37, letter dated Oct 1951: From D. W. J. (Durward) Cruickshank of Leeds University [1924–2007, chemist], requests a copy of the Mark I manual; in his reply, Turing says he cannot as his supplies are running very short. "Possibly I could let you have one when the new ones come out some time next year".; (e) Item 47, letter dated 22nd Nov 1951 from F. M. Colebrook, National Physical Laboratory, asking to come [to Manchester] on Monday 17th (& Tues) Dec. AMT's reply on 27/11/51: "We let Ferrantis have the use of the machine on Mondays".

[4]Turing, A.M. 1950. Computing Machinery and Intelligence. *Mind* 59: 433–460.

[5]Turing, A.M. 1952. The Chemical Basis of Morphogenesis. *Philosophical Transactions of the Royal Society* B237 (641): 37 ff.

equations concerned lead in some cases to stationary waves. The appearance of Fibonacci numbers in phyllotaxis may be explained in these terms".

In March 1951 Alan Turing was elected a Fellow of the Royal Society. He died tragically on 7th June 1954.

Newman also had the administrative and teaching responsibilities attached to being the Head of the Mathematics Department at Manchester in addition to pursuing his main research interests of topology and mathematical logic. In 1949 Newman took the lead, with W. V. D Hodge (of Cambridge) and J. H. C. Whitehead (of Oxford), in launching the British Mathematical Colloquium which held its first meeting in Manchester from 8th to 10th September 1949. The Colloquium later became the largest pure mathematical conference to be held annually in the UK. It seems doubtful whether the world inhabited by Newman, Hodge and Whitehead had any connection with the world inhabited by Williams and Kilburn. It is true to say that Newman's interest in computers faded after 1948.

The Ferranti Mark I was expected to arrive by June 1950 but it was not actually delivered to the University until 12th February 1951.[6] The delay must have been a frustrating time for Alan Turing. In correspondence dated 16th May 1950 with S. F. (Stan) Frankel, Turing wrote: "The delivery of the machine is very much behind what was expected in February [1950]. I feel fairly confident it will be available in the middle of June, but doubt whether I will be able to give much attention then. I expect looking after the machine will be rather hectic then. I think you would be wise to postpone your visit …"[7] Frankel was an important player. He had joined the Manhattan (atomic bomb) Project at Los Alamos in 1943, had learned how to program the ENIAC computer and, by 1950, was Head of a new digital computing group at the California Institute of Technology working on what became known as Monte Carlo analytical methods of computation. Frankel's equipment at Caltech was inadequate for the task and he had asked to use the superior computing power available at Manchester. Frankel eventually came to use the Ferranti Mark I in August/September 1951.[8]

Even by December 1950 Turing, in answer to a query from the University of Aberdeen, wrote that "I am not able to hold out much encouragement. It would certainly not be possible for you to use the machine within the next six months …".[9] By the spring of 1951, Turing was able to be more positive. Professor

[6]Hodges, Andrew, *Alan Turing: The Enigma.*

[7]Items 9 and 31: correspondence with S. P. (Stan) Frankel, Los Angeles, 16th May 1950. Item 31 (14th Sept 1951) says: "Dr. Frankel from Cal tech has been working with us here for a couple of months".

[8]Items 9 and 31: correspondence with S. P. (Stan) Frankel, Los Angeles, 16th May 1950. Item 31 (14th Sept 1951) says: "Dr. Frankel from Cal tech has been working with us here for a couple of months".

[9]Item 11: Letter dated 2nd Dec 1950 from Sheila Macintyre, Dept. of Mathematics, University of Aberdeen. On 11th Dec 1950 AMT replies: "I am not able to hold out much encouragement".

E. G. Cox, of the Department of Chemistry at Leeds University, wrote on 20th April 1951 asking whether D. W. J. (Durward) Cruickshank could use the computer for X-ray structure calculations.[10] By the end of 1951 Cruickshank had started to use the machine (see later).

Notwithstanding his frustration at the delayed delivery, Turing set to work writing a 110-page comprehensive *Programmers' Handbook*, the first edition of which appeared in March 1951.[11] Details of the Ferranti computer's programming conventions and sample code is given in Chap. 15. But be warned! The representation of the instruction set was based on the 5-track teleprinter code, so users had to become familiar with the binary form of 32 alphanumeric symbols. Furthermore, digital quantities were represented in 'backwards binary', meaning that the least-significant digit was at the left-hand end of a string. This suited the engineers, for whom the time-representation of signals in a serial computer flowed from left-to-right. To Alan Turing, who had spent the war years immersed in 5-bit teleprinter codes as a cryptanalyst at Bletchley Park, backwards binary presented no difficulty and was efficient. He justified his choice.[12] The programming conventions were to be made easier within a couple of years, when Ferranti Ltd. introduced the Ferranti Mark I* (*Mark One Star*) computer as described in Chap. 4 and Chap. 15, which adopted the so-called *Radix 32* system of programming.

Let us leave the University in its state of anticipation and return to Ferranti's busy Moston factory.

2.2 The Moston Team Gets to Work

In the beginning Ferranti's computer group was small. To quote from[13]: "This team, which consisted of four or five people, started its work at the beginning of November, 1948, by attending a series of [4] lectures at Manchester University on this new Computer, and the ideas surrounding it. Then, a number of the members of

---

[10]Item 18. Letter dated 20th April 1951, from E. G. Cox, Chemistry Dept. University of Leeds, mentioning D. W. J Cruickshank.

[11]*Programmers' handbook for Manchester Electronic Computer Mark II.* 110-page typed foolscap manual in buff card covers. Anon and undated, but known to have been written by Alan Turing. An issue-date of about March 1951 can be deduced from inserted errata sheets, two of which were issued and respectively dated 13th and 28th March 1951. The qualification 'Mark II' in the title was used to distinguish this computer from its research predecessor at the University of Manchester. The nomenclature 'Mark II' refers to the production version which was later called the Ferranti Mark I computer. The manual contains an Appendix (on pp. 85–97) entitled *The Pilot Machine (Manchester computer Mark I).* This gives a description of the research prototype which was used at the University between April 1949 and August 1950.

[12]*Local programming methods and conventions.* A. M. Turing. Proceedings of the Manchester Computer Inaugural Conference, p. 12. The Proceedings were produced and printed locally and circulated by the University. A copy is at NAHC/MUC/2/D3.

[13]Pollard, B.W. 1957. The Rise of the Computer Department. *Ferranti Journal* 15 (3): 21–23.

the team worked for some considerable time with the University research staff, to gain further detailed knowledge". The four lectures were given by Tom Kilburn on the afternoons of 8th, 9th, 10th and 12th November 1948.[14]

At that stage most of the team's members were hardware people. The exception was the physicist D. G. (Dietrich) Prinz who had worked on servomechanisms. He had visited the USA in 1948 and learned about the state of computing development —(see Chap. 1). By 1950 Prinz was programming the prototype Manchester University Mark I (*MADM*), tabulating values of the Laguerre function in connection with Ferranti's design of guided weapons. We shall introduce more Ferranti programmers in Chaps. 4 and 17, where we'll see that Ferranti's software efforts were concentrated on applications and were therefore related to sales. Ferranti's centre of gravity of applications and sales gradually moved southwards to London in the mid-1950s.

During the production of the Ferranti Mark I the de facto head of Ferranti's computer efforts was Eric Grundy, the Instrument Department's Manager. Grundy became the Manager of the whole Moston site on the 15th of March 1949 and from that date all the Moston departments were integrated into one entity, known as 'The Moston Factory',[15] situated in St. Mary's Road, Moston. The premises no longer exist.

Above Grundy in the Ferranti hierarchy was his Managing Director, J. D. Carter. Above Carter was Sir Vincent de Ferranti. Under Grundy, the principal design engineers were B. W. (Brian) Pollard and Keith Lonsdale and their Moston colleagues. It has proved impossible to name all those who contributed to the Mark I's hardware but the list in Chap. 17 goes some way in recognising individual contributions. Whilst this list may seem to be only of parochial interest, it is sad that academic contributors get their names on publicly-available scientific papers whereas their industrial colleagues usually remain anonymous. In collaborative projects such as the Ferranti Mark I this is a pity. Chapter 17 is an attempt to correct the matter.

Brian Pollard remembers[16] that "to begin with, the general work [at Moston] was divided into two sections; one section, which consisted of the majority of the electronic circuit development, was the responsibility of the Radio Department, whilst the other section, which was responsible for the precision mechanical equipment, such as the magnetic-storage drums and the paper-tape devices, was the responsibility of the Instrument Department. However, after a very short time these two sections combined together into one group, the Computer Group, in Instrument Department.

"The work of manufacturing this first Computer went on during 1949 and 1950, and in parallel with this work a number of small computer-like projects were tackled. The chief one of these was the manufacture, for a Government Department [GCHQ],

[14]Ferranti Internal memo dated 1st November 1948 from J. D. Carter to Grundy, Aitken, Searby, Prinz and Lunt. NAHC/FER/B2.

[15]News item on page 60 of the Ferranti Journal, Vol. 7 No. 2 May 1949.

[16]Pollard, B.W., The Rise of the Computer Department.

Fig. 2.3 An assembly line at Ferranti's Moston factory in about 1950. The four women in the main row are soldering components onto four chassis for a Ferranti Mark I computer

of a storage system [called Revolver, see Lavington[17]] consisting of a magnetic drum and its associated electronics. [This and other GCHQ/Ferranti projects are described in Chap. 10]. During 1950, the greater part of the Computer was being constructed in the Experimental Workshop at Moston, and testing was going forward as quickly as possible". Drath,[18] quoting J. D. Carter, records that "During this period and up to the

[17]Lavington, Simon. 2006. In the Footsteps of Colossus: A Description of Oedipus. *IEEE Annals of the History of Computing* 28 (2): 44–55.

[18]Drath, Paul. 1973. *The Relationship Between Science and Technology: University Research and the Computer Industry 1945–1962*. Ph.D. thesis, University of Manchester. Drath quotes a conversation he had with J. D. Carter on 27th February 1973.

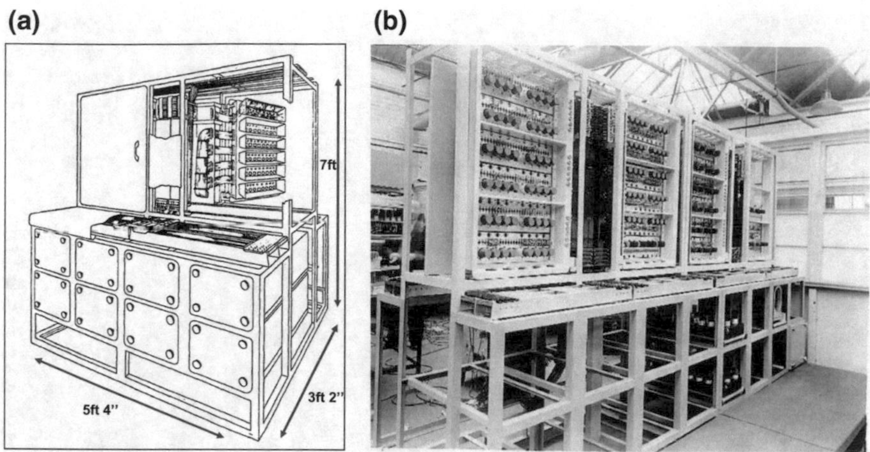

Fig. 2.4 The arrangement of racking to house the logic and storage units of a Ferranti Mark I computer. Figure **a** gives the dimensions as 7 ft (2.130 m) high, 5 ft 4" (1.625 m) wide and 3 ft 2" (0.965 m) deep. Photo **b** shows two racks side-by-side. Three racks side-by-side formed a complete bay (16 ft long) and two such bays made up the complete computer. The Williams/Kilburn CRT storage tubes of the RAM primary store were housed in some of the lower compartments. The magnetic drum secondary store occupied one end of a modified rack

installation of this computer, the number of Ferranti graduate engineers engaged on the project rose to twelve, of whom two were especially detailed to work continuously at the university and were largely responsible for transmitting the university researchers' specifications and designs to the rest of the team".

Geoff Tootill had transferred from the University to Ferranti in the autumn of 1949.[19] He was the main channel through which the logical design and detailed circuit specifications of the university's computer were conveyed to Ferranti. Tootill wrote up the specifications, consisting of 49 typed pages and many diagrams, in November 1949.[20] Tootill's report contained an Appendix written by Alan Turing on the electronic generation of random numbers. Alex Robinson, who had been responsible for the design of the University Mark I's multiplier, transferred to Ferranti in April 1950—by which time Tootill had left Ferranti and joined the Royal

[19]Geoff Tootill interviewed by Thomas Lean. British Library, National Life Stories, Ref. C1379/02. Dates of interview: various, between 3rd December 2009 and 9th April 2010. Transcript available from the British Library, London NW1 2DB.

[20]*Informal report on the design of the Ferranti Mark I computing machine.* 30 typed pages and many diagrams. Text dictated by G. C. Tootill, November 1949. See NAHC/MUC/2/C4. The opening introduction of this Report states: "This report not only endeavours to explain in detail the principle of operation of the machine but also to explain the reasons for various design decisions which have been taken. It will be assumed that the reader is broadly familiar with the computing machine at the University, which will be referred to as the existing machine".

Fig. 2.5 The first production Ferranti Mark I computer nearing completion at Moston in 1950. The magnetic drum store is contained within the nearest rack at the end of the right-hand bay. In the far distance in the photo, the unit on wheels in the centre is an oscilloscope and test trolley

Military College, Scrivenham, complaining that Ferranti were not willing to pay him enough![21]

By November 1950 the first production Mark I was showing signs of life at Moston: "We hope to finish it by the end of the year … and to install it in a new building which is being erected to house it in the University".[22] By mid-December "simple computations had been performed" and, since the Ministry of Supply (MOS) contract was due to run out the following April, "discussions have already been opened between Ferranti and Lockspeiser about the future of the work".[23] Follow-on contracts were indeed forthcoming, as is discussed in Chap. 4.

[21]Geoff Tootill interviewed by Thomas Lean. British Library, National Life Stories, Ref. C1379/02. Dates of interview: various, between 3rd December 2009 and 9th April 2010. Transcript available from the British Library, London NW1 2DB.

[22]Bowden, B.V. 1950. *The Ferranti High Speed Digital Computer*. Ferranti Ltd., typed internal report produced in November 1950. Contains five photos. See NAHC/FER/C17.

[23]*Ferranti computing machine*. NRDC internal memo dated 18th December 1950, from H. J. Crawley to Lord Halsbury. NAHC/NRD C7/3.

2.3 The New Computing Machine Laboratory

Back in the autumn of 1948, whilst the Ministry of Supply's contract for Ferranti to build a computer was being formalised, Manchester University urgently sought suitable accommodation for the machine. It was decided to build a new structure of about 3000 ft^2 floor area, "on the site of a hut abutting the metallurgy Department on Coupland Street".[24] The site is indicated in Figs. 2.1 and 2.2. It was planned that the building should be ready for occupation by June 1950. For the longer-term future, it was decided that the Department of Electrical Engineering and, within it, the Computing Machine Laboratory, should occupy the first of several new buildings planned for a proposed Science Centre near the junction of Dover Street and Rumford Street. The new Dover Street building was ready by the summer of 1954. Figures 2.1 and 2.2 show the location of these structures in relation to other University buildings.

The University's new (but temporary) Computing Machine Laboratory was financed by the unspent equipment portion (£20,000) of a 1946 grant to Professor Max Newman. The grant came from the Royal Society and the background to this is described in Chap. 14. The architects were J. W. Beaumont & Sons of Spring Gardens, Manchester. Special Building Permits, necessary in post-war Britain with steel shortages, were obtained in July 1949 and the premises were more or less ready by the end of 1950. Staff moved in mid-January.[25] The specially-designed two-storey structure, incorporating electro-magnetic screening for the computer room, is shown in Figs. 2.6 and 2.7. It must have been one of the earliest, if not *the* earliest building to be custom-built for running a computing service on a modern digital computer. As we shall see, this service was made available not only to academics but to Ferranti Ltd. and to outside government and business organisations.

The Computing Machine Laboratory was planned to accommodate the following staff: Alan Turing, Tom Kilburn, two mathematical Research Assistants, two engineering Research Assistants, two data-prep persons and one secretary. The total estimated salary bill was £10,000 per annum. To put this staffing in context, the initial (October 1949) establishment had been Alan Turing, Tom Kilburn and the newly-arrived Cicely Popplewell. Popplewell, a Cambridge mathematics graduate with experience of punched cards used in the compilation of housing statistics, was employed to help generally with system programming. She became a fondly-remembered stalwart of the Computing Service at Manchester, retiring in the 1960s.

[24]A File of letters, reports and working papers (some in Williams' own hand) (1948–50), concerning a new computer building. See especially University of Manchester Buildings, general projects, Sub-Committee meeting of 15th December 1948. See NAHC/MUC/1/C2.

[25]Broadbent, T.E. 1998. *Electrical Engineering at Manchester University: The Story of 125 Years of Achievement.* Published by The Manchester School of Engineering, University of Manchester. ISBN 0-9531203-0-9.

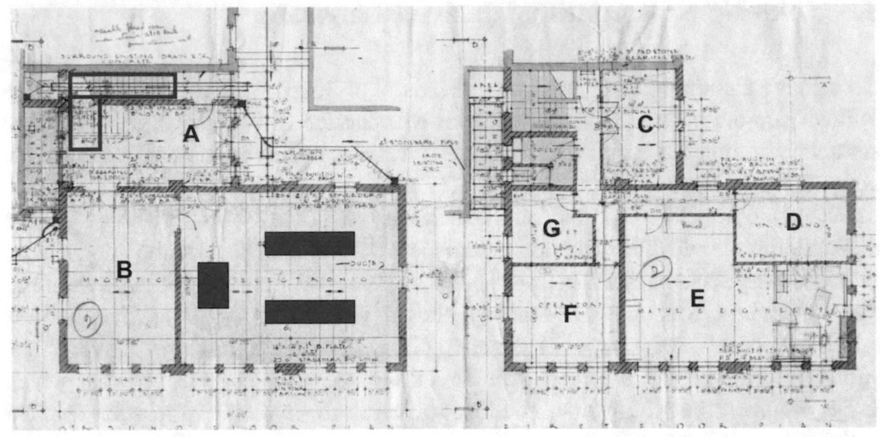

Fig. 2.6 Architect's original plan of the ground floor (*left*) and first floor (*right*) of the new Computing Machine Laboratory. The control desk and two bays of the Ferranti Mark I are superimposed as blue blocks in the largest of the ground-floor rooms, whose measurements are 32 ft 6" by 25 ft. The rooms are named on the original plan as follows: A = Workshop; B = Magnetic; C = Dr. Kilburn; D = Mr. Turing; E = Mathematicians and Engineers; F = Operators; G = Typist. The exact position of a ground-floor interior wall and doors in area A had changed slightly by the time the building was completed. It is believed that the rotating machinery (motor-generators) and the power supply control panel were in the blue-bordered areas in room A. Some 5-track data-preparation equipment was in room B

2.4 The Giant Arrives

Ferranti's marketing brochure[26] described the new machine as *The Manchester Universal Electronic Computer,* proudly stating that it "differs in detail from all other machines, as it makes use of two new and improved 'memories' both of which were developed in Manchester. It is the first machine of this type to be built by an engineering firm and to be commercially available". Continuing the hype, the installation was said to contain "about 4000 valves, 2500 capacitors, 15,000 resistors, 100,000 soldered joints and six miles of wire". It consumed 27 kW of power and was "continuously ventilated by means of a re-circulating air system with heat exchanger connected to an external cooling unit". Clearly, Ferranti was proud of what was, at the time, a significant engineering achievement.

As can be seen from the plan in Fig. 2.6 and the photographs of Figs. 2.8, 2.9 and 2.10, the main electronics of the Ferranti Mark I filled a large ground floor room measuring 32 ft 6 in. by 25 ft (about 10 m by 7.6 m). About a third as much space again was taken up with the rotating machinery (motor-generator sets) and

[26]*The Manchester Universal Electronic Computer*. 5-page illustrated glossy brochure, Ferranti Ltd., List DC1, August 1952.

Fig. 2.7 The Computing Machine Laboratory *(the two-storey flat-roofed building to the left of centre)* at Manchester University in about 1990, by which time the premises had been put to other uses in the University

control panels associated with the power supplies. The electrical details are described in Chap. 16. To this list of equipment should be added the 5-track paper tape data-preparation units.

Not surprisingly, this giant creation took some time to settle down. The Ferranti engineer Brian Pollard wrote[27]: "There was, however, a lot of work still to be done on this computer as it had not been possible to develop the appropriate maintenance techniques. In point of fact, no one knew what the problems of maintenance would be, and the solution of these problems, together with that of improving the general efficiency, gave rise to a considerable amount of difficulty". The engineer Allan Ellson, who joined Ferranti in May 1951, recalls that: "New engineers for commissioning were taught their job by Ianto Warburton. Ianto was the rough Welsh wizard who had no formal rank but was endlessly knowledgeable about all aspects of the digital circuitry of the Mk1*. He was the unsung hero of the early Ferranti Computer Department and we all relied on him to teach us the job".[28]

After delivery on 11th February 1951 there followed three or four months of installation and commissioning work, during which no useful user-programs could

[27]Pollard, B.W., The Rise of the Computer Department.

[28]*Ferranti MkI* commissioning and installation: some recollections.* Allan Ellson. Two-page typed document, 14th October 2015.

Fig. 2.8 One of the two bays making up the Ferranti Mark I at Manchester University. The rack (or compartment) to the right and nearest the camera houses the magnetic drum—compare Fig. 2.5. An electro-magnetic (EM) screening mesh can be seen over the window on the extreme right. It was known that CRT memory tubes were susceptible to EM radiation. The user's control desk or *console* is in the left background of the photo

be attempted. The computer was at last in a good state when it was officially unveiled at an Inaugural Conference from 9th to 12th July 1951, attended by 169 delegates including 13 from overseas. 14 formal papers were presented. Bernard Swann remembers[29] that "The Manchester University Computer was formally opened by Sir David Brunt on July 9th 1951 with a very large attendance of scientists, Government officials and interested businessmen. It was clear that a revolution had begun, though it was to take some years before schemes first suggested at that meeting were to yield results".

The reliability of the computer was not good during the early autumn of 1951. The hardest faults to fix were, naturally, the intermittent ones. As Turing remarked in July 1951,[30] these "would at first occur at rare intervals of time. This meant that

[29]Swann, B.B. 1975. *The Ferranti Computer Department: A History*, 98 p personal memoire. This 1975 document was initially circulated privately and marked 'confidential'. A copy is held at the NAHC, catalogue number FER/C30.

[30]*Local programming methods and conventions.* A. M. Turing. Proceedings of the Manchester Computer Inaugural Conference, p. 12. The Proceedings were produced and printed locally and circulated by the University. A copy is at NAHC/MUC/2/D3.

Fig. 2.9 Part of one bay and the operator's console of the Ferranti Mark I in the Computing Machine Laboratory at Manchester University. A view of the console from further back and showing the opposite bay is given in Fig. 2.8

serious work could not be done, but that it was impossible [for the engineers] to service the machine because the faults were too infrequent. In the course of an hour or two the faults would become more frequent and the machine could be serviced. In the intermediate period, when the machine was only occasionally faulty, it was possible to use it for correcting programmes [i.e. on-line debugging]."

There were uncertainties about the electrical supply in Manchester at that time. Olaf Chedzoy, a Ferranti programmer, remembers that "We were in an era of undersupply of power and the University was but a mile or so away from Old Trafford, where all the big engineering manufacturers were located, BTh and AEI. They used a lot of heavy electrical equipment, and during the day the power would drop from 230 V to 210 or lower—once it dropped below 190 V. That played havoc with the electronic storage system—we used to get 'clods'—that is the spots on the screen [of the CRT memory system] all turned from '0' to '1'."[31]

Perusal of the computer's Log Books[32] shows that early users battled heroically on. Things had got better by December 1951 when, for example, the entry for the

[31]*Starting work on the world's first electronic computer: 1952 memories.* Olaf Chedsoy. Notes prepared for a lecture to local residents at Kilve, Somerset, c. 1995. 18 typed pages, including photos. Copy of notes sent to Simon Lavington in November 2015.

[32]Some surviving log books from the Manchester University's Ferranti Mark I computer, 1951–58, are preserved in: NAHC/MUC/2/C6.

Fig. 2.10 The Ferranti Mark I's console with *(left-to-right)* Keith Lonsdale (Ferranti), Brian Pollard (Ferranti) and Alan Turing (Manchester University). This was one of several publicity photos taken in 1951, just before the computer's Inaugural Conference in July

night of 13th/14th December read: "6:0 p.m.–7:0 a.m. machine working like a bomb". Some hardware fault statistics are given in Chap. 16.

2.5 The Laboratory Prepares for Serious Users

In October 1951 R. A. (Tony) Brooker arrived in Manchester from Cambridge, where he had been working on EDSAC following a spell at Imperial College on the design of a relay sequence-controlled calculator called ICCE.[33] Brooker took over from Turing the responsibility for software development and systems organisation at the Computing Machine Laboratory. Brooker produced a second, and third,

[33]Lavington, Simon. 1980. *Early British Computers*. Manchester University Press. ISBN 0-7190-0803-4. This is out of print but has helpfully been made available at: http://ed-thelen.org/comp-hist/EarlyBritish.html.

editions of Turing's programming manual.[34] With Cicely Popplewell he set up a University Computing Service, providing help with programming and data-preparation.

A software library of subroutines was created (see Chap. 15) and was added to as time went by. To give some idea of size, in June 1952 the Canadian mathematician Kelly Gotleib obtained paper tape copies of the complete subroutine library, which at that time amounted to approximately 9000 lines of code.[35] Gotleib, from the Computation Centre at the University of Toronto, had been sent to Manchester to learn how to use the Ferranti Mark I in preparation for Toronto taking delivery of a copy of this computer. The Toronto installation, called FERUT, is described in Chap. 3.

N. E. (Nick) Hoskin and R. K. (Richard) Livesley, both graduates of the Mathematics Department at Manchester, became the first Research Assistants in the Computing Machine Laboratory in August 1952.[36] Livesley remembers that his "official duties included extending the library of subroutines, teaching programming to new users and giving advice on appropriate numerical techniques".[37]

Livesley recalls that: "From 1951 to 1954 I was one of the small group of regular users of the Mark I. This group, numbering about 20, included individuals from outside the University and staff and research students from various Departments …. During my time in the Laboratory Alan Turing was a heavy user of the Mark I, booking it regularly for two nights a week, but his research work on morphogenesis produced relatively little in the way of new ideas for software development. Nor did he show much interest in the work on Autocodes which was beginning in the laboratory at that time. I suspect that in common with many other 'professional' programmers he regarded such languages as only suitable for the casual user". Tony Brooker's development of the Mark I Autocode, and its experimental predecessor written by Alick Glennie, are discussed in Chap. 15.

Livesley continues: "At the suggestion of my Ph.D. Supervisor [Professor Louis Matheson in Civil Engineering] I began work on the Mark I by implementing some ideas on the elastic analysis of rigid-jointed structural frames put forward by J. M.

---

[34]The second edition was issued in 1952. As the Preface indicates, "much material has been taken over unaltered, or only slightly modified from the 1st Edition which was written by Dr. A. M. Turing. In addition some of the results of the first years' experience of programming for the MK II [i.e. the Ferranti Mark I] have been incorporated in the later chapters. Miss C. M. Popplewell and N. E. Hoskins of the staff of this Laboratory, and A. E. Glennie of the Armament Research Establishment, Fort Halstead, Sevenoaks, Kent, were responsible for Chaps. 3, 7 and 6 respectively". The third edition was issued by Brooker in 1956.

[35]Gotlieb, Calvin C. 1954. The Cost of Programming and Coding. *Computers and Automation* 25: 14 ff.

[36]Campbell-Kelly, Martin. 1980. Programming the Mark I: Early Programming Activity at the University of Manchester. *Annals of the History of Computing* 2 (2): 130–168.

[37]*Minimum weight design: Memories of Alan Turing*. Dr. R. K. Livesley. 7 typed pages plus an Appendix (mixed hand-written and typed) of 13 pages. Document AMT/C/33 received by the Turing Digital Archive from Dr. R. K. Livesley in 2001. See: http://www.turingarchive.org/browse.php/C/33.

Bennett [of Ferranti], who had analysed some simple frames on the EDSAC. Although by present standards the Mark I had a very small working store it had an advantage over the EDSAC in that it possessed a magnetic drum backing-store. This made it possible to construct a program for the elastic analysis of plane frames which was capable of doing serious commercial work, and in 1953 T. M. Charlton and I carried out the analysis of a 20-joint frame for a new power station. As far as I know this was the first computer analysis of a commercial civil engineering structure in this country".

Livesley continues: "Turing was fond of saying 'An ounce of mathematics is worth a ton of computing'. To my surprise and pleasure he [Turing] took up the problem [the elastic analysis of plane frames] with almost childish enthusiasm. I remember that he spent some time transposing it into a problem in game theory …." This brief encounter with Turing appears to be typical of his interaction with other users of the Mark I. As Alick Glennie recalled in the paper by Martin Campbell-Kelly,[38] "Turing's manner of thought was so different from most other people's that his ideas both irritated and invigorated. You had to be mentally prepared for all sorts of unusual ideas where Turing was".

Later on, other Manchester academic users included D. C. (Dennis) Gilles (numerical analysis), Bernard Richards (Turing's research student, working on morphogenesis), F. H. (Frank) Sumner (molecular Chemistry), and Derrick Morris (civil engineering). All went on to become Professors of Computer Science in UK universities.

The Ferranti Mark I at Manchester University was finally switched off just before Christmas 1958. The machine had been moved to a new Electrical Engineering building in mid 1954 and, by the time of its final switch-off, a new Ferranti Mercury computer had been installed in the same building and had taken over as the University's main computer. We will describe the Manchester and Ferranti computing scenes from 1954 onwards in Chap. 13. For now, let us remain in the early days and return to life in the Computing Machine Laboratory in 1952.

2.6 External Users Arrive

At this stage (1952) the Ferranti Mark I provided a computing power far in excess of the University's own needs and so outside bodies came to make use of the facility. This was officially encouraged by the Ministry of Supply, the nominal owners of the machine.

There was some debate about how much per hour the external (e.g. industrial) users were to be charged. In a letter from Professor Blackett to Lord Halsbury dated 9th May 1952, Blackett says that "Williams has, in effect, two alternative types of

[38]Campbell-Kelly, Martin., Programming the Mark I: Early Programming Activity at the University of Manchester.

agreement: (i) £20 per h of production time; (ii) £10 per h of production plus preparation time". Later on in the letter, Blackett remarked on a casual conversation he had had with Tony Brooker, who was of the opinion that "for every problem that needs the whole of the Williams machine, there are 100 which need only one tenth of it". It was becoming clear to the National Research Development Corporation that, in addition to 'supercomputers' such as the Ferranti Mark I, there was also a need for smaller and cheaper machines. This is a theme to which we will return later.

For some potential users, the large Ferranti machine was indeed attractive. On 30th November 1950 Williams received a letter from Lord Portal, Controller of Production (Atomic Energy) at the Ministry of Supply, requesting use of the computer. He writes[39]: "The British atomic energy project has now reached a stage when it is necessary to do a series of very lengthy calculations. I am told that these calculations are too elaborate to be done by hand-machines and that your machine at Manchester University is the only one in the country which can handle them expeditiously and with the accuracy required. I am writing to you to enquire whether you would be willing to give us permission to use your machine on our work. The programming would be done by two of our staff and there would be no need to worry about security because the calculations would involve only strings of numbers. You will see from the attached memorandum from Dr. J. Corner, who would be in charge of the work, that we would require about 200 solutions each of which might take the machine one hour. The Ministry of Supply would be glad to pay for the services of this machine and any operating staff that would be required for running the machine on our problems .... I assure you that the matter is one of the highest importance".

Lord Portal was actually talking more about atomic *weapons* than *energy*. To put this in context, the Armaments Research & Development Establishment at Fort Halstead and the projected Atomic Weapons Research Establishment at Aldermaston had been looking at computers for a year or so before Lord Portal's letter, as described in Chaps. 11 and 12. Both organisations were eventually to acquire their own Ferranti Mark I* computers.

Since Ferranti did not have its own computer at Moston, Ferranti programmers came to use the University machine on a regular basis, specifically at least on every Monday.[40] In some respects, the programming objectives of Ferranti staff overlapped with those of the academics. In March 1953 John Bennett wrote a report on *Uses for the Manchester University computer*.[41] He identified two groups of users: "Firstly, there is the mathematical staff of the Royal Society Computing Laboratory

[39]Letter from Viscount Portal of Hungerford to Prof. F. C. Williams, 30 November 1950. Document NAHC/MUC/1/B1c.

[40]Item 47, letter dated 22nd Nov 1951 from F. M. Colebrook, National Physical Laboratory, asking to come [to Manchester] on Monday 17th (& Tues) Dec. AMT's reply on 27/11/51: "We let Ferrantis have the use of the machine on Mondays".

[41]Bennett, J.M. 1953. *Uses for the Manchester University Computer*. Six typed foolscap pages. See document NAHC/FER/C10a.

[i.e. the Computing Machine Laboratory at the University], and users whose contact with the machine is through this organisation; most of these users come from Universities or research establishments. The second group is a section of the Computer Department of Ferranti Ltd., whose function is to construct demonstration programmes and to carry out such routine computations as industrial users may require".

At the start of 1951 Ferranti's Moston programmers probably numbered four: two 'home grown' and two 'imported'. The home members of the team were Dr. Dietrich Prinz, who we met in Chap. 1, and the mathematician Cyril Gradwell. The first imported member of the team was Audrey Bates (later Clayton). Audrey had graduated in 1949 with a first-class mathematics degree from Manchester and had become Turing's first research student. She obtained her M.Sc. in October 1950, the title of her thesis being *on the mechanical solution of a problem in Church's lambda calculus.* She thereupon left to join Ferranti rather than pursuing university research because, she remembered, Professor Newman had remarked: "Miss Bates, I don't think anyone will ever get a Ph.D. involving computers". She found the move to Ferranti convenient because "I had been working alongside Ferranti people while still a research student".[42] The senior 'imported' person was Dr. John Bennett, an Australian structural engineer who had joined the EDSAC team at Cambridge as its first research student in 1947. On obtaining his Ph.D. he moved to Ferranti. We shall meet John Bennett again in Chap. 4.

Ferranti set about recruiting more programmers. By April 1953 the total had reached 15, of whom over half were women. More details of this remarkable group of skilled people are given in Chap. 17. It was, at the time, the largest such programming group in the UK's emerging computing industry. To accommodate them on what was, after all, a re-furbished 19th century factory site, Ferranti installed a special prefabricated building, known by all as the *Tin Hut*, next to the Moston factory. There is more about this fondly-remembered building in Chap. 17.

One Ferranti programmer, Olaf Chedzoy, remembers that: "We all were given small jobs (usually research oriented) to try out to see the potential, and occasionally larger jobs (with a commercial bias) where we had close contact with the management of commercial organisations … The largest job which I was given with a commercial organisation was to liaise with British Railways, to see if a system could be developed to improve use of their rolling stock …. A much more interesting job was to work with two members of the British Iron and Steel Research Authority, BISRA, who had written a program for the Mark I to assess the linear correlation coefficients relating to the production of steel".[43] Other external organisations with which Ferranti worked were the Shirley Institute (the British Cotton Industry Research Association's headquarters in Manchester) and the

[42]M. A. (Audrey) Clayton (neé Bates), e-mail exchanges with Simon Lavington, spring 2010.

[43]*Memories part 1*. Olaf Chedzoy. November 2015. Three-page typed note sent to Simon Lavington.

aircraft manufacturer A. V. Roe (Avro) at Chadderton, Manchester. Avro eventually acquired its own Ferranti Mark I* machine in 1956, as described in Chap. 6.

2.7 Applications in Science, Engineering and Commerce

The Ferranti Mark I at Manchester was used to investigate a great variety of problems, most of them in science and engineering. At the July 1951 Inaugural Conference F. C. Williams had anticipated that "in the immediate future, the following items are among those which will be investigated:

(a) Partial differential equations arising out of biological calculations.
(b) Simultaneous linear differential equations and matrix algebra, and their applications to the cotton and aircraft industries, and electricity distribution.
(c) Tabulation of Laguerre functions.
(d) Design of optical systems.
(e) Fourier synthesis for X-ray crystallography.
(f) Design of plate fractionating towers.
(g) Chess problems.

A start on areas (c) and (d) had already been made in 1950 using the University's prototype computer. Area (e) commenced late in 1951 when the Leeds University Ph.D. student Farid R. Ahmed and his supervisor, Dr. D. W. J. (Durward) Cruickshank used the Ferranti Mark I to investigate molecular structures by analysing X-ray diffraction patterns. Their programming methods, which drew inspiration from earlier work carried out by John Bennett and John Kendrew on the Cambridge EDSAC computer,[44] are described[45] Area (e) remained a preoccupation with Manchester University research chemists for some years, peaking in the early 1960s with the implementation of an on-line X-ray Diffractometer connected to a large computer—(see Chap. 13).

Area (g) was perhaps suggested as a marker for some non-numeric and theoretical problems for, as G. H. Hardy said, "Chess problems are the hymn tunes of mathematics". Certainly, the prototype computer *MADM* had been used to investigate Mersenne Primes and the Riemann-Zeta function. A later Ferranti marketing brochure,[46] after giving a long list of engineering applications, coyly notes that "From time to time ad hoc problems arise which have little connection with real life. Two enumeration problems in pure mathematics are being coded". This contrasts starkly with Max Newman's 1946 vision of a Manchester computer that

---

[44]Bennett, J.M., and J.C. Kendrew. 1953. The Computation of Fourier Syntheses with a Digital Electronic Calculating Machine. *Acta Crystallographica* 6 (10): 109–116.

[45]Ahmed, F.R., and D.W.J. Cruickshank. 1952. Crystallographic Calculations on the Manchester University Electronic Digital Computer (Mark II). *Acta Crystallographica* 5: 765–769.

[46]*The Manchester Universal Electronic Computer.* 5-page illustrated glossy brochure, Ferranti Ltd., List DC1, August 1952.

would be primarily used for research into pure mathematics itself. Indeed, we shall see in Chap. 4 that some of the Mark I's more unusual instructions originally specified by Newman and Turing were dropped in the formulation of the Ferranti Mark I*'s rigorously-curtailed instruction set. Lest these comments seem rudely mechanical, just wait until Sect. 2.7 when a couple of comfortingly human applications are revealed.

Returning to the theme of science and engineering, a snap-shot of the work being carried out on the computer as at October 1953 was given by John Bennett.[47] It is estimated that, by 1953, about 50 people had been trained to use the Mark I computer.[48] Here is a summary of the activity, arranged according to John Bennett's original classification.

At the University's Computing Machine Laboratory:

Applications involving the solution of differential equations included: rotating shaft problems; the mass-ratio of binary stars; properties of fibres under 'ballooning' forces during spinning; electron density in unit-cell metal crystals; weather forecasting; vorticity of fluids emerging from a hole; compressible fluid flow over a delta wing.

Applications involving matrix methods included: modes of vibration of organic molecules (benzene, naphthalene). A general program dealing with up to 63 simultaneous equations had also been developed. When used for 31 unknowns, the input time was four minutes, the computational time was four minutes and the output time was 10 minutes. There were clearly lessons to be learned about the slowness of the computer's input/output equipment!

At Ferranti's Moston programming group:

Applications involving the solution of differential equations included: cotton spinning; auto-pilot performance; aircraft stability; guided missile trajectories; surge generator calculations; flux distribution in sheet steel; potential distribution in transformer bushing; photo-elastic photograph reduction.

Applications involving matrix methods included: electrical network analysis; bending moments in structural networks; statistical problems in steel production; modal frequencies and flutter problems in aircraft.

Other Ferranti work labelled by Bennett as 'miscellaneous' included: Ordnance Survey traverse reduction; gear train ratios; electrical transmission line losses; production schedules in factory machine shops; payroll calculations; English Life Tables based on data from the 1951 Census; market survey tabulations; monthly price index numbers for the Board of Trade; mathematical model of a highway intersection for traffic control studies; readership survey analysis.

Apart from the payroll example, it may be deduced that Moston had not done much work involving realistic volumes of commercial data-processing. This was

---

[47]Bennett, J.M., *Uses for the Manchester University Computer*.

[48]Campbell-Kelly, Martin., Programming the Mark I: Early Programming Activity at the University of Manchester.

not so much that this market area was being ignored but more a realisation that the Ferranti Mark I's paper tape based input/output equipment and its lack of magnetic tape archival storage made it ill-suited for commercial data processing. These were issues that were being actively addressed by Ferranti Ltd. from 1953 onwards. Liaisons were being sought with Powers Samas, the punched card equipment company, and Ferranti engineers were developing and conducting trials of various forms of magnetic tape equipment. A lineprinter and its buffer store was being attached to the machine. We will return to these data-processing issues in Chaps. 4 and 5. In the light of hindsight we may observe that Ferranti was slow to make inroads into the business world—just as the UK's business world itself was slow to venture into the unknown territory of automatic digital computers.

Applications of the Ferranti Mark I at Manchester that emerged after 1954 are discussed in Chap. 13, where we will describe the move of this machine to a new building and the impact upon users of the Mark I Autocode programming language. At the same time the engineering collaborations between the University and Ferranti Ltd. continued. The theme was the design of follow-on high-performance computers, culminating in the delivery of a Ferranti Mercury to the first customer in August 1957 and the inauguration of the first Ferranti Atlas computer in December 1962. These were but two of Ferranti's many other computer products that came to market from 1956 onwards.

2.8 Music, Games and Romance

Amongst the light-hearted demonstrations of the Ferranti Mark I was the playing of melodies on the console's loudspeaker. It was possible to create crude music because certain Mark I instruction caused a pulse to be sent to the loudspeaker. Sequences of pulses were heard as notes. Although the ability to 'hoot' had been incorporated into the Small-scale Experimental Machine (aka Baby) in 1948, the first use of this facility for melodies—we may hardly call it *music*—occurred in the Computing Machine Laboratory in the autumn of 1951.[49] These were not the world's first computer-generated melodies: that accolade should go to the Australian computer CSIRAC which, according to the memories of those who worked on it, first generated *Colonel Bogey* and other tunes sometime between April and August 1951 (though maybe a little earlier).[50]

[49](a) Link, David. 2016/17. God Save the King—An Early Musical Program on the Ferranti Mark I. *Resurrection, The Journal of the Computer Conservation Society* 76: 11–16. A re-creation of the sound generated by Christopher Strachey's program may be found at www.computerconservationsociety.org/images/gstk.mp3.; (b) Lavington, Simon. 2017. Reflections on the Hoot. *Resurrection* (77): 13–20.

[50](a) Doornbusch, Paul. 2004. Computer Sound Synthesis in 1951: The Music of CSIRAC. *Computer Music Journal* 28 (1): 10–25.; (b) Doornbusch, Paul. 2017. Early Computer Music Experiments in Australia and England. *Organised Sound* 22 (2): 297–307.

The Manchester musical pioneer was Christopher Strachey, a historically fascinating person who we'll meet again in Chaps. 3 and 4. Strachey, then a mathematics master at Harrow School, had obtained a copy of the Mark I programmer's manual. The exact sequence of what followed is a little unclear but here is a plausible version of the story.

Strachey had written a program to play draughts (checkers) and wrote to Turing, whom he had known as a student at Cambridge, and asked if he could visit the Computing Machine Laboratory during a half-term holiday to run his code. Alan Turing agreed but persuaded Strachey to write a different program to start with, namely an interpretive trace program (or *debugging* environment) roughly equivalent to the machine simulating itself. (The draughts program was run later in the summer of 1952, see below). It is supposed that Turing was trying to be helpful in suggesting a project that would educate Strachey into the mysteries of the Ferranti Mark I before attempting a game-playing task.

Anyway, Strachey duly complied and sent his debugging program, which he called *Checksheet*, to Manchester for punching onto paper tape. It was about 20 pages long (over a thousand instructions) and the naivety of a first-time user attempting anything this length caused much amusement in the Laboratory.

From the computer's log book it seems that Strachey arrived at the Computing Machine Laboratory for the first time on 25th September 1951. He remembers that Turing "came in and gave me a typical high-speed, high-pitched description of how to use the machine" and left me alone that evening with the computer. "I sat in front of this enormous machine, with four or five rows of twenty switches and things, in a room that felt like the control room of a battleship".[51] He was undeterred.

After a couple of sessions spent in fixing program faults and dodging intermittent machine hardware errors, it seems likely from Strachey's entries in the log book that the first full performance of the Mark I's rendering of *God Save the King* took place early in the morning of 27th September 1951.

Certainly, the anthem had been heard and admired by many in the Computing Machine Laboratory by the start of October. A result was that Turing recommended Strachey as an exceptional programmer to the National Research Development Corporation. Strachey was interviewed in November 1951 and joined NRDC formally in June 1952 (Fig. 2.11).

Strachey's music routines were copied and modified by the Ferranti Mark I's maintenance engineers and other programmers, to use as party tricks when demonstrating the computer to visitors. It was during one such visit, probably on 7th December 1951, that derivatives of Strachey's music were recorded by BBC *Childrens' Hour* sound engineers. B. V. Bowden from Ferranti observed that: "In December 1951 the BBC broadcast the machine's performance of *Jingle Bells*, *On a One Horse Open Sleigh* and *Good King Wenceslas* …. On the whole it seems

[51] *The word games of the night bird.* An interview with Christopher Strachey by Nancy Foy. Computing Europe, 15 August 1974, pp. 10–11.

Fig. 2.11 Christopher Strachey (1916–1975) was a complex and brilliant character who made his mark on the development of Computer Science. After working for the National Research Development Corporation (NRDC) from 1952 to 1959 and then as an independent computer consultant, he entered academia. In due course he became Oxford University's first Professor of Computer Science

probable that this technique will have no other applications than the amusement of the programmers and their friends at Christmas time".[52]

Music aside, the Mark I was inevitably tried out on simple games. For example it was programmed to make an opening bid for Bridge using the Acol system.[53] The interest in getting the Manchester computer to play board games such as chess and draughts with a human in the period 1951–1953 is carefully described in Chap. 5 of a contemporary book *Faster Than Thought*.[54] In February 1951 Christopher Strachey had written the first version of his draughts (checkers) program, which he ran on the Pilot ACE computer at the National Physical Laboratory (NPL) in July 1951. However the Pilot ACE's memory capacity proved inadequate for the task. Strachey finally ran his re-coded program on the Ferranti Mark I in June/July 1952. The code is analysed in detail by Link.[55] Strachey's sense of fun was lurking in the program. If the user (the human player) made an illegal move or input nonsense information, the machine became increasingly angry until it eventually printed the message: I REFUSE TO WASTE ANY MORE TIME. GO AND PLAY WITH A HUMAN BEING.

Finally, Christopher Strachey made a light-hearted use of the Ferranti Mark in 1953/54 for 'creating' love-letters according to a menu of pre-determined

[52]Bowden, B.V. (eds). 1953. *Faster than Thought.* Published by Pitman.

[53]Olaf Chedzoy, e-mail dated 6th November 2015 to Simon Lavington.

[54]Bowden, B.V. (eds)., *Faster than Thought.*

[55]Link, David. 2012/13. Programming ENTER: Christopher Strachey's Draughts Program. *Resurrection The Journal of the Computer Conservation Society* (60).

syntactical layouts and words.[56] The choice of which words to use for an individual letter was decided by the computer's hardware random number generator, so the results were often curious! (The Ferranti Mark I was one of the very few—possibly the only—production computers to have a special instruction for generating random numbers. More comments on this are given in Chap. 15).

Here's an example output from Strachey's original *love letter* program, as reproduced in mid-1950s editions of the popular reference book *Pears Cyclopedia*:

Darling Sweetheart,
You are my avid fellow-feeling. My affection curiously clings to your passionate wish. My liking yearns to your heart. You are my wistful sympathy; my tender liking.
Yours beautifully,
M.U.C.

MUC stood for 'Manchester University Computer'.

Perhaps Strachey was producing the first tiny example of computer-generated romantic literature? Anyway, here's another not-so-successful example,[57] proving that one should be somewhat suspicious when Artificial Intelligence comes on too strongly:

Honey dear,
You are my seductive love. My lovable eagerness attracts your erotic infatuation. You are my devoted liking: my sweet sympathy: my anxious rapture.
Yours keenly,
M.U.C.

But enough of this fun and frolic!

[56](a) Zielinsky, Siegfried, and David, Link. 2006. There Must Be an Angel: On the Beginnings of the Arithmetics of Rays. In *Variantology 2, On deep time relations of Arts, Sciences and Technologies*, 15–42. Published Cologne, Konig. See: http://www.alpha60.de/research/there_must_be_an_angel/DavidLink_MustBeAnAngel_2006.pdf. (Translated from the German by Gloria Custance). This careful paper gives Strachey's original code and some detailed background information.; (b) In 2009 David Link produced a travelling art installation that uses displays evocative of the original Mark I together with behind-the-scenes processors, to re-create the love letters. See: http://www.alpha60.de/art/love_letters/.

[57](a) Zielinsky, Siegfried, and David, Link., There Must Be an Angel: On the Beginnings of the Arithmetics of Rays.; (b) In 2009 David Link produced a travelling art installation that uses displays evocative of the original Mark I together with behind-the-scenes processors, to re-create the love letters. See: http://www.alpha60.de/art/love_letters/.

2.9 Getting Serious

Back at Moston, things were getting serious. By the end of 1951 Ferranti's factory was completing a second Mark I computer which was, after some uncertainties, exported to Canada. At the same time, the Ministry of Supply started making definite approaches to Ferranti for the purchase of a modestly-enhanced version of the Mark I. Finally, the National Research Development Corporation, which had assumed responsibility for the computer patents emanating from Manchester University, had begun to put pressure on Ferranti to produce and sell more computers. The wider world beckoned.

The company took tentative steps to face the market place. In 1950 Ferranti had hired B. V. (Vivian) Bowden to be a roving ambassador and technical salesman for its Computer Department's products. Bowden was a good choice in the early days of the digital revolution. He was an electronics engineer who had worked at TRE during the war. His qualifications included a Physics Ph.D. from the University of Cambridge. Bowden perceived that someone with a more commercial and administrative background was needed to promote sales. He therefore recruited B. B. (Bernard) Swann in June 1951. Swann was a Civil Servant with an accountancy background. He had risen to be Assistant Secretary of the Statistical Division of the Board of Trade. In Chap. 5 we'll see that he led Ferranti's marketing efforts in the commercial data-processing sector.

In the next chapter we'll see that all was not plain sailing. Britain, and indeed the whole of Europe, was still in the period of post-war financial stress. The Ministry of Supply, which had originally promoted Ferranti's computer-building efforts, was planning to install Ferranti computers in three of its crucial Establishments but the sorting out of priorities and the allocation of necessary funding meant that firm orders took time to arrive. Specifically, the three formal MOD contracts did not arrive until, respectively, autumn 1951, autumn 1952 and autumn 1953. Fortunately, meanwhile, computing activity in Canada was blossoming. Chapter 3 gives us the opportunity of telling the fascinating story of Canada's pioneering computing efforts and the role played by a Ferranti Mark I computer in putting Toronto firmly on the historical map. There is also an intriguing sub-plot of close technical collaborations between Manchester and Canada.

References[58]

Ahmed, F.R., and D.W.J. Cruickshank. 1952. Crystallographic Calculations on the Manchester University Electronic Digital Computer (Mark II). *Acta Crystallographica* 5: 765–769.

Bennett, J.M. 1953. *Uses for the Manchester University Computer*.

Bennett, J.M., and J.C. Kendrew. 1953. The Computation of Fourier Syntheses with a Digital Electronic Calculating Machine. *Acta Crystallographica* 6 (10): 109–116.

Bowden, B.V. 1950. *The Ferranti High Speed Digital Computer*. Ferranti Ltd.

Bowden, B.V. (eds). 1953. *Faster than Thought*. Published by Pitman.

Broadbent, T.E. 1998. *Electrical Engineering at Manchester University: The Story of 125 Years of Achievement*. Manchester: Published by The Manchester School of Engineering, University of Manchester. ISBN 0-9531203-0-9.

Campbell-Kelly, Martin. 1980. Programming the Mark I: Early Programming Activity at the University of Manchester. *Annals of the History of Computing* 2 (2): 130–168.

Doornbusch, Paul. 2004. Computer Sound Synthesis in 1951: The Music of CSIRAC. *Computer Music Journal* 28 (1): 10–25.

Doornbusch, Paul. 2017. Early Computer Music Experiments in Australia and England. *Organised Sound* 22 (2): 297–307.

Drath, Paul. 1973. *The Relationship Between Science and Technology: University Research and the Computer Industry 1945–1962*. Ph.D. thesis, University of Manchester.

Gotlieb, Calvin C. 1954. The Cost of Programming and Coding. *Computers and Automation* 25: 14 ff.

Hodges, Andrew. 1983. *Alan Turing: The Enigma*. Published by Burnett Books. ISBN 0-09-152130-0.

Lavington, Simon. 1980. *Early British Computers*. Manchester University Press. ISBN 0-7190-0803-4.

Lavington, Simon. 2006. In the Footsteps of Colossus: A Description of Oedipus. *IEEE Annals of the History of Computing* 28 (2): 44–55.

Lavington, Simon. 2017. Reflections on the Hoot. *Resurrection* 77: 13–20.

Link, David. 2012/13. Programming ENTER: Christopher Strachey‘s Draughts Program. *Resurrection The Journal of the Computer Conservation Society* (60).

Link, David. 2016/17. God Save the King—An Early Musical Program on the Ferranti Mark I. *Resurrection, The Journal of the Computer Conservation Society* 76: 11–16.

Pollard, B.W. 1957. The Rise of the Computer Department. *Ferranti Journal* 15 (3): 21–23.

Swann, B.B. 1975. *The Ferranti Computer Department: A History*, 98 p personal memoire.

Turing, A.M. 1950. Computing Machinery and Intelligence. *Mind* 59: 433–460.

Turing, A.M. 1952. The Chemical Basis of Morphogenesis. *Philosophical Transactions of the Royal Society* B237 (641): 37 ff.

Zielinsky, Siegfried, and David, Link. 2006. There Must Be an Angel: On the Beginnings of the Arithmetics of Rays. In *Variantology 2, On deep time relations of Arts, Sciences and Technologies*, 15–42. Published Cologne, Konig. See: http://www.alpha60.de/research/there_must_be_an_angel/DavidLink_MustBeAnAngel_2006.pdf.

[58]*Note:* NAHC is the National Archive for the History of Computers, held in the University of Manchester Library.

Chapter 3
Canada Calling: Toronto Gets a Mark I

Britain's post-1948 defence requirements for new computing facilities, channelled through the contracts sections of the Ministry of Supply (MOS), caused lots of paperwork to fly round before firm priorities and firm orders for equipment were established. Ferranti, as the UK's earliest supplier of the new type of automatic digital computer, was naturally drawn into the discussions. By the late summer of 1951 Moston was faced with a potential difficulty: a second Mark I was nearing completion for which no buyer had been fixed because the MOS still had not made up its mind. Fortunately, the University of Toronto had become very interested in acquiring new computing facilities. After some debate, an order was placed by the Canadian National Research Council in the autumn of 1951.

This gives us the opportunity of telling the fascinating story of Canada's pioneering computing efforts and the role played by a Ferranti Mark I computer in putting Toronto firmly on the historical map. There is also an intriguing sub-plot of close technical collaborations between the UK and Canada.

3.1 The Ministry of Supply Needs Some Computers

Even as the first production Mark I computer was being readied for its inauguration at Manchester University in July 1951, Ferranti's Moston factory was working on a second machine. The funding for Moston's first computer and for initial work on the second had come from the Ministry of Supply. The MOS was to place more contracts during the early 1950s, intended to equip three specific defence-related sites with Ferranti computers.

One of the MOS sites, the Armaments Research Development Establishment at Fort Halstead in Kent, appeared to be first in the queue. Brigadier G. H. Hinds had become Director of Weapons Research (Defence) in the MOS in 1950. He was an early convert to digital computers and was actively promoting Fort Halstead's case. The enthusiastic Brigadier gave Ferranti a Letter of Intent to purchase a Mark I

S. Lavington, *Early Computing in Britain*, History of Computing,
https://doi.org/10.1007/978-3-030-15103-4_3

computer for Fort Halstead in the spring of 1951 without, it seems, the necessary full strategic approval from higher authority.[1] Brigadier Hinds knew that crucial atomic weapons research was initially being carried at Fort Halstead, though the longer-term plan was to move the work to a new and dedicated site at a redundant RAF base at Aldermaston in Berkshire, once the infrastructure had been improved.

In the event, the first MOS-related site to get a Ferranti computer was the Government's secret Communications Headquarters (GCHQ) at Cheltenham. GCHQ's machine was delivered in the autumn of 1953 (see Chap. 10). Fort Halstead was next, followed closely by Aldermaston—as is described in Chaps. 11 and 12.

By the summer of 1951 Brigadier Hinds' Letter of Intent had been withdrawn. The delay placed on Brigadier Hinds' ambitions may have been influenced by two factors:

(a) the UK government's financial problems;
(b) GCHQ's relatively late enthusiasm for acquiring a Ferranti computer quickly.

As for government finance, this was a time of uncertainty. The general election of February 1950 left Clement Atlee's Labour government with a tiny majority of only five seats. June 1950 saw the start of the Korean War, whose cost had a major impact on the British economy. There was a balance of payments crisis in the summer of 1951 which led to speculation against sterling. With the Government looking increasingly unstable, Attlee called a general election in October 1951 which was won by the Conservatives.

As for GCHQ, it was so secret that its name does not appear either in Ferranti's surviving Computer Department records or in those of the National Research Development Corporation. Indeed, we'll see in Chap. 10 that deception was part of GCHQ's strategy. And whereas Brigadier Hinds' group could arrange for initial atomic weapons calculations to be carried out on the Ferranti Mark I at Manchester University, there was no chance of GCHQ allowing its highly-classified cryptanalytic calculations to be performed elsewhere. In short, GCHQ's needs trumped Brigadier Hinds' needs.

Notwithstanding these uncertainties, Ferranti's Moston factory continued to construct a second Mark I computer in the expectation that a purchaser could be found, whether or not it turned out to be the MOS.

As an aside on the MOS's remit, Canadian folk-law has sometimes asserted[2] that the UK's atomic energy authority wished to purchase the second Ferranti Mark I

[1]Swann, B.B. 1975. *The Ferranti Computer Department: A History*. (personal memoire). This 1975 document was initially circulated privately and marked 'confidential'. A copy is held at the NAHC, catalogue number FER/C30.

[2]Williams, Michael R. 1994. UTEC and FERUT: The University of Toronto's Computation Centre. *IEEE Annals of the History of Computing* 16 (2): 4–12. This paper quotes original documents, including two University of Toronto Computation Centre Progress Reports, respectively covering the periods 1st October 1951 to 31st December 1951 and 1st April 1952 to 30th June 30 1952.

computer. This is perhaps due to a deliberate UK false rumour or possibly to a confusion between atomic *energy* and atomic *weapons*. The UK Atomic *Energy* Authority's principle establishment at Harwell set up a Computer Section in 1948. Harwell used a mixture of electro-mechanical manual calculators augmented, from 1951, by an in-house designed sequence-controlled calculator using relays and decatron valves. (This machine has been restored and is working at The National Museum of Computing at Bletchley Park.) It was not until February 1958 that Harwell obtained its first full-scale stored-program computer, a Ferranti Mercury.[3]

The uncertainty surrounding MOS orders persisted until at least May 1952, by which time Moston had completed the second Mark I computer and had moved on to start a revised design called the Mark I* (*Mark One Star*). It was this revised design that attracted the active interest of GCHQ.

We now describe events in Canada that did eventually lead to the sale and installation of the second Ferranti Mark I computer in Toronto.

3.2 The Canadian Computing Scene

Whilst the first Mark I computer was being completed for Manchester University, Ferranti had begun to look for sales prospects for follow-on copies. Vivian Bowden, who visited America and Canada in the period December 1950–January 1951, returned with some encouraging impressions of the Canadian computing scene. We will pick up the thread of Bowden's other marketing endeavours in Chap. 5. For now we describe the background to what turned out to be Ferranti's first sales success: the University of Toronto. This University already had a lively Computation Centre when Bowden arrived.

A Committee on Computing Machines had been set up at the University of Toronto during the academic year 1945/46 by representatives of the Mathematics and Electrical Engineering Departments. The aim was to make the case for establishing a Computation Centre at the University. The Committee visited several American research groups and produced a budget "from figures that had been given to them by John von Neumann when the Committee had visited Princeton in the summer of 1946".[4] The Committee's report pointed out that no electronic computing machine was at that time available but that $75,000 (about £26,000 at that time) should be set aside for copying a design when one had been produced elsewhere. At first the Committee had in mind copying the Bell Labs Model V Programmable Calculator which used about 9000 electro-mechanical relays. This idea was quickly dropped in favour of waiting for an *electronic* stored-program

[3]Hopgood, F.R.A. 2015. *Early History: Harwell Computing*. 3rd Sept 2015. This three-page typed document was written by Bob Hopgood in response to a query from Simon Lavington concerning the early use of digital computers at Harwell. Bob Hopgood spent his working life at Harwell and has taken a leading role in documenting Harwell's history of computing.

[4]Williams, Michael R. UTEC and FERUT: The University of Toronto's Computation Centre.

Fig. 3.1 Ferranti engineers commissioning the second Mark I computer at the Moston factory at the end of 1951 or start of 1952. This is the machine that went to Toronto. The engineers shown in the photo are (*left-to-right*): Bill Wallace and David Wilde looking at the monitor (oscilloscope); Brian Welby in the background; Alan Ellson crouching; Jock Wilson standing at right, holding a logic door

computer to come along. It was estimated that if an electronic machine was to be developed from scratch at Toronto University then the in-house development costs would exceed $300,000 (about £105,000)—certainly an ambitious vision. To begin with, the Centre's less ambitious goal was to hire personnel and equip them with conventional calculating devices.

Toronto's Computation Centre got off the ground towards the end of 1947 when two computing assistants, J. P. (James) Stanley and B. H. (Beatrice, or Trixie) Worsley were hired to work initially with a Madas Calculating Machine and IBM electro-mechanical punched card equipment (a 601 Multiplying Punch, a 405 Accounting Machine and 031 Punch). Some of this equipment has been pictured in Chap. 1. The IBM 601 was replaced in July 1949 by an IBM 602A.[5] C. C. (Kelly) Golieb of the Physics Department managed the Centre's day-to-day activities. Then in late 1948 Stanley and Worsley were sent to Cambridge University to study

[5]Campbell, Scott M. 2006. *The Premise of Computer Science: Establishing Modern Computing at the University of Toronto (1945–1964)*. Unpublished Ph.D. thesis, Graduate Department of the Institute for the History and Philosophy of Science and Technology, University of Toronto, 2006.

something completely different—the EDSAC project. They remained in Cambridge for some time and naturally learned to program EDSAC, an experience that surely paid off later when Worsley helped to develop Transcode (see below) at Toronto.

Trixie Worsley also spent time in the Mathematics Department at Manchester University and it seems that Alan Turing may have acted as a second supervisor for her Ph.D. research. Certainly, Worsley kept up an on-going correspondence with Turing. The first surviving letter is dated 23rd December 1950, when she wrote: "As you will see by the enclosed letter, I have been advised to take a prolonged change from academic work. This is therefore to inform you that I am making plans to return home [to Canada] as soon as possible. However, I hope to return as soon as I am fit and complete the work I have undertaken at Manchester".[6]

In Toronto the Computation Centre's ambition to provide a first-rate national computing facility was to be realised by the end of 1952, but not quite in the manner that Toronto University had initially imagined.

Meanwhile, an interest in digital computers had been growing within Canadian defence circles. The principle example is DATAR (digital automated tracking and resolving).[7] This was a target-tracking and command-integration project for the Canadian Navy which was conceived at the very end of 1948. Ferranti Canada had been in at the start of DATAR and became the prime contractor. In 1949 Arthur Porter (see later) was recruited by Vincent de Ferranti to lead Ferranti Canada's DATAR developments, which included a small parallel computer containing 3800 thermionic tubes (valves), some electrostatic primary storage and a magnetic drum backing store. The DATAR command-and-control system was successfully demonstrated in the autumn of 1953. Whilst a novel command-and-control system was also being developed concurrently for the Royal Navy by the Elliott company in England,[8] DATAR was certainly world-class. Being a classified project, detailed information about DATAR was hidden from Toronto University.

[6]A collection of Alan Turing's correspondence between April 1949 and June 1954 is held at the University of Manchester Library, Reference GB 133 TUR/Add. The listing can be seen at: https://archiveshub.jisc.ac.uk/search/archives/4f6c3f0c-9a70-33c5-bd03-df331fb06146?terms=%22BBC%22; Of particular relevance to Chap. 3 are these items, which cover correspondence with Beatrice Worsley: Tur/Add/13, 23rd December 1950; Tur/Add/45, 14th November 51; Tur/Add/91, nine letters covering the period 9th January 1952 to 26th November 1952.

[7]Vardalas, John. 1994. From DATAR To The FP-6000 Computer: Technological Change in a Canadian Industrial Context. *IEEE Annals of the History of Computing* 16 (2).

[8](a) Lavington, Simon. 2011. *Moving Targets: Elliott-Automation and the Dawn of the Computer Age in Britain, 1947–67*. Springer. ISBN 978-1-84882-932-9.; (b) Boslaugh, David, Peter Marland, and John Vardalas. 2017. The Information Age & Naval Command and Control. In: *Presented at the McMullen Naval History Symposium, at USNA Annapolis*, 14–15 Sept 2017. Possibly to be re-published by USNA.

3.3 UTEC: Toronto's Home-Grown Computer

In parallel with Toronto's acquisition of punched card equipment, Professor V. G. Smith of the Electrical Engineering Department had become interested in designing and building circuits for automatic computers. In May 1948 three recent graduates, Alfred Ratz, E. Doeringer and Josef Katz (later Kates), were recruited by Smith to begin working at Toronto on ideas for a locally-designed electronic computer. By 1949 they had decided to use electrostatic storage techniques similar to those developed by F. C. Williams at Manchester—(see Chaps. 1 and 16). Their thoughts on computer architecture were strongly influenced by John von Neumann's computer project at the Institute for Advanced Study, Princeton. In March 1949 Professor Smith's group visited the IAS project.

The funding for the Toronto electronic computer project was met from a grant of about $50,000, of which the Department of National Defence's Defence Research Board (DRB) provided $30,000 and the Canadian National Research Council (NRC) $20,000. This computer was to be named UTEC. F. C. Williams' group at Manchester became aware of the work and he visited the University of Toronto's Computation Centre in 1949.[9] Williams later received a copy of a July 1950 report on UTEC.[10] In the summer of 1950 Maurice Wilkes, leader of the EDSAC team at Cambridge University, also visited the project.

UTEC was unusual at the time in being a parallel (rather than serial) computer. In this respect it followed the general structure being developed by von Neumann's IAS group. The first, prototype, version of UTEC was deliberately kept simple. It had a 12 bit word, of which three bits were assigned for the instruction (i.e. op code) and nine to the operand address. By the end of March 1950 a parallel binary adder/subtracter had been demonstrated for this prototype; by June a control unit and one bit of storage were working. By the end of September 1950 the storage system was nearing completion and undergoing tests, but the input-output equipment was still on order. By the end of June 1951 the storage, control, arithmetic and input/output components had still not been fully integrated.[11] However in a letter dated 14th November 1951 Trixie Worsley wrote to Alan Turing to say "Toronto has just completed a pilot model called UTEC. It does the basic instructions Add, Subtract, Transfer, Input, Output and has CRT storage operating in parallel".[12]

---

[9]Williams, Michael R. UTEC and FERUT: The University of Toronto's Computation Centre.

[10]*University of Toronto Digital Computer.* Seven-page typescript bearing the names of A. Ratz and J. Katz and dated 28th July 1950. NAHC FER/C, Ferranti Ltd. Reports, Working Papers, etc., C2.

[11]Campbell, Scott M. *The Premise of Computer Science: Establishing Modern Computing at the University of Toronto (1945–1964).*

[12]A collection of Alan Turing's correspondence between April 1949 and June 1954 is held at the University of Manchester Library, Reference GB 133 TUR/Add. The listing can be seen at: https://archiveshub.jisc.ac.uk/search/archives/4f6c3f0c-9a70-33c5-bd03-df331fb06146?terms=%22BBC%22; Of particular relevance to Chap. 3 are these items, which cover correspondence with Beatrice

Fig. 3.2 UTEC pictured in 1952, when the 12-bit word prototype was nearing completion. The primary store, consisting of 12 Williams CRTs, are in two rows towards the top centre of the racking. Joseph Katz (later Kates) is the centre of the three people

Possibly there may have been doubts about certain aspects of Toronto's storage endeavours. In a letter to F. C. Williams, Vivian Bowden says in January 1951 that he "found it extremely difficult to assess what Kates has really done".[13] Also, in a letter to F. C. Williams dated 14th September 1951, Arthur Porter had said that "we came to the conclusion, as you did, that his [Josef Katz's] theory was haywire".[14] Arthur Porter, about whom more details are given below, was Head of Ferranti Canada's Research Department.

Trixie Worsley wrote again to Alan Turing on 9th January 1952, reporting that "unfortunately Toronto is letting UTEC pass away without giving it much of a work-out. The engineers are concentrating on the design of a much more ambitious

Worsley: Tur/Add/13, 23rd December 1950; Tur/Add/45, 14th November 51; Tur/Add/91, nine letters covering the period 9th January 1952 to 26th November 1952.

[13]Papers of Professor Sir F. C. Williams (1911–1977); B4 Correspondence (1951–69) with Ferranti Ltd. NAHC/MUC/1.

[14]Papers of Professor Sir F. C. Williams (1911–1977); B4 Correspondence (1951–69) with Ferranti Ltd. NAHC/MUC/1.

machine, which seems rather a short-sighted policy to me. They refuse to hire enough mathematicians to give them a run for their money".[15]

Although reliability was at first questionable, nevertheless the 12-bit word UTEC prototype was apparently running satisfactorily with 512 words of storage by the summer of 1952.[16] This was a significant achievement for a university research group. But the writing was on the wall. By the autumn of that year UTEC was discontinued and dismantled due, fundamentally, to impatience on the part of funders and potential external computer users who wielded a lot of influence. News of this impatience filtered back to Manchester.

3.4 Toronto Orders a Ferranti Computer

Arthur Porter, Head of Ferranti Canada's Research Department, was an early enthusiast for Canada to invest in computers. Porter had a good pedigree.[17] He had worked with Douglas Hartree on Differential Analysers at Manchester University from 1933–36, had then spent two years as a Fellow at MIT, returning to the UK to spend the war years working on computations, particularly at the Air Defence Research Establishment at Malvern. Another strong influence on Canadian investment was the distinguished scientist W. B. Lewis FRS, who had become Chief Superintendent of TRE at Malvern in 1945. In 1946 Lewis was appointed Director, Division of Atomic Energy Research, at Canada's new nuclear research facility at Chalk River, Ontario. By 1951 Chalk River's computational needs were growing rapidly.

Both Porter and Lewis were on good terms with F. C. Williams, having shared experiences of Manchester and of Malvern—see for example the social references to their respective families in.[18] Letters were exchanged which kept Williams, and hence Ferranti at Moston, abreast of Canadian developments. In February 1951 Williams was to write to Porter saying that Lewis "gives me the impression that he

[15]A collection of Alan Turing's correspondence between April 1949 and June 1954 is held at the University of Manchester Library, Reference GB 133 TUR/Add. The listing can be seen at: https://archiveshub.jisc.ac.uk/search/archives/4f6c3f0c-9a70-33c5-bd03-df331fb06146?terms=%22BBC%22; Of particular relevance to Chap. 3 are these items, which cover correspondence with Beatrice Worsley: Tur/Add/13, 23rd December 1950; Tur/Add/45, 14th November 51; Tur/Add/91, nine letters covering the period 9th January 1952 to 26th November 1952.

[16]*University of Toronto Digital Computer*. Seven-page typescript bearing the names of A. Ratz and J. Katz and dated 28th July 1950. NAHC FER/C, Ferranti Ltd. Reports, Working Papers, etc., C2.

[17]Robinson, Tim. 2008. *Computer History Museum, Oral History of Arthur Porter*. Interviewed by Tim Robinson, 8th Mar 2008. See http://archive.computerhistory.org/resources/access/text/2015/06/102658245-05-01-acc.pdf.

[18]Papers of Professor Sir F. C. Williams (1911–1977); B4 Correspondence (1951–69) with Ferranti Ltd. NAHC/MUC/1.

would like to see a commercially made computer available in Canada".[19] This was also the impression gained by Vivian Bowden during his sales tour in 1950/51.

In short, the signs were encouraging that Canada was eager to obtain realistic electronic computing facilities. Although the progress towards a full-scale UTEC was disappointing, Toronto University was certainly one of the country's major research establishments. Could its Computation Centre be equipped with a large modern computer built elsewhere? Were there, indeed, any companies offering such computers for sale? By the autumn of 1951 Ferranti was one company, perhaps one of only of two companies, producing computers for sale.

By the end of 1951 the Canadian government was proposing that Toronto should acquire a "Ferranti Machine", for which the DRB and NRC would each contribute $150,000 towards the estimated $300,000 cost. It appears that the Canadian funding authorities had decided to re-deploy the money previously earmarked for developing UTEC. The inference was that "this use of the money would provide everyone with an electronic computer much sooner than would be the case if the full scale machine [UTEC] were constructed by the Computation Centre".[20] On 11th January 1952 the University of Toronto accepted the DRB/NRC proposal.[21]

This was the background from which Ferranti at Moston was able to report on 28th August 1951 to NRDC that: "We have been asked to quote the National Research Council of Canada, acting on behalf of the Canadian government, for a High Speed Digital Computing Machine of the type which we have supplied to Manchester University … Whilst we have not yet worked out the actual selling price, we do not think it would be less than 225,000 dollars, that is, about £80,000". This letter implies that Ferranti's final quote to Canada, required by the end of September, would not be much more than the $225,000 mentioned above—(the exact figure is not stated). For comparison, the cost of the subsequent Ferranti Mark I* computers was about £90,000. It is not known whether shipping costs were extra to the figures quoted above. Most probably they were extra, judging from the known costs of the Ferranti Mark I* computer that went to Rome in 1955—(see Chap. 9).

Meanwhile, Moston continued to assemble the second production version of the Ferranti Mark I that was destined for the Computation Centre at Toronto University. It was ready for despatch by the end of February 1952. The machine was named FERUT (Ferranti computer at the University of Toronto) by Trixie Worsley, once it had arrived. According to a former Ferranti Mark I maintenance engineer, *FERUT* was pronounced *Ferret*.[22]

---

[19]Papers of Professor Sir F. C. Williams (1911–1977); B4 Correspondence (1951–69) with Ferranti Ltd. NAHC/MUC/1.

[20]Williams, Michael R. UTEC and FERUT: The University of Toronto's Computation Centre.

[21]Campbell, Scott M. *The Premise of Computer Science: Establishing Modern Computing at the University of Toronto (1945–1964).*

[22]*Ferut: Canada's first commercially available general purpose electronic digital computer.* Brian (Simon) Jeffrey. See: http://ferut.ca/. Brian joined the joined the NRC Structures Laboratory on Montreal Road in Ottawa in 1958, where he became responsible for maintaining Ferut.

Fig. 3.3 One section of the Ferranti Mark I's central processor being lifted into the Computation Centre at the University of Toronto in the summer of 1952. This computer was called FERUT *(pronounced Ferret)*. In the photo, the unit is suspended with its upper edge pointing to the left. Two logic doors are shown in place but the lower framework (to the right in the photo) is as yet unpopulated

By way of preparation for FERUT, Kelly Gotlieb was sent to Manchester in late April 1952 for six weeks to learn how to use a Ferranti Mark I. "During his stay, he was given instruction in programming, and wrote a few routines to solve simple problems. More importantly, he was able to obtain 'the complete subroutine library at Manchester' by making physical copies of the paper tapes and bringing them home. This amounted to approximately 9000 lines of code, which represented a sizable amount of experience and knowledge that could easily by transported".[23] Before that, Trixie Worsley had been sent a Ferranti Mark I programming manual by Alan Turing in about December 1951.[24]

[23]Gotlieb, Calvin C. 1954. The Cost of Programming and Coding. *Computers and Automation* 25: 14.

[24]A collection of Alan Turing's correspondence between April 1949 and June 1954 is held at the University of Manchester Library, Reference GB 133 TUR/Add. The listing can be seen at: https://archiveshub.jisc.ac.uk/search/archives/4f6c3f0c-9a70-33c5-bd03-df331fb06146?terms=%22BBC%22; Of particular relevance to Chap. 3 are these items, which cover correspondence with Beatrice

3.5 FERUT's Sea Voyage from Manchester to Toronto

The installation of FERUT put a considerable strain on Moston's fledgling Computer Department, especially since the Director of Toronto's Computation Centre, Kelly Gotlieb, had invited the Association for Computing Machinery (ACM) to hold their annual conference in Toronto in September 1952 in the hope that FERUT would be operational by then. The ACM had been founded in America in 1947 and was, by 1952, the principle learned society for anyone seriously interested in automatic digital computers.

FERUT certainly had its teething troubles. The computer's tribulations clearly caused concern in high places, causing Brian Pollard to send a detailed post hoc explanation to F. C. Williams in November 1952.[25] Here are the main events in Brian Pollard's account of FERUT's arrival. They give some flavour of the uncertainties faced by installation engineers of that era.

End-Feb 1952: after factory tests at Moston, the computer is dismantled.
On March 29th: crated & ready for shipment by the *Manchester Pioneer* direct from Manchester docks to Toronto. [This vessel, completed in 1952, was built specially for trans-ocean trade to the American Great Lakes and was probably the first ocean-going ship to make the direct connection].[26]
On 25th April: *Manchester Pioneer* docks in Toronto & computer [packed in 15 crates] is unloaded.
On 5th May: after customs clearance, computer delivered to University of Toronto.
By 19th May: computer unpacked, racks positioned, crates removed.
By 10th June: computer erected.
By 21st June: inter-rack wiring connected; wiring from generator room to power supply in place; most of air conditioning and ducting in place;
By 23rd July: console and temporary DC supply etc. connected.
By 28th July: 3-phase 60 Hz supply connected to AC generator set;
By 1st Aug.: testing of computer begun, with temporary DC supply.
5th Aug.: final DC generator set arrived.
By 8th Aug.: final DC generator set up and working.
By 20th Aug.: component parts of the drum arrived in Toronto [from Manchester].
By 1st Sept.: "virtually the whole of the machine had been tested and was in operation". At this point the outstanding tasks were: (i) commission more drum

Worsley: Tur/Add/13, 23rd December 1950; Tur/Add/45, 14th November 51; Tur/Add/91, nine letters covering the period 9th January 1952 to 26th November 1952.

[25]Pollard, Brian. 1952. *The Installation of the University of Toronto Computer* [Ferranti]. Two-page typed report dated 11th Nov 1952, sent to F. C. Williams. NAHC/MUC B4: Papers of Professor Sir F. C. Williams (1911–1977); B4 Correspondence (1951–69) with Ferranti Ltd.

[26]The Toronto Globe and Mail, 29th April 1952, p. 3 [as quoted in reference Scott M. Campbell, *The premise of Computer Science: Establishing modern computing at the University of Toronto (1945–1964)*].

tracks; (ii) install the paper tape punch, the random number generator and the hooter.
From 2nd to 7th Sept.: investigating serious trouble that had developed with the AC generator for the heater supplies; machine not available.
From 8th to 9th Sept.: "temporary measures" introduced to bring the computer into service for the duration of the [ACM] Conference. "No breakdown occurred during the two days of the Conference, although computing errors occurred about every 30 min during the first day of the Conference, but no errors were observed during the second day".
By 2nd Oct.: the operators "were allowed to begin using the machine during the afternoon" whilst the installation of further equipment was carried out.
31st Oct.: acceptance tests were successfully carried out and the computer was formally handed over to the University of Toronto.

Pollard was later to elaborate from personal experience[27]: "The installation engineers, who at that time were also the design engineers, had to work right through the heat of a Toronto summer. Fortunately the computer was kept cool by a large refrigeration system, and some relief could be obtained by actually getting inside the computer itself. The attempt to get the machine ready for the Conference was successful and it was opened for demonstration only one hour late on schedule. During this hour, the doors to the Computer Room had to be bolted and barred, so great was the interest, and there was a surge of people into the room when finally the doors were opened. To our consternation, the Computer suddenly failed after about half-an-hour of successful demonstration, but it was discovered that a visitor, who must be nameless, was so interested in the type of valves used that he had actually taken one out of the computer during operation!"

In a letter to Alan Turing dated 19th November 1952, Trixie Worsley said: "By now you will probably have heard that our new Computer, the FERUT, has passed its acceptance trials, and is working well enough to keep us more than busy. We are certainly glad to have all the experienced help we can get, our most recent addition being Peter Bandler [a former Manchester research student]. Mr. Strachey has promised to tell you about the INPUT and DIRECTORY scheme we have devised when he returns to England at the end of the month. We are all looking forward to having Miss Popplewell join us early in the new year".[28] The technical background to the INPUT and DIRECTORY scheme is given in Chap. 15.

---

[27]Pollard, Brian. 1957. The Rise of the Computer Department. *Ferranti Journal* 15 (3): 20–23.

[28]A collection of Alan Turing's correspondence between April 1949 and June 1954 is held at the University of Manchester Library, Reference GB 133 TUR/Add. The listing can be seen at: https://archiveshub.jisc.ac.uk/search/archives/4f6c3f0c-9a70-33c5-bd03-df331fb06146?terms=%22BBC%22; Of particular relevance to Chap. 3 are these items, which cover correspondence with Beatrice Worsley: Tur/Add/13, 23rd December 1950; Tur/Add/45, 14th November 51; Tur/Add/91, nine letters covering the period 9th January 1952 to 26th November 1952.

Fig. 3.4 FERUT's console and operator's desk at the University of Toronto. High on the wall above the console is a guide to interpreting characters punched on 5-track paper tape. The cabinets housing the main computer are in two rows behind the operator's chair, of which part of the left-hand row can be seen

Reliability continued to be an issue at Toronto, causing Ferranti to re-examine the specification of Acceptance Tests, a topic to which we'll return later. Of FERUT, Christopher Strachey [NRDC] said[29]: "After experience with the Toronto machine, which passed its acceptance tests on November 1st and broke down very seriously two days later, I feel that some reconsideration should be given to the form of the acceptance tests required for future machines. The first point is that any overall test is not really satisfactory as it may easily be passed while the machine is far from satisfactory. The second is that the tests (whatever form they take) should first be passed *in the factory* and then again when the machine is assembled. (If this had been done with the Toronto machine, several weeks would have been saved.)"

Partly as a result of the Toronto experience, a detailed specification for stringent Acceptance Tests was produced for the Ferranti Mark I* computer that went to Shell's Amsterdam Laboratory in 1954—see Chap. 8.

---

[29]Memo, Christopher Strachey [NRDC] to H. J. Crawley [NRDC], 10th December 1952. NAHC/NRD C7/5: Computers, Manchester University/Ferranti Development, Vol I. Reports, etc, 1951–52.

3.6 Continuing Links with Manchester

The ACM meeting at Toronto in September 1952 had attracted several Ferranti and Manchester University attendees, amongst whom the following were authors or co-authors of papers: John Bennett, Dietrich Prinz, Mary Lee Woods, Christopher Strachey and Alec Robinson.[30] Cecily Popplewell (from Manchester University's Computing Machine Laboratory) and Audrey Bates (from Ferranti Moston) were loaned temporarily to Toronto from Manchester to help with programming and teaching others how to program. Popplewell spent the first six months of 1953 in Toronto and helped Strachey with the Saint Lawrence backwater calculations (see below). Audrey Bates remained and married Ken Wallis, one of Ferranti's former Moston engineers.[31] Audrey Wallis served at one point as the FERUT librarian in Toronto's Computation Centre.

This is the point to recall the connection that Trixie Worsley developed with Manchester. It will be remembered that Worsley joined the Toronto Computation Centre in 1948 and was sent to Cambridge University in the autumn of 1949, where she registered as a Ph.D. student. There is evidence that in 1950 Trixie Worsley was also in contact with Alan Turing, at that time the Deputy Director of the Computing Machine Laboratory at Manchester University.[32] It seems that Turing became an informal advisor for Worsley's research, though the formal supervisor continued to be Professor Douglas Hartree of Cambridge. Worsley returned to Canada for health reasons. By November 1951 she writes to Turing saying she will not return to the UK, but instead will complete her studies at Toronto. She regrets that she won't be able to work on the Ferranti Mark I at Manchester but asks for a copy of the Mark I programming manual (written by Turing). Worsley received her Ph.D. in 1952, the title of her thesis being *Serial Programming for Real and Idealized Digital Calculating Machines.* She then received a research fellowship from Canada's National Research Council, Turing supporting her application. Worsley remained an active force in Toronto's Computation Centre until she moved to Queen's University in Kingston, Ontario, in 1965.

[30]The following papers appeared in the Proceedings of the 1952 ACM National meeting, Toronto. Published by ACM Press, 1952. (i) Bennett, J.M., D.G. Prinz, and M.L. Woods. 1952. *Interpretative Sub-routines,* 81–87.; (ii) Strachey, Christopher. 1952. *Logical or Non-mathematical Programmes*, 46–49.; (iii) Robinson, A. 1952. *The testing of cathode ray tubes for use in the Williams type storage system,* 42–45.

[31]Campbell, Scott M. *The Premise of Computer Science: Establishing Modern Computing at the University of Toronto (1945–1964).*

[32]A collection of Alan Turing's correspondence between April 1949 and June 1954 is held at the University of Manchester Library, Reference GB 133 TUR/Add. The listing can be seen at: https://archiveshub.jisc.ac.uk/search/archives/4f6c3f0c-9a70-33c5-bd03-df331fb06146?terms=%22BBC%22; Of particular relevance to Chap. 3 are these items, which cover correspondence with Beatrice Worsley: Tur/Add/13, 23rd December 1950; Tur/Add/45, 14th November 51; Tur/Add/91, nine letters covering the period 9th January 1952 to 26th November 1952.

Fig. 3.5 Beatrice (Trixie) Worsley at FERUT's console. She has rightly been called "a pioneering computer scientist and the first female in Canada to make significant contributions to the field". Trixie was a key member of the Computation Centre at the University of Toronto from 1948 until 1965. She died in 1972

There is an interesting set of correspondence between Worsley and Turing covering the period 9th January 1952–26th November 1952, in which Worsley's originality and imagination shine out.[33] Her ideas include remote access to a computer, RISC instruction sets, and spontaneity as a property of the human mind that could not be duplicated by Turing's *Imitation Game*. Trixie Worsley was a skilled mathematician whom history has rightly called "a pioneering computer scientist and the first female in Canada to make significant contributions to the field".[34] She died aged 51 in 1972, of heart complications.

[33]A collection of Alan Turing's correspondence between April 1949 and June 1954 is held at the University of Manchester Library, Reference GB 133 TUR/Add. The listing can be seen at: https://archiveshub.jisc.ac.uk/search/archives/4f6c3f0c-9a70-33c5-bd03-df331fb06146?terms=%22BBC%22; Of particular relevance to Chap. 3 are these items, which cover correspondence with Beatrice Worsley: Tur/Add/13, 23rd December 1950; Tur/Add/45, 14th November 51; Tur/Add/91, nine letters covering the period 9th January 1952 to 26th November 1952.

[34]Campbell, Scott M. 2003. *Beatrice Helen Worsley: Canada's Female Computer Pioneer* (University of Toronto). *IEEE Annals of the History of Computing* 25 (4): 51–62.

3.7 The Saint Lawrence Seaway Calculations

We saw in Chap. 2 that the brilliant programmer Christopher Strachey was recruited by the National Research Development Corporation (NRDC) in June 1952. Strachey was to become very familiar with FERUT during work on the Saint Lawrence Seaway project. Connecting the Great Lakes to the Atlantic for large sea-going vessels had been a dream since the 1890s. Various minor improvements to waterways were made over the years but the Canadian and American governments had consistently failed to agree on details of a full-scale project. By the end of the Second World War the possibilities for linking the Seaway project to hydro-electric schemes had heightened the Canadian's sense of frustration and in September 1951 Canada announced that it intended to construct a Seaway alone. It was acknowledged that this would necessitate massive amounts of engineering calculations for the 'backwaters'. These backwaters comprised the area of rivers and islands upstream from a proposed hydro-electric dam, subsequently called the Moses-Saunders Powerhouse, near Cornwall, Ontario, on the Canadian/USA border. The construction of the dam and the deepening of the St. Lawrence navigational passages was estimated to result in significant changes in water levels and surface profiles along the entire length of the backwater. When various options for the navigational route and various climatic scenarios had been taken into consideration, 99 backwater cases were identified for investigation.[35] Each case required much computation before its effects could be defined.

By the autumn of 1952, treaty negotiations with the United States were under way to determine the exact route that the Seaway would take. Clearly, it would be advantageous if Canadian engineers were able to support their preferred route by numerical results. Estimates had indicated that it would take almost 20 years to compute the required results using IBM electro-mechanical punched card equipment. FERUT's arrival was seen by the Canadians as a civil engineer's dream come true.

The reputations of the University of Toronto, Ferranti and NRDC—not to mention the Canadian Government—all stood to gain from FERUT producing a timely set of results for the Seaway. NRDC agreed to second Christopher Strachey to the University of Toronto to do the programming—an excellent choice because Strachey had proved himself an outstanding programmer of the Ferranti Mark I at Manchester. As previously mentioned, Strachey arrived in Toronto in time to attend the September ACM Conference. He quickly made contact with Professor W.H (William) Watson of the Department of Physics, who from 1952 was Chairman of the University's Computation Centre. Watson had made notable contributions to radar antenna theory during the war and had then spent from 1946 to 1951 under W. B. Lewis at Chalk River as the Head of Theoretical Physics and leader of the computing section.

[35]Campbell, Scott M. *The Premise of Computer Science: Establishing Modern Computing at the University of Toronto (1945–1964).*

Fig. 3.6 FERUT's console in 1958, after the computer had been moved to the National Research Council's Structures Laboratory in Ottawa

Although Strachey was the principle programmer for the backwater calculations, Kelly Gotlieb guided the entire project and other Computation Centre staff including Trixie Worsley helped with various parts of the program. Cicely Popplewell, on an extended visit from Manchester, also assisted. Scott Campbell[36] has deduced that: "The program relied on the library of subroutines from Manchester, but the rest of it was new …. The final program contained about 2000 instructions, and the data tape ran nearly 2.5 km in length …. As a final verification that the computer and computer program were producing proper results, several of the individual backwater cases and one entire run were computed by hand on a desktop calculator … From this it was estimated that manual computations would have taken 20 man years of time, compared to the 500 h of machine time over eight months consumed by FERUT".

Strachey spent all of October and November and part of the following spring on the Saint Lawrence Seaway project. He finally returned to the UK in March 1953. The results from Strachey's program, codenamed *Backwater*, were decisive: the USA accepted the Canadian plans for the Seaway because, it is deduced, the

---

[36]Campbell, Scott M. *The Premise of Computer Science: Establishing Modern Computing at the University of Toronto (1945–1964).*

Americans appeared not to have had access to equipment for doing the calculations for an alternate route. Construction on the St. Lawrence Seaway and Power Project began in August 1954. The Seaway was officially opened in June 1959 in the presence of the Canadian Prime Minister John Diefenbaker, US President Dwight D. Eisenhower and H. M. Queen Elizabeth.

Back to computing matters. During his stay in North America, Christopher Strachey visited several research laboratories and computer manufacturers in the US, making a detailed study of the instruction sets (order codes) of different machines. This experience enabled him to contribute with some authority to NRDC's discussions with Ferranti over the instruction sets for the Ferranti Mark I* and, especially so, for the Ferranti Pegasus computer—topics to which we'll return in later chapters.

3.8 FERUT in Action

The success of the backwater calculations brought in $35,000 to Toronto's Computation Centre in 1953, their first income in addition to government grants. News of the backwater project eventually led to customers such as the Dominion Observatory, major Canadian insurance companies such as Manufacturer's Life, and industrial corporations such as A. V. Roe, Eastman Kodak and Imperial Oil paying the Computation Centre for work. However, the largest percentage of time on FERUT was used by the NRC, the DRB, and Atomic Energy of Canada Limited. The NRC and DRB contributed to an annual grant of $50,000.

Writing in 1954, a user of FERUT described the set-up at Toronto thus: "The Centre had a staff of eight to ten mathematicians and physicists and in addition Atomic Energy of Canada Ltd., the National Research Council of Canada and the Defence Research Board each maintain a staff at the Centre to deal with specific problems".[37] The author then goes on to give this simple example of FERUT's impact on the computing facilities available to researchers: "Values were required of a function of two variables. Using FERUT, 400 values were computed in three minutes, whereas an estimate of the time required using an ordinary electrically-operated desk machine is about one hundred hours".

Because the University of Toronto had the only large-scale computer in Canada for several years, consulting companies used FERUT for demonstrations before their clients would commit to an investment of their own. By the end of 1955 at least ten computers were on order for various Canadian government, business, or academic organisations.[38] Regrettably, none of them was a Ferranti machine.

[37]McDonald, Jean K. 1954. *FERUT, Canada's Electronic Computing Machine. Journal of the Royal Astronomical Society of Canada* 48 (5): 176–184.

[38]Campbell, Scott M. *The Premise of Computer Science: Establishing Modern Computing at the University of Toronto (1945–1964).*

Fig. 3.7 One row of logic cabinets of FERUT, as re-installed in the NRC's Structures Laboratory in Ottawa. A similar row of cabinets was placed on the opposite wall. The doors have been removed from the right-most unit in the photo to reveal the magnetic drum backing store

After the St. Lawrence Seaway backwater calculations, one of the largest programs written for FERUT was a series of self-consistent field calculations carried out by Trixie Worsley. "The goal was to develop a program for FERUT that could be used to calculate atomic wave functions using the Hartree-Fock formulation. Douglas Hartree himself consulted on the project in Canada until his death in 1958".[39]

The staff at Toronto's Computation Centre had done more than give programming tuition and assistance to users. They had ventured into systems software and had produced a much more user-friendly programming system than Turing's original system based on 'backwards binary' teleprinter symbols. In the autumn of 1953, possibly inspired by John Backus's *Speedcode* system for the IBM 701, Trixie Worsley and J. N. Patterson Hume from the Physics Department, were asked to create an automatic coding system for FERUT. Their language, called TRANSCODE,[40] was ready by the end of 1954. TRANSCODE was immediately

---

[39]Campbell, Scott M. *The Premise of Computer Science: Establishing Modern Computing at the University of Toronto (1945–1964).*

[40](a) Hume, J.N.P., and Beatrice H. Worsley. 1955. T4R13a: Transcode: A system of Automatic Coding for Ferut. *Journal of the Association for Computing Machinery* 2 (4): 243–252.; (b) Hume,

Fig. 3.8 Close-up of one of the logic doors of Fig. 3.7. There are six sub-chassis, each with up to 8 EF50 thermionic tubes (valves) and a small 1400 Hz heater transformer. Various other components cannot be seen in the photo

popular with users. It incorporated meaningful four-letter mnemonic instructions such as ADDN, SUBT, MULT, LOOP, PRNT, etc. The whole subject of programming languages for the Ferranti Mark I and Mark I* is reviewed in more detail in Chap. 15.

One ground-breaking example of remote use of FERUT took place in December 1955. To quote[41]: "In a test conducted on 22nd December last, FERUT, the high-speed Electronic Digital Computer supplied to the University of Toronto in Canada in 1952, gave the answer to a question presented to it by the University of Saskatchewan, 1700 miles away. The problem, normally a 100 man-hour job, was punched direct on to tape at Saskatoon, fed through a transmitter over telegraph lines to Toronto with the co-operation of the Canadian National and Canadian Pacific telegraph companies, and the answer provided in thirty minutes. As a result of the

J.N.P. 1994. Development of Systems Software for the FERUT Computer at the University of Toronto, 1952–1955. *IEEE Annals of the History of Computing* 16 (2): 13–19.

[41] *Long distance test of the University of Toronto computer*. Ferranti Journal, Vol. 14, No. 1, Spring 1956, page 31.

Fig. 3.9 Close-up of the magnetic drum in Fig. 3.7

experience gained, it is hoped that all Canadian Universities will eventually be connected to FERUT".

3.9 The End of FERUT

By 1958 FERUT was obviously outclassed by other commercially available machines. Toronto decided to replace it with an IBM 650 which, though of similar speed, was more reliable and had a much wider software base. In 1962 the 650 was replaced by Canada's first IBM 7090.

In 1958 FERUT was passed to the National Research Council in Ottawa, where it served the needs of the NRC's Mechanical Engineering Section (in the Structures Laboratory). The Structures Laboratory became part of the National Research Council's (NRC) National Aeronautical Establishment, where aircraft structures were tested. The date of FERUT's final switch-off has not been discovered: perhaps it was towards the end of 1966? It is known that in July 1967 there was a proposal that the remains should be "turned over to the National Museum".[42] As far as can be ascertained, no significant pieces of FERUT's hardware survive today.

In the next chapter we return to England in 1951/2, where discussions were going on between Ferranti, Manchester University, NRDC and potential users about possible minor alterations to the Ferranti Mark I's instruction set. By the end of 1951 sufficient programmer experience had been obtained to make such discussions realistic—and probably necessary from the viewpoint of future marketing opportunities.

References[43]

Bennett, J.M., D.G. Prinz, and M.L. Woods. 1952. *Interpretative Sub-routines,* 81–87.

Boslaugh, David, Peter Marland, and John Vardalas. 2017. The Information Age & Naval Command and Control. In: *Presented at the McMullen Naval History Symposium, at USNA Annapolis*, 14–15 Sept 2017. Possibly to be re-published by USNA.

Campbell, Scott M. 2003. Beatrice Helen Worsley: Canada's Female Computer Pioneer (University of Toronto). *IEEE Annals of the History of Computing* 25 (4): 51–62.

Campbell, Scott M. 2006. *The Premise of Computer Science: Establishing Modern Computing at the University of Toronto (1945–1964).* Unpublished Ph.D. thesis, Graduate Department of the Institute for the History and Philosophy of Science and Technology, University of Toronto.

Gotlieb, Calvin C. 1954. The Cost of Programming and Coding. *Computers and Automation* 25: 14.

Hopgood, F.R.A. 2015. *Early History: Harwell Computing*. 3rd Sept 2015.

Hume, J.N.P. 1994. Development of Systems Software for the FERUT Computer at the University of Toronto, 1952–1955. *IEEE Annals of the History of Computing* 16 (2): 13–19.

Hume, J.N.P., and Beatrice H. Worsley. 1955. T4R13a: Transcode: A System of Automatic Coding for Ferut. *Journal of the Association for Computing Machinery* 2(4), 243–252.

Lavington, Simon. 2011. *Moving Targets: Elliott-Automation and the Dawn of the Computer Age in Britain, 1947–67*. Berlin: Springer. ISBN 978-1-84882-932-9.

McDonald, Jean K. 1954. FERUT, Canada's Electronic Computing Machine. *Journal of the Royal Astronomical Society of Canada* 48 (5): 176–184.

Pollard, Brian. 1952. *The Installation of the University of Toronto Computer* [Ferranti].

Pollard, Brian. 1957. The Rise of the Computer Department. *Ferranti Journal* 15 (3): 20–23.

Robinson, A. 1952. *The Testing of Cathode Ray Tubes for use in the Williams Type Storage System,* 42–45.

[42]Letter dated 18th July 1967 from J. V. Scott, Senior Staff Scientist at Centennial Centre of Science & Technology, Don Mills, Ontario, to Crown Assets Disposal Corporation, Ottawa. See http://ferut.ca/the-mystery/.

[43]NAHC denotes documents held in the National Archive for the History of Computing, at the John Rylands Library at the University of Manchester.

Robinson, Tim. 2008. *Computer History Museum, Oral History of Arthur Porter*. Interviewed by Tim Robinson, 8th Mar 2008. See http://archive.computerhistory.org/resources/access/text/2015/06/102658245-05-01-acc.pdf.

Strachey, Christopher. 1952. *Logical or Non-mathematical Programmes*, 46–49.

Swann, B.B. 1975. *The Ferranti Computer Department: A History*. (personal memoire).

Vardalas, John. 1994. From DATAR To The FP-6000 Computer: Technological Change in a Canadian Industrial Context. *IEEE Annals of the History of Computing* 16 (2).

Williams, Michael R. 1994. UTEC and FERUT: The University of Toronto's Computation Centre. *IEEE Annals of the History of Computing* 16 (2): 4–12.

Chapter 4
A Star Is Born: Ideas and Upgrades

By the spring of 1952 Ferranti's team at Moston was in a position to reflect on the computer product that they had launched. Twelve months' of maintenance experience had revealed a few weak points in the Ferranti Mark I's hardware's reliability and nine months of end-user experience had revealed the strengths and weaknesses of its instruction set and programmers' interface relative to facilities on offer elsewhere. There was a feeling amongst certain Moston programming staff, and notably John Bennett, that beneficial modifications could be made to the Mark I's software without disturbing the computer's general architecture. Inevitably, this chapter will get a bit technical towards the end as we summarise the transition from Mark I to Mark I*.

To begin, though, we take a more human look at the Moston programmers who were well placed to suggest changes to the Mark I's facilities. What was it like to be employed as a programmer at Moston?

4.1 The Tin Hut

In 1951 Ferranti had a problem finding suitable accommodation for the team of 15 or so programmers that was being recruited. The team clearly had to work on the Moston site, in order to be close to the Computer Department where the hardware was being designed and assembled. But the factory's internal spaces were filled with noisy equipment, ranging from machine tools such as lathes to specialist installations where valves and TV tubes were manufactured—frankly not a tranquil environment. The external aspects were also far from uplifting. Figure 4.1a, b give the picture. Since a significant proportion of the programmer recruits were female mathematics graduates from universities in the south-east of England, Ferranti's northern industrial image could appear rather forbidding. Another issue was gender-division: programmers were salaried 'white collar' staff but before 1950

S. Lavington, *Early Computing in Britain*, History of Computing,
https://doi.org/10.1007/978-3-030-15103-4_4

practically all such Ferranti staff were male. Industrial employment norms of the time indicated that females could expect a lower salary than their male colleagues.

The Moston management went some way towards solving the image problem by housing the programmers in a special prefabricated building partly made from aluminium alloy trusses, sheathing and panels. In the immediate post-war years the British government had encouraged the development of a range of prefabricated structures, from small houses to larger administrative buildings, in order to provide rapid accommodation where it was most needed. Ferranti acquired a prefabricated AIROH (Aircraft Industries Research Organisation on Housing) building, possibly manufactured by the Bristol Aircraft company. The building, affectionately called the Tin Hut, was placed amongst rare patches of green vegetation on the perimeter of the main Moston factory—see Figs. 4.2 and 4.4.

Mary-Lee Woods (later Mary Berners-Lee) attended a recruitment interview in the Tin Hut in the autumn of 1951. She recalls taking a bus from the middle of Manchester, northwards along the Oldham Road and through an industrial area surrounded by cotton mills. "I thought: Never, never, never will I live in this ugly place!" However, after a successful interview Mary-Lee Woods accepted the job. Other young people were recruited and she remembers that: "It was a very good team—very exciting! And we were a good group, too, socially".[1]

Mary-Lee went on to marry a Ferranti mathematician (Conway Berners-Lee) (Fig. 4.3). She stayed with the company until a few weeks before her first son, Tim, was born in June 1955. Both Mary and Conway are fondly remembered by former Ferranti colleagues and of course their son Sir Tim Berners-Lee deserves to be equally remembered by all who use the World Wide Web.

There was a special reason why the Tin Hut women have cause to remember Mary-Lee Woods. In 1952 she took the lead in a campaign to get Ferranti to adopt a policy of equal pay for male and female programmers. Mary-Lee had an advantage: at the time she was sharing a flat with the Assistant Personnel Manager for Women (Joy Badham). "I knew what they were thinking in the Personnel Department; and I was the person who was selected to go and see the management and make the case—and we won".[2]

There was also a special reason why the Mark I engineers had cause to remember Mary-Lee Woods. In about 1953 she was struggling to develop a program to invert

[1]Mary-Lee Berners-Lee: An Interview Conducted by Janet Abbate for the IEEE History Center, 12 September 2001. Interview #578 for the IEEE History Center, The Institute of Electrical and Electronic Engineers, Inc. See: https://ethw.org/Oral-History:Mary_Lee_Berners-Lee#About_Mary_Lee_Berners-Lee. The IEEE History Center has a collection of more than 800 oral histories in electrical and computer technology which can be accessed via http://ethw.org/Oral-History:List_of_all_Oral_Histories.

[2]Mary-Lee Berners-Lee: An Interview Conducted by Janet Abbate for the IEEE History Center, 12 September 2001. Interview #578 for the IEEE History Center, The Institute of Electrical and Electronic Engineers, Inc. See: https://ethw.org/Oral-History:Mary_Lee_Berners-Lee#About_Mary_Lee_Berners-Lee. The IEEE History Center has a collection of more than 800 oral histories in electrical and computer technology which can be accessed via http://ethw.org/Oral-History:List_of_all_Oral_Histories.

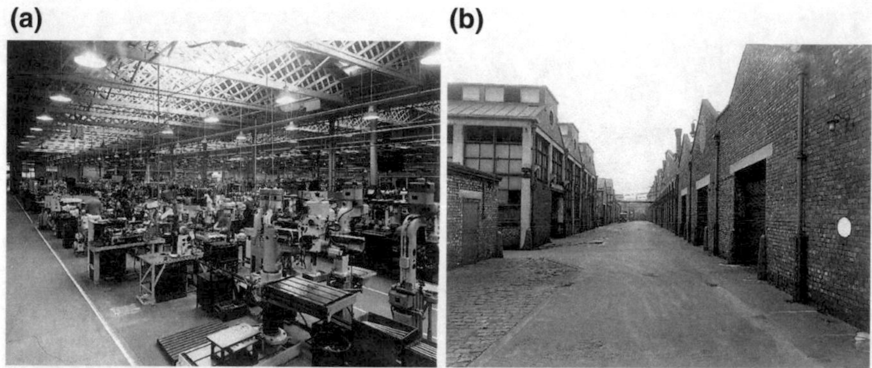

Fig. 4.1 **a** and **b** Ferranti's Moston factory. Photo **a** is the machine shop, showing an array of lathes and other tools. Photo **b** is an internal access road on the site

Fig. 4.2 Part of the Moston site in 1984 looking south, with the Tin Hut ringed in red

Fig. 4.3 Mary-Lee Woods and Conway Berners-Lee in about 1954. These two well-remembered Ferranti programmers were married about the time this photo was taken. Their first son became Sir Tim Berners-Lee of World-Wide Web fame

large matrices, as part of a technique used for the solution of simultaneous linear (differential) equations. This required long error-free computing runs—of three hours or more, depending upon the number of equations. Mary remembers: "I just couldn't get the program through, because the machine always made a mistake before we got to the end …. So I found myself very much in the middle of this big row between the engineers and the mathematicians as to what constituted a really working machine. And I got very tired over it. One time I went along to see a professor at the university about this, the great Tom Kilburn, to state my case. He was very polite to me. He was very courteous—and at the time I felt resentful, because I thought he'd been courteous to me instead of arguing with me because I was a woman … I resented this, and I had resented it for years; but when it came to the 50th anniversary [in 1998] of the Manchester [Baby] computer, and we went back, and there was a really good dinner, and I'd had enough wine, I went and tackled the great man and told him this. And he said, "Oh, no, no, no! I wouldn't have been like that. I didn't argue with you because I knew you were right!".[3] Mary-Lee's matrix inversion program became part of Ferranti's standard Acceptance Tests for new computers.

[3]Mary-Lee Berners-Lee: An Interview Conducted by Janet Abbate for the IEEE History Center, 12 September 2001. Interview #578 for the IEEE History Center, The Institute of Electrical and

Fig. 4.4 The Tin Hut in the early 1950s, as photographed from the main factory buildings looking towards St. Mary's Road

As for the Tin Hut itself, an approximation to the internal layout is shown in Fig. 4.5. John Bennett was in charge of the team of about 15 programmers, most of whom inhabited a large open area with a smaller office for Dietrich Prinz and Cyril Gladwell. Another large open area contained the Ferranti Computer Departments' design and development engineers. In between was a small office for John Bennett and another office for the Sales Staff which at that time consisted of Vivian Bowden and Bernard Swann.

This is the point to re-introduce Moston's senior programmer John Bennett, who was the initial driver for modifications to the Ferranti Mark I. Bennett, an Australian who had had experience of radar during wartime service, came to Cambridge in 1947 as the EDSAC team's first Ph.D. student. He used EDSAC in the Cambridge Mathematical Laboratory for structural engineering calculations during his research. He then joined Ferranti at Moston in 1950 and stayed with the company until 1956. Bennett was clearly influenced by Cambridge programming conventions, as set down in the seminal book.[4]

Electronic Engineers, Inc. See: https://ethw.org/Oral-History:Mary_Lee_Berners-Lee#About_Mary_Lee_Berners-Lee. The IEEE History Center has a collection of more than 800 oral histories in electrical and computer technology which can be accessed via http://ethw.org/Oral-History:List_of_all_Oral_Histories.

[4]Wilkes, M.V., D.J. Wheeler, and S. Gill. 1951. *The Preparation of Programs for an Electronic Digital Computer.* Published by Addison-Wesley.

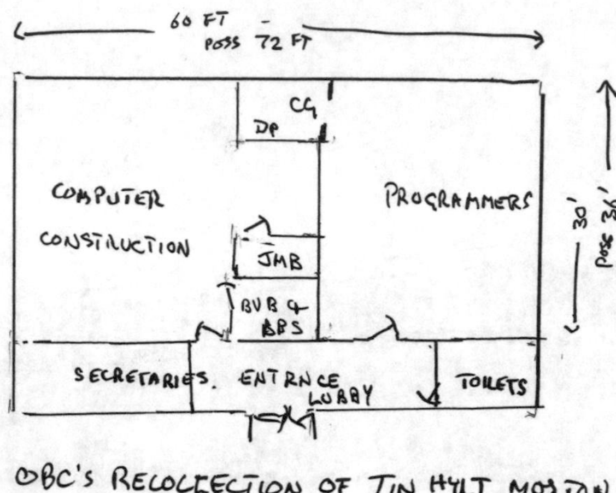

Fig. 4.5 Olaf Chedzoy's sketch of the approximate layout of the Tin Hut. The initials identify the following staff: CG = Cyril Gladwell; DP = Dietrich Prinz; JMB = John Bennett; BVB = Vivian Bowden; BBS = Bernard Swann. There were approximately a dozen programmers in the large open office on the right. The space marked *Computer Construction* was used for experimental circuitry and not for assembly of the full-sized computers. From Fig. 4.4 it seems probably that Olaf has under-estimated the dimensions

The debate about how to modify the Ferranti Mark I, and indeed whether a completely new computer design should be attempted, went back and forth for about a year. Below we set out the story chronologically, starting with Bennett's early suggestions and the feedback he attracted from Manchester University, then moving on to Christopher Strachey's radical input on behalf of NRDC. The debate had two outcomes: firstly a set of agreed modifications to the Mark I's instruction set; secondly some novel ideas from which new computer designs were later to emerge. One of the new designs originated in Ferranti's London premises, thereby heralding the start of a north/south culture divide within the company. This was to have far-reaching consequences—as we demonstrate in Sect. 4.6.

4.2 John Bennett Calls for Change

John Bennett first circulated his ideas to the University in a discussion document received by Manchester on 21st March 1951. This was surprisingly early: clearly before much real applications programming had been undertaken. Bennett's paper resulted in a meeting held at the University on 13th April to discuss *'proposals for* *the Ferranti Mk II computer'.*[5] At this stage, the *Mk II computer* in the meeting's title just indicated some hypothetical advance on the Mark I. In the light of hindsight, the 21st March document described more than minor enhancements. It also contained some ambitious ideas that did not in fact see the light of day until Ferranti had launched other computer projects towards the end of the 1950s. But in the short term the document did influence the modifications that became the Ferranti Mark I* computer—modifications that were commenced in Moston once the FERUT machine had been despatched to the University of Toronto in Canada in the spring of 1952. FERUT, the *Ferranti* Mark I computer for the *University of Toronto*, is described in Chap. 3.

In his 1951 discussion document, Bennett set out his stall as follows: "This preliminary report is meant to be an *Aunt Sally*, in that it proposes a complete machine, the logical design and instruction code of which will be found to include many controversial features. In designing the code, I have endeavoured to make reasonable complication of hardware subservient to greater versatility of instructions, on the assumption that most users, we hope, will be much less specialised than the machine maintenance engineers; moreover, many instructions have been chosen specially to facilitate modes of operation which, experience has shown me, are expected from a general-purpose computer concerned with engineering and physical problems e.g. floating-point operations, etc."

One may imagine the curiosity, not to mention surprise, in the University's Computing Machine Laboratory when John Bennett's document arrived before their Ferranti Mark I had had a chance to prove itself.

4.3 Reactions from Chorlton-on-Medlock and Cheltenham

Whilst Professor Williams' team at Manchester University were certainly consulted about proposed modifications to the Ferranti Mark I computer, surviving documents indicate that the evolution from Mark I to Mark I* was entirely controlled by Moston. There were two main reasons for this. Firstly, in 1951 the University's hardware researchers had started work on the design of a new computer called Meg, which was to be about ten times faster, to have floating-point hardware and consume half the power of the Mark I. Meg, short for *megacycle machine*, had a clock-rate of 1 MHz. (The Ferranti Mark I's clock rate was 100 kHz). Meg first ran

[5]*Meeting to be held at Manchester University, Friday 13th April 1951 to discuss proposals for the Ferranti Mk II computer.* This document consists of a one-page notice of the meeting followed by John Bennett's proposal consisting of 17 typed foolscap pages which appear to have no formal heading or date. From the pencilled marginal notes, this is Tom Kilburn's copy. Evidently there was an accompanying letter that has not survived. The document is preserved as NAHC/MUC/2/C/8.

a program in May 1954 and the design was later taken by Ferranti to become the Ferranti Mercury.

Secondly, Alan Turing had only just issued his *Programmer's Handbook,* which quickly became the computer users' Bible. Turing's programming scheme thereupon dominated the University's software activities. He was not the sort of person to change direction at the suggestion of a commercially-oriented organisation. Audrey Bates, who had joined Ferranti at Moston, wrote: "Dr. Turing had been considering the proposed changes mainly from the point of view of their use in a computer sold by Ferranti Ltd. to other organisations for scientific use, and not for a second university machine—in which case the problem of invested man-hours on library subroutines with the existing machine becomes important".[6] Other academic programming staff, and in particular Tony Brooker, took their cue from Turing.

Both Tom Kilburn and Alan Turing commented upon the Bennett paper. We have Kilburn's pencilled notes on the document[7] and some transcribed comments from Turing.[8] For example, adjacent to Bennett's Preface Kilburn had scribbled: "Question the type of machine indicated; particular watch on size of machine should be kept". And when Bennett writes: "At the moment, facilities only exist for 40 digit arithmetic and logical operations. However, in many problems involving the reduction of data and low order of accuracy tabulated information, arithmetic operation in the 20 digit regime is desirable". Kilburn simply scribbled: "Agree to general principle".

Much of Bennett's document is actually taken up with a critique of the existing instruction set, in some cases with a bit-by-bit analysis of possible alternative forms. Both Kilburn and Turing indicate that some of Bennett's suggestions seem to be minor tweaks but others are indeed worthwhile. Turing particularly approved of the suggestions for rationalising the 'magnetic' instructions and the shift instructions

---

[6]A collection of Alan Turing's correspondence between April 1949 and June 1954 is held at the University of Manchester Library, Reference; GB 133 TUR/Add. The listing can be seen at: https://archiveshub.jisc.ac.uk/search/archives/4f6c3f0c-9a70-33c5-bd03-df331fb06146?terms=%22BBC%22; Of particular relevance to Chap. 4 is this item: Tur/Add/19, April/May 1951: Exchange between Audrey Bates (Ferranti Ltd.) and Turing concerning his advice on technical aspects of the Ferranti Mark I. Includes copies of a meeting between Turing and Ferranti staff on 20th April 1951 (2-page report) and Turing's additional comments of 30th April 1951 (2 pages).

[7]*Meeting to be held at Manchester University, Friday 13th April 1951 to discuss proposals for the Ferranti Mk II computer.* This document consists of a one-page notice of the meeting followed by John Bennett's proposal consisting of 17 typed foolscap pages which appear to have no formal heading or date. From the pencilled marginal notes, this is Tom Kilburn's copy. Evidently there was an accompanying letter that has not survived. The document is preserved as NAHC/MUC/2/C/8.

[8]A collection of Alan Turing's correspondence between April 1949 and June 1954 is held at the University of Manchester Library, Reference; GB 133 TUR/Add. The listing can be seen at: https://archiveshub.jisc.ac.uk/search/archives/4f6c3f0c-9a70-33c5-bd03-df331fb06146?terms=%22BBC%22; Of particular relevance to Chap. 4 is this item: Tur/Add/19, April/May 1951: Exchange between Audrey Bates (Ferranti Ltd.) and Turing concerning his advice on technical aspects of the Ferranti Mark I. Includes copies of a meeting between Turing and Ferranti staff on 20th April 1951 (2-page report) and Turing's additional comments of 30th April 1951 (2 pages).

and for emphasising fractional rather than integer arithmetic. Both Kilburn and Turing approved of Bennett's scheme for stepping on the store addresses between CRT tubes—(see Chap. 15 for further details).

The Moston team also listened to the views of potential customers, once these had started to come forward. In the spring of 1952 the Ministry of Supply was still holding the developmental purse-strings. And we now know that MOS was poised to order up to three Ferranti computers for the use of its own establishments. Tony Brooker remembers that the changes from the Mark I to the Mark I* "were probably demanded by GCHQ Cheltenham, before they would take delivery".[9] Whilst the surviving Ferranti and NRDC documents are silent about the specific requirements of GCHQ as a customer, recently-released GCHQ documents indicate that detailed discussions and meetings between Moston and GCHQ took place between early November 1951 and early March 1952—see Chap. 10.

On Ferranti's side, the company started by presenting their thoughts on the Mark I*'s instruction set. This triggered GCHQ into making a series of requests for changes in the computer's facilities, each item being countered either by agreement or by a statement that the requested item was too complicated to be included. The list of GCHQ requests and Ferranti responses is detailed in Chap. 10. The final outcome for this particular customer was a Ferranti Mark I* with some special extra facilities that did not make their way into other production Ferranti Mark I* installations. So in this sense, Cheltenham's ideas had no noticeable effect on other users. This was, indeed, in line with GCHQ's stated aim of keeping their acquisition of a digital computer a closely-guarded secret.

4.4 Christopher Strachey's Rocket

What we do know is that by March 1952 Moston issued an internal document entitled *Ferranti High Speed Digital Computer No. 3* which proposed a revised form of the original Mark I's instruction set.[10] It can be inferred that the author was John Bennett. This document was sent to NRDC, who naturally sent it to Christopher Strachey for comment. We saw in Chap. 2 that Strachey had proved himself an exceptionally skilful programmer, as a result of which he was to take up a post with NRDC in June 1952 as soon as he was able to leave his current employment as a mathematics master at Harrow School. In May 1952 Strachey reported back to NRDC on John Bennett's proposed instruction set as follows.[11]

---

[9]R. A. Brooker, telephone conversation with the author, 1st October 2015.

[10]*Ferranti High Speed Digital Computer No. 3.* Document DC6, Ferranti Moston, March 1952. 6 typed pages. No author is given but one may infer that it was written by J. M. Bennett. See NAHC/NRDC/C5/7: Computers, Manchester University/Ferranti Development, vol. I. Reports, etc., 1951–52.

[11]Letter dated 15th May 1952 from Christopher Strachey (whilst still a master at Harrow School) to H. J. Crawley of NRDC. NAHC/NRDC/C5/7.

"Thanks for the draft specification. This is the first time I have seen the details of the list of instructions for the MOS machine, and I had no idea it was so different from the Manchester one.

"My first impression is that this has been designed by someone who has been trained on the EDSAC and has not fully appreciated the facilities of the Manchester machine. As a result, I think the proposed list of functions is definitely a retrograde step. It includes a few of the necessary functions, but it leaves out so many others that I think the programming of anything other than purely mathematical operations would be considerably harder".

Strachey then gives some detailed criticisms, including: the lack of functions dealing with the least significant half of the accumulator; number representation and the sign digit; shift instructions; B instructions to be available both with, and without, B-modification; the lack of relative transfers; accumulator tests; output instructions; coding of the absolute stop instruction; lack of clarity with magnetic instructions 13–16. He ends by saying:

"The design of the MOS machine is clearly tending towards a strictly mathematical machine only. This is quite contrary to the way I have been thinking about the development of programming, so I don't much like it. It seems to me that we must be prepared to investigate the commercial and other non-mathematical uses of these machines if we are to make full use of them or to tap the whole available market – to say nothing of the possibility of amplifying the programming of mathematical problems. In order to do this, we should have more, not less, logical functions available, at any rate in some machines".

Strachey clearly felt passionately about the subject. His letter concludes: "Would it be a good idea if I came down to see you on Tuesday afternoon? I'm afraid I can't manage any sooner". We can deduce that this meeting did take place from Strachey's next letter to Crawley dated 21st May 1952.[12] In this next letter, Strachey says: "After our talk yesterday, I made a list of the functions I should like to see in the new Ferranti machine …. The functions are not all of equal importance by any means, and I should be quite prepared to see some of them go. I think this list would form a better basis of discussion than the MOS list—it certainly includes all the functions there".

Strachey's *list of functions* attached to his 21st May letter is interesting and ambitious. It includes:

(a) Instructions where the operand is a literal (i.e. constant) rather than an address, so as "to facilitate the planting of instructions".
(b) Indexing instructions, where a match is sought between a comparand held in the accumulator and the contents of successive drum tracks, thus giving a form of on-the-fly searching.
(c) Provision for *two* program counters: "There are to be two separate controls. A digit in the instruction indicates which control line is concerned. This provision is to allow subroutines to be entered and left easily".

[12]Letter dated 21st May 1952 from Christopher Strachey to H. J. Crawley (NRDC). NAHC/NRDC/C5/7.

(d) Provision for half of the random-access memory to be implemented in a non-volatile storage technology rather than by the use of the volatile Williams Tubes. This non-volatile section would "contain permanent or semi-permanent information such as constants, routine changing sequence and very frequently used sub-routines".

Some of Strachey's suggestions, notably points (b) and (d) above, carried significant hardware development implications—hence causing potential production delays and cost increases. We note in passing that features (b) and (c) did not appear in the Ferranti line of computers, or those of its successor company ICL, until much later. Feature (b), on-the-fly searching of a spinning store, was sometimes part of special-purpose equipment designed for organisations such as GCHQ. The feature found a more general manifestation in the ICL Content-Addressable File Store[13] which was available in the late 1970s. Feature (c) in the form of *three* program counters was included in the Ferranti Atlas computer, available from 1962.[14]

Brian Pollard, the Ferranti engineer leading the developments at Moston, wrote back to John Crawley (NRDC) on 29th May 1952 saying only that "We have read Strachey's instruction list with interest and Bennett is sending off a separate letter to you about this, outlining the principles upon which our instruction list is based. I do agree however that a meeting should be called, when Strachey is available, to discuss the list in detail".

Crawley and Pollard met in London on 6th June[15] but their discussion covered the more general matters such as total size of memory, specification of the power supplies and form of the acceptance tests. The final specification of the Mark I*'s instruction set, together with various other details such as the form of the Acceptance Tests, appeared in a document dated 18th July 1952[16]—by which time the construction of the first Mark I* had already begun.[17] It was to take another 12 months or so before this machine was delivered to GCHQ Cheltenham.

Meanwhile, Christopher Strachey was about to take his revolutionary ideas elsewhere. He was to play a significant role in the design of a completely different computer that became the Ferranti Pegasus. This was another star in the firmament of novel computer architectures, as described later in Sect. 4.6.

---

[13]Maller, V.A.J. 1980. Information Retrieval Using the Content Addressable File Store. In *Information Processing 80, The Proceedings of IFIP Congress*, Tokyo/Melbourne, 187–192. Published by North-Holland.

[14]Kilburn, T., D.B.G. Edwards, M.J. Lanigan, and F.H. Sumner. 1962. One Level Storage System. *IRE Transactions on Electronic Computers* EC-11 (2): 223–235.

[15]Internal NRDC file note by H. J. Crawley, dated 6th June 1952. See: NAHC/NRD/C7/5.

[16]*Schedule 1: Specification for computing machines*. Final draft, 18th July 1952. Nine typed foolscap pages, no author indicated. See NAHC/NRD/C7/5.

[17]*Note for File from the Managing Director (NRDC): Electronic Computers*. 1st May 1952. This is a two-page typed foolscap document, summarising a long discussion that had taken place between Halsbury Hennessey, Crawley and Bowden on 29th April 1952. See NAHC/NRD/C9/1a.

4.5 Moston's Star Twinkles

Brian Pollard and the Moston engineers had not been idle whilst the programmers were debating instruction sets. There were some more pressing Mark I hardware loose ends to tidy up. The magnetic drum at the University's Computing Machine Laboratory had been giving lots of trouble since its installation in February 1951 so the drum design was re-visited and a new version produced (see below). The whole subject of power supply provision was revised, as described below and in Chap. 16. The new power supply systems were employed for all Ferranti Mark I* installations from 1953 onwards. Finally, the machine's cooling system was re-designed. Looking ahead, the Moston engineers were beginning to address the challenges of the commercial data-processing market, with their need for punched card bulk input/output and archival storage based on magnetic tape decks. There was also a need for faster printing of results. The first Bull lineprinter, along with its Williams/Kilburn tube buffer store, was attached to the Manchester University machine during the few weeks following 11th August 1953.[18] Several of the Mark I* installations also had Bull lineprinters.

All this engineering activity, intimately bound up with the planning for a Mark I*, was putting pressure on space at the Moston factory. Ferranti therefore took steps to re-activate a lease which had been negotiated for a redundant nearby cotton mill on the outskirts of Oldham. Operations at Gem Mill got under way in 1954. The Ferranti engineer Allan Ellson remembers visiting Gem Mill before the move. "It was a grim, empty building with the line shafting still in place overhead. Before we moved in they had cleared it out, cleaned it and painted the iron columns blue".[19] Initially Gem Mill was just used for the testing and commissioning of new computers. Full production of Mark I* machines was moved from Moston to Gem Mill in about 1955. Research and Development remained at Moston.

So, what of the revised instruction set? The next few paragraphs inevitably get rather technical. Readers with no stomach for detail can skip to Sect. 4.6 without losing the historical thread.

A full comparison of the instruction sets for the Mark I and Mark I* is given in Chap. 15. From the end-user's perspective, the main differences are:

- Numbers in the Mark I* are written down and displayed in conventional binary, rather than the 'backwards binary' of the Mark I;
- The Mark I distinguished between unsigned and signed arithmetic operations. In the Mark I*, only signed arithmetic is performed. Both machines use the two's complement representation for negative numbers. In the Mark I, the binary point for signed quantities is one place from the end; in the Mark I* it is two places from the left -hand (more-significant) end, so that data is held in fractional form.

[18]Some surviving log books from the Manchester University's Ferranti Mark I computer, 1951–58, are preserved in: NAHC/MUC/2/C6.

[19]Allan Ellson, e-mail of 21st September 2015 to Simon Lavington.

Fig. 4.6 John Bennett standing at the console of the Ferranti Mark I Computer at Manchester University in about 1951. This was one of several publicity photos taken shortly before the computer's Inaugural Conference in July 1951

- The number of B instructions has been rationalised, so as to eliminate duplicates, etc.
- When addressing operands in an ascending sequence, the Mark I* increments addresses from page to page throughout the memory, whereas the Mark I treats ascending addresses as modulo the page of the first-occurring address in the sequence.

Besides the reversion to standard binary representation, the most obvious change from the programmer's view was the reduction of the total number of meaningful instructions from 39 to 32. This meant that each instruction could now be represented by a single 5-bit character. From this simplification sprang the so-called *Radix 32* Input and Organisation Scheme, on which all Ferranti Mark I* software was founded. At a stroke, there was to be incompatibility and divergence between software written for the Ferranti machines at Manchester and Toronto Universities and software written for the seven Mark I* computers that were eventually delivered to other sites.

On the hardware side, the main changes were improvements to the drum technology, buffering provision for a relatively fast Bull lineprinter, a re-designed operator's console and a change to the heater supply for all the thermionic valves.

Fig. 4.7 An engineer commissioning a Ferranti Mark I* computer in about 1954. This was about the time when commissioning activity was being transferred from the Moston factory to Gem Mill, a couple of miles away. On the open logic door can be seen the three larger dark heater transformers which may be compared with the Mark I's smaller transformers shown, for example, in Fig. 3.8

These last two modifications give visual clues to the interpretation of contemporary photographs. Firstly, there were significant differences in console layout (compare Figs. 4.6 and 4.7). Secondly, whereas the Mark I's logic doors featured compact heater transformers operating at 1400 Hz supplied by a separate motor-generator set, the Mark I* used larger transformers operating at the standard 50 Hz (see Fig. 4.7). In all other respects, the outward appearance of the two types of machine were virtually indistinguishable.

A word should be said about Ferranti's drum technology because drum faults were an annoying feature of the first two Mark I's and also the first Mark I* computer. In April 1952 Vivian Bowden gave an account of "the recent troubles of the Manchester machine on the magnetic side".[20] Bowden was of the opinion that Ferranti "know the reason for this, which is that they are trying to work on the limit of resolution of the drum. An increase in the diameter of the drum from 6" to 10"

[20] *Note for File from the Managing Director (NRDC): Electronic Computers*. 1st May 1952. This is a two-page typed foolscap document, summarising a long discussion that had taken place between Halsbury Hennessey, Crawley and Bowden on 29th April 1952. See NAHC/NRD/C9/1a.

will, they feel sure, dispose of these difficulties once and for all". Bowden was being too optimistic, it later transpired, and Moston's drum improvements for the Mark I* continued until about 1954.

Moston engineers also made improvements from time to time to the logic circuits. Here's an example, again taken from Bowden's April 1952 comments on the reliability of the Manchester Mark I. Bowden says: "The difficulties with the cathode ray tube storage have been traced to the marginal failure of a valve in the common time base circuit. A number of instances of these failures have occurred due apparently to running EF50 pentodes at 12 mA, which appears to be too close to their rating of 15 mA. By replacing approximately 100 valves in the machine with EF55s they [Ferranti] hope that these troubles will also disappear. The Toronto machine [FERUT] will probably not suffer from a number of teething troubles, as those in Manchester, because EF55s will be installed, there will be a 10" diameter drum and one or two other modifications". Again, Bowden was later proved too optimistic but it should perhaps be said that all early valve computers would, by today's standards, have been considered unreliable.

4.6 London's Star: A Different Galaxy

The Mark I* was not, however, the only computer product being considered by Ferranti in the early 1950s. Quite separately from Moston and Manchester University, a new line of development was starting 200 miles south in London. This line had had a somewhat confusing start—not in London but in nearby Hertfordshire. In August 1950 NRDC had made contact with the Research Laboratory of Elliott Brothers (London) Ltd. at Borehamwood in Hertfordshire. NRDC was attracted by rumours of advanced modular (i.e. easily-replaceable, packaged) digital circuits that had been developed by the Borehamwood Laboratory for a secret Admiralty gunnery-control computer.[21]

NRDC went on to negotiate a series of contracts with Borehamwood, aimed at producing a computer that would address the perceived need for a smaller, cheaper machine that would balance the *supercomputer* Ferranti Mark I at the costly end of the market. The final contractual specification of a small-scale machine, which was to become the Elliott 401, was agreed on 12th December 1952. By then, the Computing Division at Borehamwood had satisfactorily tested a range of new

[21]Lavington, Simon. 2011. *Moving Targets: Elliott-Automation and the Dawn of the Computer Age in Britain, 1947–67*. Springer. ISBN 978-1-84882-932-9.

Fig. 4.8 Gem Mill, at Chadderton, on the outskirts of Oldham, as it appeared in the 1980s. This cotton mill was built in 1901. In 1915 it was recorded as having textile machinery with a total of 115,000 spindles, powered by a 1700 hp stationary steam engine. Ferranti's Computer Department took over Gem Mill in 1954. The Mill was demolished in 2008 to make way for housing

packaged circuits and a new nickel memory technology based on the magnetostrictive effect in nickel delay lines.[22]

It is intriguing, and relevant, to reveal that clause 2 of the 11-page 401 computer contract has a specific provision: "Elliotts shall arrange that so long as Mr. William Sydney Elliott is in the employment of Elliotts he shall during the continuance of this Agreement be employed upon the project and shall devote so much of his working time thereto as shall be necessary for the proper furtherance of the project". W. S. (Bill) Elliott, who was not related to the company's 18th Century founder, was at that time the Head of the Computing Division at Borehamwood. He seems to have been an ambitious, far-sighted, player with an eye to the main chance. He had caught wind of a down-turn in the Elliott company's finances and had feared (unnecessarily, as it turned out) for the future of the Computing Division. As if to strengthen Bill Elliott's fears, the Head of the Borehamwood Laboratories, John

[22]In this form of delay-line memory, electronic pulses were converted into shock waves (vibrations) that travelled along a piece of nickel wire and were detected and re-converted into electronic pulses at the receiving end. Two forms of shock waves could be employed: longitudinal or torsional, the latter being preferred in later versions of the memory system. The original research is described here: (a) Millership, R., R.C. Robbins, and A.E. De Barr. 1951. Magnetostriction Storage Systems for High-Speed Digital Computers. *British Journal of Applied Physics* 2 (10): 304.; (b) De Barr, A.E. 1953. Digital Storage Using Ferromagnetic Materials. *The Elliott Journal* 1 (4): 116–120. See also Borehamwood Internal Research Reports numbers 309 dated 30th December 1952 and 325 dated 22nd February 1953.

Coales, had formally left the company in April 1952[23] after disagreements with senior management based at the company's main factory south of the Thames at Lewisham.

The Elliott 401 computer first became operational at Borehamwood in March 1953. It was publicly demonstrated at the Physical Society Exhibition held in April 1953 in London (Fig. 4.9). Christopher Strachey, from NRDC, took the lead in writing the demonstration programs. Interestingly, the Elliott 401's debut took many by surprise. M. H. (Harry) Johnson, who had already gained practical programming experience on the Cambridge EDSAC, the NPL Pilot ACE and the Ferranti Mark, I recalled that 401's performance at the Exhibition "amazed everyone involved in computer activity, particularly as most of us had been completely unaware of any Elliott Brothers interest in computers".[24]

However, all was not quite as it seemed. By the spring of 1953 NRDC became aware that Bill Elliott had given notice of his intention to resign from Borehamwood and that he wished to take several members of the computer design team with him. Bill Elliott and others eventually joined Ferranti, as described below. Meanwhile, to cut a long and sometimes acrimonious story short, NRDC removed the 401 computer from Borehamwood. On 14th June 1953 Maurice Wilkes gave the computer a temporary home at the Mathematical Laboratory at Cambridge, where an NRDC consultant and former Borehamwood engineer, Harry Carpenter, continued work to refine the machine's facilities. In March 1954, the 401 was transferred to the government's Rothamsted Agricultural Experimental Station in Hertfordshire, eventually being retired in July 1965 after a long and useful life. The machine lay in pieces in a Science Museum store in London for some years. Restoration by volunteers from the Computer Conservation Society began in a small way in 1993, then languished for a few years, then resumed only to be terminated due to a change in Science Museum policy. The 401 finally became a static display in a new Mathematics Gallery which was opened in 2016.

Back at Borehamwood in 1953, Andrew St. Johnston, a leading figure in the 401's original design, became Head of the Computing Division where the successful Elliott 402 and 405 series of computers, and several others, were produced —(see also Chap. 5).

So what of Bill Elliott? After making overtures to a few companies, he joined Ferranti Ltd. with effect from 1st November 1953. Within a year, the key former Borehamwood employees Charles Owen, Hugh Devonald and George Felton had joined him. In June 1954, NRDC gave Bill Elliott (by then based at Ferranti Ltd., 18 Manchester Street, London) a Report and complete set of engineering drawings

---

[23]Lavington, Simon, *Moving Targets: Elliott-Automation and the Dawn of the Computer Age in Britain, 1947–67.*

[24]M. H. (Harry) Johnson, My work with computers: from the Ferranti Mark I to the ICT 1900 (1952–1966). 84-page typed manuscript, circulated privately, June 2002. By 1953 Harry Johnson was working for the Operational Research Group of the Ministry of Civil Aviation. He had already gained practical programming experience on the Cambridge EDSAC, the NPL Pilot ACE and the Ferranti Mark I.

Fig. 4.9 The Elliott/NRDC 401 computer in 1953. This machine was designed to be reliable, easily manufactured in quantity and to cost about a third of the price of a Ferranti Mark I. These days it is bemusing to recall that this 'small' computer weighed over a ton, had only 4 kbytes of memory and a clock-rate of only 0.33 MHz

of the 401 computer. Suitably equipped, he was determined to build a re-designed 401 in London. The project was initially known as the Ferranti Packaged Computer, FPC1, to emphasise its construction from a limited set of carefully-designed packages (circuit modules). The aim was to produce a cheap and reliable computer that was easy to manufacture and maintain. The FPC/1 was soon re-named the Ferranti Pegasus.

When the Pegasus project finally matured, its architecture was radically different from that of the Elliott 401. The Pegasus instruction set, primarily conceived by Christopher Strachey, was probably the first in the world to use a general set of registers.[25] For ease of programming, Pegasus proved to be an outstanding machine. George Felton led the software development team.

Charles Owen, a first-class circuit designer from Borehamwood, developed improved 401-type modules for Pegasus (the FPC/1), based on commodity double-triode valves. The computer used an improved version of the Borehamwood

[25]Bell, C.G., and A. Newell. 1971. *Computer Structures: Readings and Examples*. McGraw-Hill. This classic book carefully analyses the architectures of a wide range of computers designed before 1970.

nickel delay-line storage for central registers and for its primary store. The primary memory was implemented as many single-word delay-line packages so that, since the computer was serial, the effect was of random-access at the word level. A magnetic drum backing store was provided. Ian Merry, recruited from the BBC's sound engineering laboratories, led the drum design team.

The Pegasus circuits, like their 401 predecessors, worked at a conservative clock-rate of 333 kHz—a frequency originally determined in 1949 by the limits of reliable operation of Borehamwood's anticipation-pulse CRT storage system used in the Admiralty's fire-control computer (the Elliott 152).[26] This clock-rate persisted in all the Elliott 400 series computers and in the Ferranti Pegasus and Perseus computers to the end of the 1950s (Fig. 4.10).

With Bill Elliott in charge, the engineering team at Ferranti's London Laboratory grew from about 10 people to 30 people between mid-1953 and early 1956.[27] News of these developments obviously reached Brian Pollard, who headed Moston's engineering team. There was a long-running clash of personalities between the two men, who had been aware of their respective career-paths for some years. Misunderstandings about project costs arose, which alarmed NRDC who intended to underwrite the process of bringing Pegasus to market. To make matters worse, there was an initial dispute between NRDC and Sir Vincent de Ferranti about certain clauses in NRDC's contract. This sorry tale is told in Hendry[28].

Meanwhile the hardware design team for the FPC1/Pegasus continued to operate at Ferranti's London Laboratory relatively independently from Moston. Besides the animosity between Pollard and Elliott, there were two more contributing reasons for this independence. Firstly, there were those within Ferranti such as Bernard Swann, the Computer Marketing Manager, who felt that the company ought sooner or later to have more of a presence in the capital. Secondly, Bill Elliott had recruited Pegasus engineers who already lived in or near London.

Almost independently of the growth of Bill Elliott's Engineering Laboratory, in 1954 Ferranti had planned to set up a high-profile base in London from which to develop software applications and at which a Ferranti computer could be demonstrated to prospective customers. An elegant Georgian house, 21 Portland Place, just north of Oxford Street, was converted into Ferranti's London Computer Centre. The house had been built in 1777. One of its first notable occupants had been General Sir Henry Clinton, who in 1778 had commanded the British forces in America. 21 Portland Place was carefully modified—for example the floor of its first-floor ballroom was strengthened to accommodate a Ferranti Mark I* computer. In the event, the first Ferranti Pegasus was installed here and began to provide a

[26]Lavington, Simon, *Moving Targets: Elliott-Automation and the Dawn of the Computer Age in Britain, 1947–67*.

[27]Lavington, Simon. 2000. *The Pegasus Story; A History of a Vintage British Computer*. Published by the Science Museum, London. ISBN 1-900747-40-5.

[28]Hendry, John. 1989. *Innovating for Failure: Government Policy and the Early British Computer industry*. Published by the MIT Press.

Fig. 4.10 W. S. (Bill) Elliott, pictured in 1993 at a Computer Conservation Society meeting at the London Science Museum. The photo shows an original magnetic drum store from the Elliott/NRDC 401 computer. A team led by Chris Burton, shown in the background, was restoring the 401 in 1993

service to external users from April 1956. The first of many Pegasus programming courses had been held in Portland Place the previous autumn.

As for Bill Elliott, he had had enough of Ferranti. He joined IBM in 1956 to head up IBM's new UK Laboratories at Hursley in Hampshire. He persuaded four key engineers of the Pegasus development team (Charles Owen, Ian Merry, Harry Metcalf, and later John Fairclough) to follow him. This essentially marked the end of Ferranti's London-based hardware design activities, though in 1956 Ferranti opened a facility in Bracknell, Berkshire, which went on to produce several interesting computer designs. After a spell at Cambridge University from 1961–66, Bill Elliott ended his working life as Professor of Computing at Imperial college, London.

As for Brian Pollard, he resigned from Ferranti at the end of 1958 under somewhat of a cloud. By that time, the company had moved the Computer Department to large new premises at West Gorton, about three miles east of Manchester city centre. Ferranti's new premises and its new products are discussed in Chap. 13.

Back in 1956 Pegasus recovered from the internal company politics and went on to be an outstanding success for Ferranti. In all, 40 Pegasus 1 and Pegasus 2 computers were sold between 1956 and 1962. The cost of each machine varied, over time and according to configuration, between £32,000 and £42,000. This was between one third and one half the cost of a Ferranti Mark I*. To put these figures in a social context, in 1962 a four-bedroom Victorian terrace house in Manchester could be bought for £1200 and a four-door family saloon car cost about £700. Specifically, the motoring extremes in 1962 were represented by an Austin Mini at £496 and a Rolls Royce Silver Cloud II at £6272.

Cost is, of course, just one factor. Facilities, reliability and performance are additional, probably more important, issues. In the next chapter we will take a look at the products of other computer manufacturers, both in the UK and in the USA, and compare their offerings. Whilst Ferranti had had the UK market to itself up to 1954, the competition was starting to build up. In May 1954 the Ferranti marketing Manager, Bernard Swann, wrote that: "Although the emergence of designs for smaller computers is having a serious effect on the demand for large computers there are outstanding several serious enquiries for our Mark I*".[29] In the event, only seven Ferranti Mark I* computers were delivered between October 1953 and October 1957 because, by the end of the 1950s, its technology was beginning to look rather out-dated and its cost-effectiveness had been superseded by the performance of more recent products from Ferranti and from other vendors.

References[30]

Bell, C.G., and A. Newell. 1971. *Computer Structures: Readings and Examples*. McGraw-Hill.

De Barr, A.E. 1953. Digital Storage Using Ferromagnetic Materials. *The Elliott Journal* 1 (4): 116–120.

Hendry, John. 1989. *Innovating for Failure: Government Policy and the Early British Computer industry*. Published by the MIT Press.

Kilburn, T., D.B.G. Edwards, M.J. Lanigan, and F.H. Sumner. 1962. One Level Storage System. *IRE Transactions on Electronic Computers* EC-11 (2): 223–235.

Lavington, Simon. 2000. *The Pegasus Story; A History of a Vintage British Computer*. London: Published by the Science Museum. ISBN 1-900747-40-5.

Lavington, Simon. 2011. *Moving Targets: Elliott-Automation and the Dawn of the Computer Age in Britain, 1947–67*. Berlin: Springer. ISBN 978-1-84882-932-9.

Maller, V.A.J. 1980. Information Retrieval Using the Content Addressable File Store. In *Information Processing 80, The Proceedings of IFIP Congress*, Tokyo/Melbourne, 187–192. Published by North-Holland.

Millership, R., R.C. Robbins, and A.E. De Barr. 1951. Magnetostriction Storage Systems for High-Speed Digital Computers. *British Journal of Applied Physics* 2 (10): 304.

Wilkes, M.V., D.J. Wheeler, and S. Gill. 1951. *The Preparation of Programs for an Electronic Digital Computer*. Published by Addison-Wesley.

---

[29]Letter dated 25th May 1954 from Bernard Swann (Ferranti) to Hennessey (NRDC). See NAHC/NRD/C9/1.

[30]*Note*: NAHC refers to the National Archive for the History of Computing, which is held in the University of Manchester's Library.

Chapter 5
Into the Market

Selling computers was not easy in the early 1950s. Ferranti had help and encouragement from the government in various guises but in truth the UK market was slow to react. And in these early days, many potential customers and applications were ruled out because of the uncertain cost-benefits. Furthermore, the computers of the early 1950s were often unreliable, with limited memory capacity and limited rates of data input/output.

B. V. (Vivian) Bowden, Ferranti's first computer salesman, recalled in 1953 that "I have interviewed a large number of people whose interests are in the fields of science, engineering and commerce, from the chief designers of aircraft firms and the presidents of insurance companies to the organisation and methods branch of the Bank of England and the Treasury. I have discussed their problems with scores of accountants, actuaries, engineers and scientists of all kinds …" He found that, though the scientific and engineering sectors were the more receptive to begin with, "there were three overwhelming objections to the use of a computer. The machine usually broke down before it finished the calculation, its memory was too small to hold all the data it needed, and, worst of all, it took so long for the mathematicians to get the program right that it was usually quicker (and cheaper) to get the calculation done by clerks …. I came to the conclusion that there were a few calculations that would be worth doing on a computer. They were neither too big nor too small … We were more worried by the shortage of mathematicians and programmer than by anything else ….". [1]

In this chapter we recount Ferranti's efforts to explore applications areas and follow up sales prospects, though this means that we jump around chronologically when describing each avenue. The final result, seven Ferranti Mark I* computers

[1]These quotations come from two sources: (a) Four-page letter written by B. V. Bowden on 8th May 1953, when applying for the post of Principal of Manchester College of Science and Technology (later UMIST).; (b) Bowden, B.V. 1970. The Language of Computers. *American Scientist* 58 (1): 43–53. This paper was originally delivered at the Brighton College of Technology as the first Richard Goodman lecture, 2nd May 1969.

S. Lavington, *Early Computing in Britain*, History of Computing,
https://doi.org/10.1007/978-3-030-15103-4_5

sold, was somewhat disappointing. But lessons were learned. By the end of the 1950s the company emerged as a stronger computer manufacturer with a wider range of products.

But first we need to set the stage and introduce the players that surrounded Ferranti in its first performance as a vendor of modern computers.

5.1 The Players and the Stage

Ferranti was not alone in developing its Mark I* computer manufacturing facilities and seeking out customers. Of course, the Ministry of Supply (MOS) had provided the initial capital. The MOS was motivated by a perceived strategic national need for high-speed automatic computers, focussed on classified applications related to the UK's defence. This eventually led to orders for three Mark I* computers, but not without a period of official indecision and delay. The need to disguise the identity of one of MOS's secret sites (see Chap. 10) does not make the surviving documentation easy to follow.

Hard on the MOS's heels came the National Research Development Corporation. The NRDC administered the Manchester University computer patents, upon which Ferranti's product depended. NRDC had a statutory responsibility to promote patent exploitation but also needed to assure the Board of Trade that unfair competitive advantage was not being given to favoured companies. Upon its establishment in 1949, NRDC was backed by a £5 million interest-bearing government loan, to be repaid within five years. So if seed-corn help was to be given to Ferranti then NRDC expected a return on its investment. NRDC first approached Ferranti in February 1951. The conditions of any loan, and the precise percentage royalty required by NRDC, was at times the subject of sharp debate with Ferranti's boss, Sir Vincent de Ferranti, who wished to maintain his family business's freedom of action. The support contract was finally agreed in December 1951.[2] We shall see much more of NRDC's involvement in promoting the Mark I* in Sect. 5.3.

Another potential player on Ferranti's side was the Brunt Committee of the Department of Scientific and Industrial Research (DSIR). The DSIR had been formed in 1916 "to finance worthy research proposals, to award research fellowships and studentships [in universities], and to encourage the development of research associations in private industry and research facilities in university science departments". Establishments for which the DSIR assumed responsibility included the National Physical Laboratory, where the Pilot ACE computer was developed. In 1949 Sir Ben Lockspeiser, who was an enthusiastic supporter of Manchester's computer developments, set up a special DSIR Committee "to keep under review

[2]Crawley, H.J. 1957. *The National Research Development Corporation Computer Project*. NRDC Computer Sub-Committee paper 132, February 1957.

the progress in the construction and use of high-speed calculating machines, to examine the fields in which they are likely to prove most useful, and to recommend the most promising types for further development". The Committee was chaired by Sir David Brunt, a distinguished meteorologist. The Brunt Committee included representatives from the DSIR, MOS, Admiralty, and three people nominated by the Royal Society (D. R. Hartree, M. V. Wilkes and F. C. Williams). Lord Halsbury, Chairman of NRDC, attended by invitation. In the event, the Brunt Committee took little part in the Ferranti-related developments, though in November 1951 it did recommend that NRDC should give financial assistance to Ferranti.[3] More generally, the Brunt Committee provided a signal that the Great and the Good of the UK's scientific establishment were available in the wings, watching Ferranti's performance with interest.

It was NRDC who attempted to open the performance. In December 1949 the NRDC invited all the UK's major electronics and punched-card machine companies to a meeting designed to foster a co-ordinated national effort towards the manufacture and sale of electronic computers.[4] Much to NRDC's disappointment, there was to be no co-ordinated national effort.

Of the ten companies approached by NRDC in 1949, Ferranti Ltd. was the first to take the commercial production of computers seriously, though Elliott Brothers, English Electric and the British Tabulating Machine Co. Ltd. (BTM) were to make a start about three years later. Another, and surprising, entrant was J. Lyons & Co. Ltd., a catering company that developed a pioneering office computer called LEO without reference to NRDC. LEO was an in-house project whose hardware did not become available to other customers until 1957, then being marketed as LEO 2. The original LEO, which was based on the Cambridge EDSAC computer with the addition of bulk input/output equipment suitable for commercial data-processing, commenced to run Lyons' regular bakeries stock-control applications from September 1951. Other Lyons applications soon followed and the LEO project was declared complete in December 1953.[5]

LEO is but one example of UK commercial computer products inspired by academic research. Figure 5.1 summarises the main UK laboratory projects being pursued during the immediate post-war years and their commercial off-springs. Within this spectrum of computer research, how fortunate for Ferranti that government agencies had arranged a well-funded connection between the company and Manchester University!

---

[3]Crawley, H.J., *The National Research Development Corporation Computer Project.*

[4]Crawley, H.J., *The National Research Development Corporation Computer Project.*

[5]*The LEO Chronicle.* T. Raymond Thompson. 23-page unpublished typescript, covering the period 25th February 1947–5th November 1954. The original is held as document NAHC/MSC/D6. Thompson was a Lyons senior manager who oversaw the LEO project. Substantial extracts from his document are quoted in the book *Early British Computers* by Simon Lavington. Published in 1980, this is out of print but has helpfully been made available at: http://ed-thelen.org/comp-hist/EarlyBritish.html.

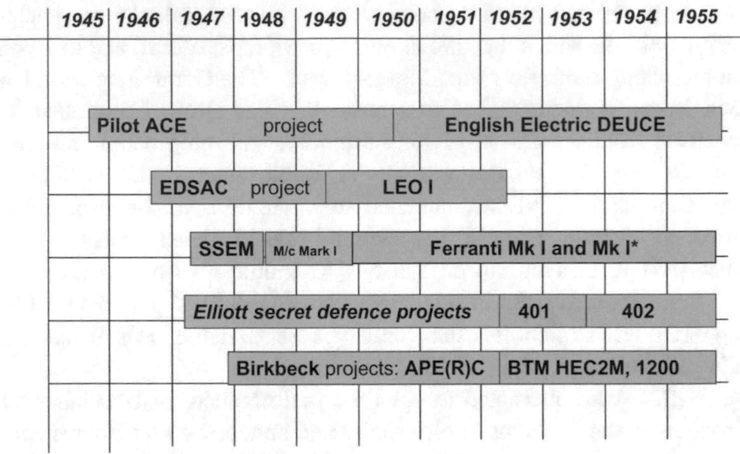

Fig. 5.1 The research origins of the first British production computers. The blue areas indicate laboratory activity and the green areas represent industrial design and manufacturing activity

The sales records of the English Electric DEUCE, the Elliott 400 series and the BTM 1200 series computers are revealed in Sect. 5.4, when describing the competition shaping up to challenge Ferranti's offerings. In the period 1951–54, however, Ferranti had a clear run with no immediate challenge from home or abroad. Furthermore, the initial impression must have been that the MOS would rapidly order the first two or three Mark I* computers and that, with this example before them, other customers would step up. Things did not turn out quite so conveniently.

5.2 Indecisions and the Three MOS Installations

The Mark I* was certainly an expensive item of equipment but in 1952 it represented the only available UK machine in what might have been called the *Supercomputer* league, if this word had been invented. The MOS judged that computers with comparatively large storage capacities and fast processing rates were needed for the calculations underpinning the UK's cryptanalysis and nuclear weapons activities.

The initial Ministry of Supply contract 6/WT/4890/CB14(d) with Ferranti Ltd., dated 19th February 1949, had only covered the construction, installation and testing of a single computer but the contract had specified the 'supply of prints of all finalised drawings of the equipments as supplied' with the intention that more examples might be built. The total amount of the contract was £113,783.[6] This is

[6]Notes on the initial MOS contract prepared by A. Ridding, the Ferranti Archivist, on 30th April 1970. The document is held in the Ferranti Archive, Science & Industry Museum, Manchester. Ridding comments that: 'All papers were transferred to ICT in 1963 when the Ferranti Computer business was sold to ICT'.

equivalent to several million pounds in 2018. A second MOS contract, 6/WT/18651/CB.15(b), was placed with Ferranti early in 1951 during the currency of the first contract; this covered servicing of the University's computer and various categories of development work[7] totalling £18,204. This contract ended on 12th February 1952. A third MOS 'servicing and development' contract ran from 1st April 1952 to 31st March 1955.

We are now in the period when the Mark I* design had been completed and production had begun. It was at this point that there was hesitation on the MOS's vision of purchasing three computers. In May 1952 Lord Halsbury, Director of the UK's National Research Development Corporation (NRDC), noted that "The [Mark I*] machine which they [Ferranti] are at present building is the second of two for the Ministry of Supply—the order for the second having been placed before the order for the first, which has still not been booked by the Ministry of Supply". Here Halsbury is referring to the suspension of Brigadier Hinds' plans in favour of a higher-priority delivery to GCHQ—though the Official Secrets Act constrained him from precise explanations. Halsbury continues: "As the Ministry of Supply are still dickering about placing an order for the second machine, it was agreed between us that Bowden [Ferranti] and Hennessey [NRDC] would go and talk to the necessary officials at the Ministry of Supply and give them the option of either placing an order now or coming 6th on the priority list, which means that they would not get delivery of the machine until perhaps October 1954".[8] At the time Lord Halsbury wrote this note, NRDC had agreed to underwrite the construction of four Mark I* computers, "which Ferranti propose to build in echelon with three on the stocks simultaneously".

After this somewhat confusing start, the three MOS computers shown in Table 5.1 were eventually delivered. The serial numbers given to these three by Ferranti (left-hand column) indicate the temporal confusion on the factory floor.

5.3 The Wider Market: NRDC's Encouragement

Wishing to promote the exploitation of the Manchester computer patents, in December 1951 NRDC proposed to sponsor the construction of four Mark I* computers subject to the agreement of the Board of Trade. Specifically, "the Corporation agreed to place a contract with Ferranti Ltd. to manufacture four machines at cost plus 7.5% profit and to act as the Corporation's agent for re-sale on a commission basis (normally 5% of the agreed selling price of £85,000 per

[7]NRDC Internal memo from Hennesey to Halsbury, dated 8th January 1954 and referring to a meeting with Brigadier Hinds. NAHC NRD C6/8.

[8]Note for File from the Managing Director (NRDC): Electronic Computers. 1st May 1952. NAHC/NRD C9 1a. This is a two-page typed foolscap document, summarising a long discussion that had taken place between Halsbury, Hennessey, Crawley and Bowden on 29th April 1952.

Table 5.1 Delivery of Ferranti Mark I* computers to Ministry of Supply sites

Ferranti's Serial no.	Government establishment	Date of order	Hardware delivered	Acceptance tests passed
DC3	GCHQ, Cheltenham	Late 1951	Sept. 1953	Approx. May 1954
DC5	Armament Research & Development Establishment, Fort Halstead, Kent	Autumn 1952	July 1954	Autumn 1954
DC7	Atomic Weapons Research Establishment, Aldermaston	Sept 1953	Early 1955	31st March 1955

More details of these three installations and their uses are given in Chaps. 10–12

computer)".[9] In effect NRDC was to act as sponsors and stockists for computers to be built by Ferranti.

At this time Lord Halsbury wrote to the Board of Trade indicated the possible destinations of these four machines, which he numbered NRDC1–4[10]:

NRDC1: "to be installed at Ferranti's on indefinite loan, to be used by them for user research on the understanding that the Brunt Committee shall have the right to nominate one half of the time of such machine for any work which they consider to be in the national interest";
NRDC2: "to be earmarked for the Ministry of Supply who are trying to make up their minds whether they can afford it or not". [This refers to the mid-1951 uncertainly over the Fort Halstead contract and the sudden appearance of GCHQ as a potentially higher-priority customer].
NRDC3 & 4: "to be earmarked for British Universities. Leeds and London might be suitable centres, as might Oxford. We shall take the advice of the Brunt Committee as to suitable sites for the installation of these machines. If the machines are sold to the universities then the cost will have to be met by DSIR ultimately".

The above wording might suggest that in December 1951 NRDC expected that Ferranti's marketing endeavours would be directed at the scientific and academic areas to the exclusion of business and commerce. This was not so. Nevertheless Lord Halsbury had to tread carefully in his letter. The Board of Trade had a duty to ensure fair market competition and NRDC did not wish to be accused of unfairly supporting Ferranti. This may account for Halsbury's mention of 'British Universities'. Actually, no British University (apart of course from Manchester) is known to have expressed any interest in acquiring a Ferranti Mark I in the early 1950s. Also, Ferranti did not borrow or acquire NRCD's first Mark I* for their own internal use, though the company hesitated for some time. In the end, a Ferranti

---

[9]Crawley, H.J., *The National Research Development Corporation Computer Project.*

[10]Letter dated 27th December 1951 from Lord Halsbury to A. ff Dakin at the Board of Trade. See NAHC/NRD/ C7/5, Computers, Manchester University/Ferranti Development, Vol. 1, reports (1951–52).

Pegasus (Fig. 5.3) was installed in the company's London Computer Centre in 1956 (see later).

In passing we note that Ferranti's hesitancy in acquiring its own computer meant that, up to the end of 1954, there was no operational demonstration model of a Ferranti Mark I* for prospective customers to view. Instead, customers could only be shown machines in part completion or under test at Moston or Gem Mill—not attractive options for business executives based in other parts of the country or from overseas.

Let us return to Lord Halsbury's list of possible destinations for NRDC's first four sponsored computers. What did happen was that, in parallel with the MOS's deliberations, a determined Dutch customer unexpectedly appeared on the scene in mid-1952. This was followed by a solid enquiry from Italy in February 1954. The result was that two Mark I* computers were exported: in July 1954 to the Amsterdam Laboratories of Shell and in January 1955 to the Instituto Nazionale per le Applicazioni Calcolo (INAC) in Rome. Section 5.5 gives the stories.

By March 1954 NRDC's summary of the state of Ferranti's sales prospects had been revised to this[11]:

NRDC No. 1: 'definitely being purchased by the MOS';
NRDC No. 2: 'definitely being purchased by the MOS';
NRDC No. 3: 'definitely being purchased by Shell';
NRDC No. 4: originally, intended for purchase by Ferranti for their London Computer Centre but now possible that this machine will be sold to Rome.

In September 1953 NRDC placed an order for another Mark I* (NRDC number 5) and in 1954 a further one (NRDC 6). NRDC's view was that "The decision to have the sixth, and final, machine manufactured was based upon the fact that the production schedules of Ferranti Ltd. for other equipment were such that a gap existed which could only conveniently be filled by the manufacture of the ninth computer [serial DC9]".[12]

No doubt NRDC was also responding to Bernard Swann who, in May 1954, had stated his reasons for hoping that one or two more Mark I*s could be sold[13]

(a) "It is the only high performance computer available in this country for delivery in a reasonable time".
(b) "The fact that we have a computer in production has considerably influenced a potential customer in the past".

---

[11]On 17/3/54, C9 1 m, Letter from Turner (NRDC) to Lane (BOT), dated 17th March 1954. See NAHC/NRD/C9/1. Turner rehearses the past history of NRDC's initial order of four machines, approved by the BOT on 6th June 1952 under sub-Sect. 2 of Sect. 4 of the Development of Inventions Act 1948.

[12]Crawley, H.J., *The National Research Development Corporation Computer Project.*

[13]On 17/3/54, C9 1 m, Letter from Turner (NRDC) to Lane (BOT), dated 17th March 1954. See NAHC/NRD/C9/1. Turner rehearses the past history of NRDC's initial order of four machines, approved by the BOT on 6th June 1952 under sub-Sect. 2 of Sect. 4 of the Development of Inventions Act 1948.

(c) "The machine has a larger store than any of the projected smaller computers and in some contexts this is very important".
(d) "The design [of the Mark I*] is the result of a great deal of experience of using this type of computer".
(e) "The production of the computer has been fully proved".

Judging from reports received by NRDC, the following organisations were amongst the sales prospects being explored by Ferranti in May 1954: the Long Range Weapons Establishment, Woomera (see Sect. 5.6); the Argentine Atomic Energy Commission; the Canadian Armaments Research & Development Establishment; Sandia Air Base, USA [the USA's principal nuclear weapons facility near Albuquerque, New Mexico]; the Tennessee Valley Authority (which had been set up to provide navigation, flood control, electricity generation and economic development to the Tennessee Valley).

By mid-1954 a sale had been made to A. V. Roe, the aircraft manufacturers (Sect. 5.6).

By December 1954 NRDC's appreciation of Mark I* sales was quoted.[14]

DC4, (NRDC 1) Now subject to a firm order (presumed Shell)
DC5, (NRDC 2) MOS has ordered this one (presumed to Fort Halstead)
DC6, (NRDC 3) Now subject to a firm order (presumed Rome)
DC7, (NRDC 4) MOS has ordered this one (presumed to AWRE)
DC8, (NRDC 5) Now subject to a firm order (presumed Avro).

No mention was made of DC9. In fact the fate of *Number 9* exercised Ferranti and NRDC for many months. The following remarks about machine *Number 9*, taken from the Minutes of NRDC committees,[15] give some indication of the scope of Ferranti's sales efforts in an age when electronic digital computers were still a novelty in many countries:

In May 1955 "Mr. Hennesey [NRDC] reported that the Mark I* machine Number 9, which is at present still unsold, had been offered to an agency of the Japanese Government for an ex-works price of £91,250".
In November 1955 machine Number 9 had still not been sold, though "the possibility that the Japanese Government might purchase the machine existed. After some discussion, it was agreed that the Corporation should be satisfied to dispose of the machine for cost plus interest and a royalty element".
By January 1956 "it was reported that an examination of the cost position had indicated that a selling price of £60,000 for the Number 9 machine would be one which would recover the Corporation's costs and provide interest and royalty

[14]Note of a meeting on 13th December 1954 attended by Pollard, Sions & Swann for Ferranti and Hennesey, Crawley for NRDC. The purpose: to consider "development recoveries in connection with commercial exploitation by Ferranti of digital equipment developed under [MOS] contract No. 6/WT/4890 and its successors". See NAHC/NRD/C9.

[15]NAHC/NRD/C1 and NAHC/NRD/C5/71/6:C1/1: Minutes and reports of the NRDC Sub-Committee on Electronic Computers: 1954–56.; c) On 14/6/55 (Minute 343):"

elements and also a margin of approximately £10,000 for contingencies. It was agreed that Ferranti should be told that they could sell No. 9 for £60,000, or more, if a higher price could be obtained".

In March 1956 "Interest in the purchase of the No. 9 machine is being expressed in Japan and Ferranti's agents there have been asked for the settling of a price".

In June 1956 "It was reported that no firm order had yet been received". Two new sales prospects were, however, reported: (i) the Portuguese Atomic Energy Authority; (ii) CDV, Admiralty, in respect of thermionic valve development".

In July 1956 Number 9 "remained unsold but that the possibility exists that the machine may be acquired by the University of Sao Paulo, Brazil. An enquiry had also been received from the Argentine Department of Marine. Mr. Swann was at the present time in Brazil".

In September 1956 "It was reported that the No. 9 Mark I* machine remained unsold, but that hope is still entertained that it would be purchased by the University of Sao Paulo".

In October 1956 it was reported that the government's Mechanical Engineering Research Laboratory (MERL), East Kilbride, had been approached and that "discussions were on foot with the Atomic Energy Authority who required a computer for one of their establishments".

In November 1956, "Lord Halsbury reported that he had had discussions with Dr. Sopwith of the Mechanical Engineering Research Laboratory and it appears that MERL will wish to buy the No. 9 Mark I* machine if the selling price can be suitably reduced".

Machine Number 9 was finally delivered to an aerospace customer in the autumn of 1957 (see Sect. 5.6). It had spent the previous year or so in storage at Ferranti's Gem Mill factory, being used periodically to test magnetic tape systems. So the overall picture of Ferranti Mark I* deliveries to non-defence customers ended up as in Tables 5.2.

Why was Machine Number 9 so hard to sell? Because by the end of 1955 Ferranti's Pegasus computer was within a few months of being delivered, Ferranti's Mercury was well under way and three other British computer manufacturers had launched products. The cost-effectiveness of the Ferranti Mark I* was being called into question. Before analysing the market competition we need to say a little more about the two 'windfall' sales, to Amsterdam and to Rome, that had happened previously.

5.4 The Dutch and the Italians Come Calling

The first real sale of a Mark I* to a non-government (i.e. non-MOS) establishment seems to have come out of the blue. Here is the story, as recounted by Swann.[16] "A sign of change came in the middle of 1952 when a team from the Shell Company

[16]Swann, B.B. 1975. *The Ferranti Computer Department: A history*. (personal memoire). This 1975 document was initially circulated privately and marked 'confidential'. A copy is held at the NAHC, catalogue number NAHC/FER/C30.

Table 5.2 Ferranti Mark I* deliveries to non-defence organisations

Serial number	Customer, site	Date of formal order	Hardware delivered	Acceptance tests passed
DC6	Shell Labs., Amsterdam	Dec. 1952	June 1954	Nov. 1954
DC4	Instituto Nazionale per le Applicazioni Calcolo (INAC), Rome	July 1954	Jan. 1955	June 1955
DC8	A. V. Roe Ltd. Chadderton, Manchester	July 1954	Early 1956	May 1956
DC9	Armstrong Siddeley Aero Engines, Ansty, Coventry	Spring 1957	Oct. 1957	Dec. 1957

More details of these four installations and their uses are given in Chaps. 6–9

visited us. They had been looking at computers in the U.S.A. and at Cambridge and the N.P.L. They heard of a machine at Manchester and, having a day available, came to see us. In November they called for a quotation and shortly afterwards decided to buy a computer". This machine was delivered to the Royal Dutch/Shell Laboratory, Amsterdam, in about June 1954. More about the life and times of the Shell Mark I* is given in Chap. 8.

It is interesting to recall that, in 1952, Shell's Amsterdam Laboratory was seemingly unable to obtain a suitable quotation from an American computer manufacturer for a machine of the power needed. Shell "started with a search for available machines, with an important criterion being that ours had to be suitable for immediate use. It soon turned out that very few machines were considered, because very few were in regular production. In America, that was the case with IBM".[17] Shell staff were aided in their search by advice from Adriaan van Wijngaarden, Head of Computing at the Mathematisch Centrum in Amsterdam.

The next sale of a Ferranti Mark I* was, in a roundabout way, connected with the economic regeneration of Italy after the Second World War. Italy in 1950 was lagging behind many of its European neighbours. For example in 1950 the country's GDP per capita was half that of Holland. Italy's leading light in the computing world at that time was Mauro Picone, Director of the Instituto Nazionale per le Applicationi del Calculo (INAC) in Rome. INAC was a division of Italy's Consiglio Nazionale delle Ricerche (CNR). After deciding not to construct his own computer, Mauro Picone applied for funding from Azienda Rilievo Alienazione Residuati (ARAR) to purchase a machine. ARAR was an Italian government entity created in October 1945 for "… the recovery, custody and disposition of material remnants of war, left behind by the Allies or abandoned by the Germans in Italy or otherwise acquired". Bernard Swann and John Bennett from Ferranti visited INAC

[17]Nouwen, Pieter. 2003. Het mirakel van Amsterdam (The Miracle of Amsterdam). *Shell's Venster Magazine* 19–23. Text (in Dutch) by Pieter Nouwen. MIRACLE was the name given to Shell's Ferranti Mark I* computer.

in early February 1954, through the good offices of Dr. Ignaxio Mottola, Ferranti's agent in Italy. In a letter to Picone dated 9th February, Ferranti quoted for the supply of "One Manchester Universal Computer". A Ferranti Mark I* was ordered by Picone and delivered in January 1955. More about the life and times of the Rome Mark I* is given in Chap. 9.

The above two exports were to countries in which, at the time, there was no indigenous competition. There were no Dutch or Italian companies actively marketing their computers at the time of the two Ferranti orders and, it seems, little or no penetration from America. We should now describe an altogether more challenging arena, where four UK manufacturers were about to launch their computer products and IBM was waiting in the wings.

5.5 The Competition

We saw in Fig. 5.1 that the post-war years produced several innovative computer projects in research laboratories. The five manufacturing companies who, in due course, benefitted directly from these laboratory innovations were: the British Tabulating Machine Co. Ltd. (BTM), Elliott Brothers (London) Ltd., English Electric Ltd., Ferranti Ltd. and Leo Computers Ltd. Figure 5.2 shows the chronology of the resulting products, indicating that the two-year period 1955–57 saw the British computer industry burst into life with a range of machines, at a range of prices and offering a variety of facilities. One small computer, the Metropolitan-Vickers MV950, has been omitted from Fig. 5.1 because only six were built, all for use within the company. The MV950, the production version of a Manchester University research prototype, was first delivered in 1956.[18] The Elliott 401 and the LEO I have also been omitted from Fig. 5.1 because neither was available for sale on the open market.

In Fig. 5.2 the number of computers of each type finally sold is shown in parenthesis. The left-hand end of each thick red horizontal line indicates the month/year when the first computer of each type was delivered.

Of course, the computers of Fig. 5.2 covered a wide range of performances and costs. Broadly speaking, the BTM 1201, the Elliott 405 and LEO II were aimed at the commercial data-processing market with the Elliott 405 at the top end. In complete contrast, the Ferranti Mark I* and the Ferranti Mercury were, when each was first delivered, high-performance scientific machines. The rest of the computers in Fig. 5.2 were middle-of-the road machines aimed at the general scientific and engineering markets, with the HEC2M at the lower end and the DEUCE II at the upper end of performance.

---

[18]Lavington, Simon. 1998. *A History of Manchester Computers*. 2nd ed. Published by the British Computer Society. ISBN 0-902505-01-8.

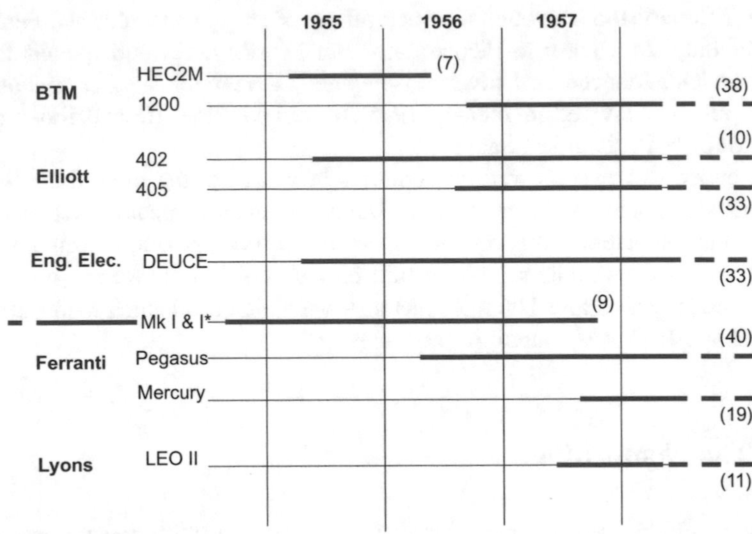

Fig. 5.2 1955–1957, the years when the British computer industry really came alive. The diagram shows the products of five manufacturers: the British Tabulating Machine Co. Ltd. (BTM), Elliott Brothers (London) Ltd., English Electric Ltd., Ferranti Ltd. and Leo Computers Ltd. See text for more explanation. The details of sales to customers and their applications are given in Chap. 18

In more detail, Table 5.3 gives the basic capabilities (memory capacity and speed of a simple *ADD* instruction) of each of the nine early British computers, along with their costs. The prices in the Table are only indicative, since they varied according to the precise configuration of each machine. Quoting the time for one isolated instruction (the fixed-point *ADD)* is only roughly indicative of processing speed, especially for the case of computers with sequential, not random-access, memories. In the case of such computers, Table 5.3 gives minimum and maximum *ADD* times. For further hardware details see.[19]

So far nothing has been said about the software that manufacturers may have provided with their computer products. Before about 1955, very little software was offered. Then gradually basic Assemblers and libraries of mathematical subroutines were provided, followed in the case of Pegasus and Mercury by primitive high-level languages (Autocodes). Compilers for high-level languages such as Fortran or Cobol were not widely available until the early 1960s. Before that time, Operating Systems as we now understand them were unknown. All of the computers in Table 5.3 spent most of their time being used by a successions of individual programmers, each taking their allocated time at the control console where they had the whole of the computer to themselves. The quality of manufacturer-provided

[19]Lavington, Simon. 1980. *Early British Computers.* This is out of print but has helpfully been made available at: http://ed-thelen.org/comp-hist/EarlyBritish.html See also the Computer Conservation Society's site http://www.ourcomputerheritage.org/.

Table 5.3 The hardware capabilities and the cost of the nine British production computers that were on the market in the period 1955–1957

Computer	Primary memory	Drum secondary memory	Fxpt Add time, min./ max. in ms	Approx. price
BTM HEC2 M	–	1K words	1.25/21.25	£20K
BTM 1201	–	4K words	1.25/21.25	£30K
Elliott 402	15 words	3K words	0.204/13.1	£25K–£35K
Elliott 405	512 words	32K words	0.306/13.1	£50K–£125K
EE DEUCE	402 words	8K words	0.064/1.064	£42K–£50K
Ferranti Mk I*	512 words	16K words	1.2	£85K–£90K
Ferranti Pegasus	56 words	5K words	0.3	£32K–£45K
Ferranti Mercury	1024 words	32K words	0.06	£100K
LEO II	1038 words	64K words	0.34	£90K

software was not much of a consideration when choosing which computer to buy. Of more consequence was the reliability of the machine.

To summarise, from 1955 onwards plenty of competition emerged that caused the Ferranti Mark I* to lose market prominence. A detailed analysis of all computer deliveries in the UK from 1955 to 1957, together with the applications for which they were used, is given in Chap. 18. One can see for example in Chap. 18 when, and where, commercial data-processing applications such as payroll processing began to be put into practice. In this field, BTM were the leaders on account of the BTM company's familiarity with existing electro-mechanical punched card equipment and the punched-card Input/Output devices provided for their BTM 1200 series computers. This was a field in which Ferranti were slow to participate. The first Ferranti computer, a Pegasus, to be fitted with punched card I/O equipment was not delivered until December 1957.

Up to 1956 no foreign-made computers were imported into the UK. October 1956 saw the first arrival of an IBM 650—a very popular small computer which was first introduced in America in 1954. By 1962 the number of IBM 650s sold world-wide was about 1800. Probably the first of the higher-performance IBM machines to reach the UK, an IBM 704, was the one installed at the Atomic Weapons Research Establishment, Aldermaston, in February 1957. By 1962 18% of all UK computer installations were IBM machines of one sort or another.[20]

[20]Campbell-Kelly, Martin. 1989. *ICL—A Business and Technical History*. Oxford: Oxford University Press. ISBN 0-19-853918-5.

Fig. 5.3 The first Pegasus computer, as installed in Ferranti's London Computer Centre in 1956. An engineer is seen extracting one of the packaged circuit boards, each roughly the size of a paperback book

It is against the aforementioned background of growing competition that we need to skip back in time to fill in details of Ferranti's efforts to achieve sales in two important areas: aerospace and commercial data-processing. Those wishing to avoid an account of the sometimes complicated and difficult sales efforts may skip to Sect. 5.8 without missing the chronological sequence of the Ferranti company's computing development after 1955.

5.6 Ferranti's Aerospace Sales Efforts

Vivian Bowden had studied the application of digital computers in the aircraft industry and had written up his conclusions in January 1953.[21] He highlighted the experiences of the Douglas Corporation in America, who "have doubled their

[21]Bowden, B.V. 1954. *The Use of Digital Computers in Aircraft Design*. 6 typed foolscap pages. See NAHC/FER/C10b.

computing staff every year since 1947 and now [1952] have 80 full-time computers and many computing machines … They spent $1 million on computation in 1952". By 'computers' Bowden meant people: mathematicians equipped with electro-mechanical desk calculators. He observed that more than seven million person-hours of computing were required in the design of the Douglas DC6 aircraft.

As an aside, a look at the 18 deliveries of IBM 701 computers between April 1953 and June 1954 reveals eight of these were sales to American aerospace companies (Boeing, Convair, Douglas (2), Lockheed (2), North American Aviation and United Aircraft). These were delivered well before the first delivery of any digital computer (IBM or otherwise) to a UK aerospace company.

Analogue computers were also in use and Bowden described an analogue machine for aircraft flutter (i.e. vibration) analysis at the UK's Royal Aircraft Establishment (RAE) at Farnborough. He commented that it was hard to maintain accuracy with analogue techniques and that "if the order of the matrix is large, digital techniques become essential". Besides flutter analysis, several other areas such as structure design and the design of rotor blades for helicopters and gas turbines seemed ripe for the application of electronic digital computers.

It appears that Vivian Bowden and John Crawley (NRDC) went to visit RAE in May 1952.[22] One of the lessons to emerge from RAE was that "They are very much taken with the idea of the [production version of the NPL] Ace machine to be manufactured by English Electric and they are thinking in terms of a cost of £35,000 of an English Electric version of the Ace". Ferranti appears to have expressed doubt about whether English Electric could deliver a production version for as little as £35,000 in a reasonable time. Nevertheless, it was decided to make an immediate effort to sell the more expensive Ferranti Mark I* to RAE. Bowden "stressed the point that the Australian bombing range [the Long Range Weapons Establishment (LRWE) at Woomera] will probably follow up the RAE and order a similar machine in order to obtain similarity of coding and that a great deal will stem from this RAE order".

In the event, Bowden was correct in fearing that NPL's Pilot ACE and English Electric's production version DEUCE would be a real threat to Ferranti's ability to sell computers to the aerospace industry. Even though the first DEUCE customer deliveries (to NPL and RAE) did not happen until May 1955, by June 1955 Halsbury told an NRDC committee that "for reasons associated with the existence of the ACE Pilot model at the NPL which is being widely used by the aircraft industry, a number of the aircraft firms had decided to buy English Electric's DEUCE machine".[23] Of the 33 DEUCE built, seven were applied to the aerospace sector. Chapter 18 gives some details.

[22]File note written by Halsbury (NRDC) on 27th May 1952, following a meeting on 26th May between Halsbury "and the computer team at Ferranti Ltd." See NAHC/FER/C9 1c.

[23]Minute 343 of committee meeting 30, 14th June 1955; comment from Lord Halsbury. See NAHC/NRD/C1: Minutes and reports NRDC Sub-Committee on Electronic Computers: 1954–56.

As for the computing needs of the LRWE at Woomera, South Australia, their problem was the huge number of person-hours taken to analyse tracking data from missiles, using electro-mechanical desk calculators. "In the two years 1955 and 1956 they had the prospect of reading 400,000 points of trajectory, velocity and attitude data, or about 800 points a day. In addition they would be calibrating six million telemetry points. Even at 100% efficiency, this was around 200,000 man-hours of work. It would take up to 200 more [human] Computers to tackle such a load manually, from a specialised labour market already drained dry".[24] LRWE also had rather special requirements for data output.

In 1950 LRWE proposed to design their own computer based on a copy of Trevor Pearcey's CSIRAC Mark I, for which the project-costs were estimated to be £40,000. This plan was later rejected, in favour of a proposal to buy a Ferranti Mark I* to be financed by the UK government at a quoted price of £95,000. In 1954 the Australians, after a visit to England to assess various options, rejected the Mark I* apparently on the grounds of maintenance difficulties and the sensitivity of the electrostatic store to interference from LRWE's large radio transmitters. After rejecting the English Electric DEUCE because it was not yet ready, the Australian team finally chose a specially-modified Elliott 402 computer at a total quoted cost of £106,625.[25] This machine, the Elliott 403 or WREDAC, included a separate and substantial Output Processor (WREDOC) which incorporated magnetic tape decks, a lineprinter and graph plotters as shown in Fig. 5.4. This output processor was nearly as big as the adjacent main computer (WREDAC, not shown in Fig. 5.4). Thanks to the use of standard Elliott 402 plug-in modules, the time between design-specification and delivery of WREDAC was only 16 months.[26] The performance of WREDAC thereafter was initially disappointing but it was giving reasonable service by about 1961 when it was replaced by an IBM 7090.

In hindsight, it was perhaps fortunate that Ferranti did not win the LRWE contract. The design specification, set by John Alan Ovenstone of LRWE, was both innovative and challenging.[27] Ovenstone, who had done doctoral research in physics at the University of Cambridge from 1949–52 before joining LRWE in 1953, was regarded by some as a rather difficult personality and Moston engineers would have found it hard to adapt the Mark I*'s architecture to suit Ovenstone's ideas. Maintaining a non-standard Mark I* in the South Australian heat would not have been easy.

---

[24] Morton, Peter. 1989. *Fire Across the Desert: Woomera and the Anglo-Australian Joint Project 1946–1980*. Canberra: Australian Government Publishing Service.

[25] Morton, Peter., *Fire Across the Desert: Woomera and the Anglo-Australian Joint Project 1946–1980*.

[26] Lavington, Simon. 2011. *Moving Targets: Elliott-Automation and the Dawn of the Computer Age in Britain, 1947–67*. Springer. ISBN 978-1-84882-932-9.

[27] Lavington, Simon., *Moving Targets: Elliott-Automation and the Dawn of the Computer Age in Britain, 1947–67*.

Fig. 5.4 WREDOC, the output processor of WREDAC *(Weapons Research Establishment's Digital Automatic Computer)*. The photo shows WREDOC at Elliott's Borehamwood Lab. just before shipment to Australia in 1956. To the left are four graph plotters (modified Fax machines) and a Bull lineprinter. Two Pye ½-in. magnetic tape decks are to the right. The WREDAC mainframe came separately

Despite the difficulties of selling the Mark I* to the aerospace industry, the company continued to keep in touch through its applications programmers, whose job was "to construct demonstration programmes and to carry out such routine computations as industrial users may require".[28] The aircraft company A. V. Roe Ltd., whose main factory was situated at Chadderton, a mile or so from Moston, were contacted. Avro's own programmers started to use the Mark I at Manchester University[29] and in 1953 were hard at work perfecting the design of a delta wing bomber to be called the Avro Vulcan. The first production Vulcan B1's flight took place in February 1955, preceded by a series of prototype test-flights which had begun in August 1952.

It did not prove particularly easy to persuade A. V. Roe to buy their own computer. It is related[30] that "Sir Arthur Vernon Roe, the proprietor, enjoyed good relations with Sir Vincent de Ferranti. So Bowden called on Sir Arthur, told him of this wonderful new machine [the Ferranti Mark I], and expressed his vision that it could be used to help design aircraft. Sir Arthur listened attentively, but without in any way committing himself. When Bowden had finished, he said, "Go and tell my

[28]Bennett, J.M. 1953. *Uses for the Manchester University Computer*. Ferranti internal document, six typed foolscap pages. See NAHC/FER/C10a.

[29]E-mail dated 11th November 2015 from Joan (Kaye) Travis to the author. Joan Travis was a programmer at Ferranti's Moston site.

[30]McGregor Ross, Hugh. 2001. Ferranti's London Computer Centre. Resurrection (*The Bulletin of the Computer Conservation Society*) (25).

Chief Designer and his senior staff about it". The designers did commit themselves: they said they could not see any use for the computer, and they did not want it. When Bowden reported this response to Sir Arthur, he retorted, "Go back and tell them they have *got* to have one!" Avro were indeed early users of the Mark I at Manchester University but it was not until 1954 that Avro finally decided to buy their own Ferranti Mark I*—proof, if proof were needed, that selling computers was a tough business.

Ferranti made one other sale to the aerospace sector when it delivered the last of the production Mark I*s (machine Number 9) to the Ansty, near Coventry, works of Armstrong Siddeley Engines Ltd. in the autumn of 1957. Both A. V. Roe Ltd. and Armstrong Siddeley were part of the Hawker Siddeley Aircraft Group and there was some sharing of computing facilities. See Chaps. 6 and 7 for the life and times of the Mark I*s at Avro and at Armstrong Siddeley.

5.7 Commercial Data-Processing

One of Ferranti's first investigations into the business and commercial uses of computers occurred during December 1951 to February 1952 when Vivian Bowden visited America and Canada. Bowden described himself informally, and with some justification, as the UK's first computer salesman. On the voyage across the Atlantic he recounts how he got into conversation with a lighthouse salesman and they debated which of the two had the more difficult job!

It has not been possible to analyse all of Bowden's American contacts but we know that he considered one of his more promising prospects to be the AC Nielsen Corporation, a global marketing research company founded in Chicago in 1923. Bowden visited Nielsens in Chicago in December 1950. In a letter to NRDC,[31] he recalled that Nielsons were keen to purchase a computer, would prefer to buy it from USA but would consider Ferranti. In the spring of 1951 Bowden had follow-up meetings in the UK with one of Nielsen's patent lawyers and with the head of Nielson's Oxford branch. His conclusion to NRDC was that the Ferranti Mark I was useful in its present form but would be even more use to Nielsen if equipped with punched card input/output. Lord Halsbury replied to Bowden with two notes of caution: (i) be careful of US anti-trust legislation; (ii) the development of input/output equipment for the Ferranti Mark I "is lagging behind the development of the rest of the machine". We will return to the question of card input/output equipment later.

Unsurprisingly, the Nielson prospect came to nought. AC Nielsen intended instead to install a Univac I. This machine was the first American computer

[31]Letter dated 7th May 1951 from Bowden (Ferranti) to Halsbury (NRDC) and reply from Halsbury dated 24th May 1951. See NAHC/NRD C7/5, Computers, Manchester University/Ferranti Development, Vol. 1, reports (1951–52).

specifically designed with commercial data processing in view, even though it did not have punched card I/O when it was first sold in 1951. Univac added some card-to-magnetic tape and magnetic tape-to-card equipment a year or two later. The first six Univac sales were to US government sites such as the Atomic Energy Commission. The first sale to an independent commercial business was probably that to the Appliance Division of General Electric in 1954.

Returning to Ferranti, promoting the Mark I computer to the business community received a boost with the arrival of Bernard Swann at the end of June 1951. Bernard Burrows Swann (Fig. 5.5) was a Civil Servant with an accountancy background.[32] He had risen to be Assistant Secretary of the Statistical Division of the Board of Trade when he was recruited by Vivian Bowden to head up Ferranti's marketing efforts in the commercial data-processing sector (Fig. 5.5).

Retrospectively, it seems as though Swann found his new job a tough one. Here are his informal (and, it seems, slightly confusing) comments, written in 1993 a few months before his death.[33] "My reason for responding to pressure [to join Ferranti] from Vivian Bowden was because I had for some years been concerned with the planning and organisation of statistical work, and it was the facilities which electronics offered which attracted. Selling computers, one could see, was best done by demonstration rather than talk and I, with the valuable help of several young men could have made a lot of money for the Company which was needed to make the machines. Unfortunately under the influence of the punched card firms we [Ferranti] were frustrated, as were customers who also wanted to compute rather than struggle with the vagaries of electronics. The policy I wanted to follow has been demonstrated in the US. Even here, it would have put Sebastian [de Ferranti, son of Sir Vincent] in the House of Lords at least. Sir Vincent also saw the light at the last visit he paid to Portland Place, but I did not ask for the Computing to be separated from the Engineering because Moston would have insisted on running it, which would have killed it as they killed the successful Pegasus". The focus of Swann's animosity was Brian Pollard, the computer engineering manager at Moston who was resistant to change and lacked vision.

From a knowledge of how Ferranti's Computer Department was to develop in the period 1952–1958, one can guess that Swann was regretting two factors: firstly that Ferranti was much too slow to introduce punched cards, despite a formal agreement with the punched-card company Powers-Samas dating from October 1954[34]; secondly, that the Moston engineers and Pollard in particular were slow to understand the potential for computers as applied to commercial data-processing.

---

[32]McGregor Ross, Hugh. 1993. *Obituary: Bernard Swann*. Resurrection (*The Journal of the Computer Conservation Society*) (8).

[33]Bernard Swann, contribution to a booklet accompanying a 1993 reunion of Mark I programmers. The reunion was organised by Olaf Chedzoy and held at Curdon Mill, Williton, Somerset on 21st April 1993. Chedzoy produced a 38-page commemorative booklet containing short contributions from 23 former Ferranti employees and their partners, plus short contributions from four others who were unable to be present.

[34]Pollard, B.W. 1955. Ferranti and Powers Samas. *The Ferranti Journal* 13 (1): 8.

Fig. 5.5 Bernard Swann, who joined Ferranti in 1951 as the new Sales Manager of the Computer Department. He was keen for Ferranti to manufacture reasonably-priced machines for commercial data processing, a plan to which the company was slow to respond

Moston received several hints about the importance of business applications and the relevance of smaller, more affordable, computers. Here, for example, are Bowden's notes of a June 1952 meeting at NRDC's London office to discuss *Future commercial policy for the development and exploitation of computing machines.* Participants included Professor P. M. S. Blackett (Manchester University) and Christopher Strachey (NRDC). Here are the main points of the meeting[35]:

(i) "Professor Blackett was anxious to be assured that Ferranti Ltd. were both able and willing to consider the development of a smaller machine that would be peculiarly suitable for commercial problems, particularly for PAYE [*Pay as you Earn,* the scheme for National Insurance], for which, he felt, the ultimate market was likely to be very big". Bowden assured the meeting that Ferranti were "well aware of the importance of this market", were proposing to demonstrate the present machine on PAYE–type problems, and would

[35]Memo from Bowden (Ferranti) to Halsbury (NRDC), dated 17th June 1952, concerning a meeting at NRDC to discuss *Future commercial policy for the development and exploitation of computing machines.* The memo is entitled *Notes of a discussion held on Monday afternoon, June* 16th *1952 at NRDC.* Present: Lord Halsbury, Mr. Hennessey, Mr. Crawley, Mr. Strachey, Prof. P. M. S. Blackett, Mr. A. Ridding, Dr. B. V. Bowden. See NAHC//NRD/C9/1.

then see by how much the operational performance of the present machine could be simplified and still be able to cope with PAYE problems.

(ii) Halsbury and Blackett then asked whether Ferranti "proposed to collaborate with British Tab [the British Tabulating Machine Co., BTM] in the use of punched cards as an input medium". Bowden replied that Ferranti should do this "at least for the time being owing to the convenience of punched cards, the wide understanding of their use and the large amount of capital which has been invested in data punched on cards and which will have to be processed in the future. On the other hand we are going to develop magnetic tape which we expect will supersede both punched tape and cards in the long run".

(iii) Blackett suggested that Ferranti "might be unable to develop a small machine in view of our other commitments, in which case he would propose to offer a contract either to Elliott Brothers or to English Electric to produce a machine for commercial use". Bowden "resisted this suggestion and countered it with the idea that our [i.e. Ferranti's] experience being so much greater than that of anyone else, we are likely to be able to do this as well as producing machines of the present type, and perhaps making one or maybe two 'prestige' machines of the high-speed type now being developed by Williams and Kilburn".

(iv) Halsbury said that NRDC had already "given a contract to Elliott Brothers to produce a machine using printed circuits and other special components and of the highest speed that they can persuade the Williams Memory Tube to work". [Elliotts later abandoned Williams Tube memory in favour of their own magneto-strictive nickel delay line storage system].

(v) Bowden told the meeting that "English Electric are being very secretive about their plans but there is no doubt that they will re-engineer the ACE almost completely and produce a machine which will be in direct competition with ours before very long". Bowden emphasised that English Electric's prices "could not be completely unlike ours [i.e. Ferranti's] in view of our own Costing Department's estimate of the price which we should have to charge for a machine of a type similar to the ACE".

Bowden's comments reveal opinions that, in the light of hindsight, we see to be unfortunately blinkered. He portrays Ferranti as the only UK company with computing expertise and scientific applications as the main target. In contrast, it seems that Bernard Swann, Patrick Blackett and Lord Halsbury were making efforts to extend the application of digital computers to data-processing applications and that Halsbury was already looking at other manufacturers. In an internal memo dated 23rd July 1952 Halsbury estimates that "if the initial cost of a computer is £80 K written off over four years, and costing £10K per annum to run, one such machine could economically replace about 40 clerical workers".[36]

[36]Internal NRDC memo by Halsbury, dated 23rd July 1952. See NAHC/NRD/C9/1 (Computers, Manchester University/Ferranti, development, vol. 2, correspondence, 1952–1967).

The surviving records of Ferranti's wider marketing activities are incomplete and mostly consist of hopes rather than definite sales deals. To put this in context, the early 1950s was a period of great uncertainty for all digital computer manufacturers. Some experts, amongst them Professor Hartree of Cambridge University, felt that there were few market opportunities. In September 1951 Hartree told Bernard Swann: "We have a computer in Cambridge, there is one in Manchester and one at the NPL. I suppose there ought to be one in Scotland, but that's about all".[37]

Fear of faults and machine errors held other experts back. At about this time Sir William (later Lord) Penney who headed the UK's Atomic Energy Authority indicated to Swann that he "would not buy one of the early machines, but would 'probably buy the sixth'; he wanted others to pioneer".[38] Cost was another barrier. The Purchasing Controller of Imperial Chemical Industries told Swann that "ICI never had, and never would, pay more than £50,000 for any one instrument". The price of a Ferranti Mark I* was initially fixed at about £85,000 which, in 2016, would be equivalent to over £2 million.

Back to early 1952. Despite some gloomy predictions, Vivian Bowden the salesman continued to be optimistic about sales prospects. In a meeting with NRDC on 29th April 1952 he suggested that Mark I* computers could be sold to ICI, British Railways, the Rootes Group (manufacturers of motor cars such as Hillman, Humber and Talbot) and the Treasury.[39] In the same meeting prospects for European sales were discussed. It was agreed that Bowden would accompany Hennessey and Strachey (both NRDC) on a visit to Germany, "to see what was going on there and what chance there was of getting some orders". There was also a suggestion that a Mark I* computer be demonstrated at a suitable venue in Europe —a plan that came to nought. The notes of Bowden's April 1952 meeting recorded that: "With regard to installation of the machine under OEEC [Organisation for European Economic Co-operation] auspices in Europe, Dr. Bowden emphasised that this project had degenerated into a racket from which all responsible members of the Committee concerned had withdrawn. It was, in his opinion, quite impractical to lend a machine as the cost of transport, erection and subsequent dismantling would be prohibitively expensive".

We will now widen our time-frame and resume the story of how Ferranti was preparing to meet the competition that grew from 1955 onwards. One strand was the planning of a much more substantial computer manufacturing facility than the one available at Moston. The site chosen was West Gorton, about three miles south-east from Manchester city centre. The progression from Moston to West Gorton is described below. This hardware-oriented move was not the only change on the agenda. The programming staff and sales staff were migrated from Moston to

---

[37]Swann, B.B., *The Ferranti Computer Department: A history.*

[38]Swann, B.B., *The Ferranti Computer Department: A history.*

[39]NRDC internal memo written by Halsbury on 1st May 1952 and summarising a meeting held on 29th April between himself and Bowden, Hennessey and Crawley. See NAHC/NRD/C9/1, Computers, Manchester University/Ferranti Development, Vol. 2, correspondence (1952–67).

London during the mid-1950s. By the late 1950s there was also a subtle diversification of cultures within the company: no longer was Manchester University to be regarded as the sole source of ideas for the company's future computer products.

5.8 The Drift Away from Moston

Under the influence of Bernard Swann, Ferranti's management had been contemplating setting up a London Computer Centre since 1952. Amongst the factors driving Swann was his observation that the leading decision-makers in government and industry were based in London. Swann, recalled a colleague, also "drew on his wartime experience of introducing punched card systems to carry out statistical work for the army in India. There he learnt that it was easier for an army man to learn punched card techniques than for a punched card expert to learn about the army's statistics".[40] Ferranti therefore needed a working computer in a central London location, with the ability to demonstrate its use on a range of applications. Additionally, end-user courses on programming should be offered on the demonstration computer.

By the spring of 1953 both NRDC and Ferranti had been exchanging "proposals and counter-proposals" for setting up a London Computer Centre and equipping it with a purchased, or rented, Mark I* but J. D. Carter, Ferranti's relevant Manager, indicated that Ferranti was still worried about the cost.[41] Curiously, in a meeting with Sir Vincent de Ferranti on 18th June 1953 it seemed as though Sir Vincent was suggesting that the company buys a Mark I* computer outright. This never happened. Instead, in January 1954 it was decided that "£102,000 of capital will be spent by Ferranti this year in the setting up of the London Computer Centre".[42] There was an associated build-up of sales staff. In September 1953 Hugh McGregor Ross and Conway Berners-Lee joined Ferranti, to be followed in April 1954 by M. H. (Harry) Johnson. All three spent a short while in Manchester before moving down to London as the core of a new team.

Harry Johnson remembers that he spent a few months in Manchester, as liaison officer between Ferranti's Moston and Wythenshawe factories on guided weapons projects. He transferred to London in about August 1954, at first into temporary accommodation in Great Ormond Street. He describes two rooms, in one of which was Bernard Swann and two secretaries and in the other were six sales engineers

---

[40] McGregor Ross, Hugh., Ferranti's London Computer Centre.

[41] Letter from J. D. Carter (Ferranti) to D. Hennessey (NRDC), 16th April 1953. See NAHC/NRD/C9/1.

[42] *Proposals for computer research and development programme, 1954*. This is a three years budget of the Ferranti computing organisation from 1st April 1954 to 31th March 1957. Two typed foolscap pages dated 28th January 1954. See NAHC/FER/C13.

including himself, Ross, Berners-Lee, Chris Wilson and Leslie Fox.[43] It was clear that more space was needed quite soon if Ferranti was not to miss opportunities. In support of this, Swann had said in May 1954 "We are told that the aircraft industry, for instance, experiences a delay of up to three months in getting work done on the ACE at the NPL".[44]

Meanwhile, more prestigious premises were being acquired and refurbished, into which staff moved in early 1955. The location was 21 Portland Place, described by Hugh Ross as: "a 1780s Nash brothers' development of a street of town palaces for the very wealthy. It had an imposing entrance with a marble floor, two large rooms downstairs (one with a fine plasterwork ceiling, which we used for presentations and programming courses), and curved stairs to the main room on the first floor. This room was very grand, with extensive plasterwork and a ceiling painted by Angelica Kaufmann with cherubs and other conceits. The floor, built on great oak beams, could cope with 50 dancers at a ball, but the architects stipulated that a false floor using steel joists had to be built to support the heavy computer" (Fig. 5.6).[45]

By the spring of 1956 it was reported[46] that Ferranti's efforts at the "sales promotion and the study and planning of computer applications" were divided between two groups:

(1) The Computer Sales Department (Mr. Swann);
(2) The Programming Research Group (Mr. Welchman).

Group (1), Sales, had a total of about 18 staff, of whom 17 were graduates. They included a Technical Application sub-group (4 senior programmers, 4 programmers), a Technical Sales sub-group (2 senior & 1 assistant), and an Industrial Applications sub-group (3 senior and 5 or 6 others). One consultant in the Insurance area was employed.

Group (2), Programming Research, consisted of 11 graduates including Stanley Gill from Cambridge and four women with first-class degrees and postgraduate experience. Their work was divided into Technical and Industrial.

Also by the spring of 1956, three 2-week training courses had been held at 21 Portland Place and a fourth was about to start. There were about 30 people per course. "The first course was arranged specially for aerodynamicists; the second course was primarily for scientists and technologists but the attendance included a small proportion of accountants; the third course was primarily technical. The attendance at these courses by representatives of commercial or industrial interests

[43]*My work with computers from the Ferranti Mark I to the ICT 1900 (1952–1966).* M. H. (Harry) Johnson. 2002. 84 typed pages, circulated privately.

[44]Letter from Swann (Ferranti) to Hennesey (NRDC) dated 25th May 1954. See NAHC/NRD/C9.

[45]McGregor Ross, Hugh., Ferranti's London Computer Centre.

[46]*Notes on sales promotion and applications of computers*, 17th January 1956. This was an Appendix to NRDC paper number 106, considered at the 13th March 1956 meeting of NRDC's Sub-committee on Electronic Computers. See NAHC/NRD/C1.

Fig. 5.6 21 Portland Place, into which Ferranti's London Computer Centre moved in 1955. Located just north of Oxford Street, this elegant Nash house was built in 1777. At one time it was the home of General Sir Henry Clinton who had been commander-in-chief of the British forces in America

has been at high level but not at Board level. Proposals for courses aimed at senior executives are under consideration".[47]

Finally by the spring of 1956[48] about 40 "memoranda and reports have been prepared describing problems which have been examined or actually put on a machine. So far, these have mainly been in terms of the Mark I machine". Commercial problems that had been examined for the Mark I included wage calculations, Board of Trade Index numbers, linear programming and the Transportation Problem, life assurance, minimisation of cutting losses in sheet material, production planning analysis, inventory control, airline seat reservation.

A somewhat smaller selection of problems had been examined for prospective Pegasus customers. "The Sales Department estimates that, on this study basis, they are in touch with about 50 prospective users" but admitted that "this aspect of the work is hampered by the fact that no machine suitable for commercial work is

[47]*Notes on sales promotion and applications of computers*, 17th January 1956. This was an Appendix to NRDC paper number 106, considered at the 13th March 1956 meeting of NRDC's Sub-committee on Electronic Computers. See NAHC/NRD/C1.

[48]*Notes on sales promotion and applications of computers*, 17th January 1956. This was an Appendix to NRDC paper number 106, considered at the 13th March 1956 meeting of NRDC's Sub-committee on Electronic Computers. See NAHC/NRD/C1.

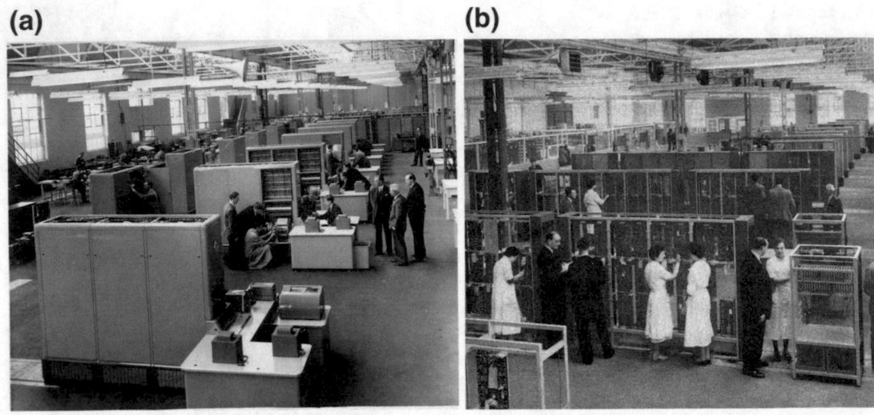

Fig. 5.7 **a** and **b** Ferranti's new factory at West Gorton, Manchester, into which the Computer Department moved in 1956. The computers being assembled in photo **a** are seven Pegasus machines; those at the other end of the floor in photo **b** are three Mercury machines. At the time, West Gorton was believed to be the largest computer manufacturing plant in Europe

available. It is hoped that a Pegasus machine with punched card and magnetic tape input and output will be available this year".[49] Pegasus 2, with cards *and* magnetic tape, was not actually available until 1959. However, before then Ferranti's Bracknell laboratory had been designing Perseus, a substantial commercially-oriented computer which used the Pegasus technology. Perseus had both magnetic tape and card facilities and an instruction set that worked directly on alphanumeric characters. Only two machines were sold, to insurance companies in Sweden and in South Africa in 1959. This was partly because, by 1959, Ferranti designers were working on two powerful new computers: Orion and Atlas (Fig. 5.7).

Whilst 21 Portland Place was getting into business, Ferranti was preparing its much more substantial computer manufacturing facility in West Gorton, about two miles south-east from Manchester city centre. The factory, formerly the premises of textile machinery manufacturer Brookes and Doxey which dated from 1859, was ready for occupation by the first group of Computer Department staff in September 1956. In due course production lines were set up for the Ferranti Pegasus and Ferranti Mercury computers. It was reckoned to be the largest computer manufacturing plant in Europe.[50]

Production of Ferranti's high-performance computers continued at West Gorton until Ferranti's mainframe computer interests were taken over by ICT in 1963. In turn, ICT was merged with the mainframe interests of English Electric, Leo

[49] *Notes on sales promotion and applications of computers*, 17th January 1956. This was an Appendix to NRDC paper number 106, considered at the 13th March 1956 meeting of NRDC's Sub-committee on Electronic Computers. See NAHC/NRD/C1.

[50] Pollard, B.W. 1957. The Rise of the Computer Department. *Ferranti Journal* 15 (3): 20–23.

Computers, Marconi and Elliott Brothers, to become ICL in 1968. Throughout this period West Gorton remained a major centre for computer design and manufacture.

We continue an account of Ferranti in the late 1950s in Chap. 13, where we will begin by picking up the story of the first Ferranti Mark I which had been moved in 1954 from the original Manchester University Computing Machine Laboratory to a new Department of Electrical Engineering in Dover Street. Here it lasted until 1958 when it was replaced by a Ferranti Mercury.

Before getting to Chap. 13 and Dover Street we should provide a proper treatment of the life and times of the seven Ferranti Mark I* computers to have been sold. This is covered in Chaps. 6–12, which now follow.

References[51]

Bennett, J.M. 1953. *Uses for the Manchester University Computer*. Ferranti internal document, six typed foolscap pages. See NAHC/FER/C10a.

Bowden, B.V. 1954. *The Use of Digital Computers in Aircraft Design.*

Bowden, B.V. 1970. The Language of Computers. *American Scientist* 58 (1): 43–53.

Campbell-Kelly, Martin. 1989. *ICL—A Business and Technical History*. Oxford: Oxford University Press. ISBN 0-19-853918-5.

Crawley, H.J. 1957. *The National Research Development Corporation Computer Project*. NRDC Computer Sub-Committee paper 132, February 1957.

Lavington, Simon. 1980. *Early British Computers.*

Lavington, Simon. 1998. *A History of Manchester Computers*. 2nd ed. Published by the British Computer Society. ISBN 0-902505-01-8.

Lavington, Simon. 2011. *Moving Targets: Elliott-Automation and the Dawn of the Computer Age in Britain, 1947–67*. Berlin: Springer. ISBN 978-1-84882-932-9.

Letter from Swann (Ferranti) to Hennesey (NRDC) dated 25th May 1954. See NAHC/NRD/C9.

Memo from Bowden (Ferranti) to Halsbury (NRDC), dated 17th June 1952, concerning a meeting at NRDC to discuss *Future commercial policy for the development and exploitation of computing machines.* The memo is entitled *Notes of a discussion held on Monday afternoon, June* 16th *1952 at NRDC.* Present: Lord Halsbury, Mr. Hennessey, Mr. Crawley, Mr. Strachey, Prof. P. M. S. Blackett, Mr. A. Ridding, Dr. B. V. Bowden. See NAHC//NRD/C9/1.

McGregor Ross, Hugh. 1993. *Obituary: Bernard Swann*. Resurrection (*The Journal of the Computer Conservation Society*) (8).

McGregor Ross, Hugh. 2001. Ferranti's London Computer Centre. Resurrection (*The Bulletin of the Computer Conservation Society*) (25).

Morton, Peter. 1989. *Fire Across the Desert: Woomera and the Anglo-Australian Joint Project 1946–1980*. Canberra: Australian Government Publishing Service.

Nouwen, Pieter. 2003. Het mirakel van Amsterdam (The Miracle of Amsterdam). *Shell's Venster Magazine* 19–23.

Pollard, B.W. 1955. Ferranti and Powers Samas. *The Ferranti Journal* 13 (1): 8.

Pollard, B.W. 1957. The Rise of the Computer Department. *Ferranti Journal* 15 (3): 20–23.

Swann, B.B. 1975. *The Ferranti Computer Department: A history*. (personal memoire).

[51]*Note:* NAHC refers to the National Archive for the History of Computing, which is held in the University of Manchester's Library.

Chapter 6
The AVRO Mark I* Installation at Chadderton

6.1 A. V. Roe at Chadderton, Manchester

The aircraft manufacturer A. V. Roe & Co., commonly known as Avro, was founded in 1910. By 1935 Avro was part of the Hawker Siddeley Aircraft Group. In 1963 the company name was changed to the AVRO-Whitworth Division of Hawker Siddeley Aviation Ltd. After a series of mergers and rationalisations within the UK's aerospace industry in the 1950s and 1960s, Hawker Siddeley (and therefore Avro) became part of British Aerospace in 1977. In due course British Aerospace became part of BAe Systems.

In 1939 Avro's large factory at Chadderton, north Manchester, was opened and it was here that, amongst other aircraft, many of the Avro Lancaster bombers were built during the Second World War. The company also had a Test and Assembly site at Woodford, 10 miles south of Manchester in Cheshire. We will begin by concentrating on Chadderton because it was here that Avro's digital computing activity began.

A Ferranti Mark I* computer was delivered to Avro's Chadderton site either very late in 1955 or (more probably) very early in 1956 and had passed its acceptance tests by about the end of May 1956.[1] Two Ferranti engineers stayed with the machine for a year until Avro staff took over the maintenance. The computer was installed in a purpose-built room.

[1]Len Hewitt, e-mail exchange with the author, September 2015. Len joined Ferranti Gem Mill in mid-1955, then became the Ferranti maintenance engineer on the Avro Mark I* at Chadderton. He stayed at Chadderton for a year and left Ferranti in mid-1957 to maintain a Ferranti Pegasus computer at ICI Dyestuffs Division, Manchester.

S. Lavington, *Early Computing in Britain*, History of Computing,
https://doi.org/10.1007/978-3-030-15103-4_6

Fig. 6.1 Ferranti engineers during the commissioning of the Mark I* at Avro Chadderton in 1956. *(Left-to-right):* Bill Wallace, Len Hewitt, Ron St. John

6.2 Avro's Computing Environment, Digital and Analogue

In Chap. 2 we described how Avro designers had started using digital computers in 1952/3 by running their programs on the Ferranti Mark I at Manchester University. Somewhat later Avro mathematicians used the Ferranti Mark I* at Fort Halstead until their own Mark I* was delivered.[2] Amongst likely tasks at that time were calculations connected with the design of an advanced specification Avro Vulcan delta wing bomber. At about this time another Avro project, the 720 Interceptor, was still in the early design stage. This interesting fighter aircraft was planned to use an Armstrong Siddeley *Screamer* rocket engine for rapid acceleration and high top speed, with a conventional jet engine for cruising flight. The fate of this project is described later.

A firm intention to purchase their own computer was made by Avro in February 1954, confirmed by July of that year. The Avro order for a Mark I* had to take its place in the queue for the limited Ferranti production facilities at Moston, behind

[2]Smith, Joan. M. 2016. *Early Computing in the Aircraft Industry: Avro's at Chadderton.* Five-page typed personal recollections, copy available from SHL. Joan died in August 2017.

Fig. 6.2 The Ferranti Mark I* at Chadderton during commissioning in 1956

orders for two Ministry of Supply computers (see Chaps. 11 and 12) and neck-and-neck with an order from the Instituto Nazionale per le Applicazione Calcolo (INAC) in Rome (see Chap. 9).

Although Avro's Mark I* was the eighth, and penultimate, machine of the Mark I/Mark I* family, it was probably the first general-purpose digital computer to be installed by a British aerospace company. The other early adopter, the Bristol Aeroplane Company at Filton, Bristol, did not receive an English Electric DEUCE computer until about June 1956. The first external customer for a Ferranti Pegasus computer was Hawker Aircraft Ltd. of Kingston-upon-Thames in October 1956.[3] An overview of all early installations in the UK's aerospace industry is given in Chap. 5.

From the summer of 1956 the Ferranti Mark I* at Chadderton was to become "responsible for all the structural, aerodynamic and control calculations for the Vulcan Mark 2. It was also used in the design of the 748".[4] The Avro 748 short-haul airliner's first flight was in the summer of 1960. "The software for the

[3]Delivery lists for all the major early British computers are given on the Computer Conservation Society's Our Computer Heritage website—see: http://www.ourcomputerheritage.org/.

[4]Hawker Siddeley. 1966. News, Aviation Edition. 3 (1): 2.

proposed Vulcan Mk III was gradually building up, as was the design knowledge gained when the results were analysed".[5]

Not all Avro's projects ran smoothly. The Avro 720 Interceptor was dealt a serious blow when the Armstrong Siddeley *Screamer* rocket engine which powered the Interceptor was cancelled by the government during 1956. Joan Smith writes that she "recalls the day when the government contract for the 720 Interceptor was cancelled [late 1956 or early 1957]. This work had been the mainstay of the drawing office. The following day it was impossible to hear yourself speak for the noise of paper being torn up. The next week the office was clear, all drawing office staff having been made redundant. A few designers remained at the end of the design office, engaged on testing early computer-assisted design (CAD) equipment, but that was all. At the other end an area was screened off for the New Projects office, to which entry was not permitted. It led to more work for the Mark I* computer".[6]

Of course Avro, in common with all other aircraft manufacturers, had been using *analogue* computers for some years. In the case of Chadderton, the in-house development of a large new analogue computer took place at about the same time as Avro was acquiring its first digital machine. Avro had a need to investigate some unexpected pitching and yawing behaviour in the new breed of fighters with highly swept wings and long narrow fuselages. The analogue computer shown in Fig. 6.3 was built by the Electronics Department at Chadderton to investigate such problems.[7] It came into operation in 1956. Each of the two outer left and right assemblies shown in Fig. 6.3 could solve four fully coupled, second order simultaneous differential equations. The solution was in real time and gave the operator an immediate "feel" for what the aircraft was doing—something that contemporary digital computers could not provide. The reason for including Fig. 6.3 at this juncture in Avro's story is to make two points: firstly, analogue and digital computing facilities existed side-by-side within the company; secondly, the company contained its own in-house electronic design skills.

---

[5]Smith, Joan. M., *Early Computing in the Aircraft Industry: Avro's at Chadderton.*

[6]Smith, Joan. M., *Early Computing in the Aircraft Industry: Avro's at Chadderton.*

[7]*The Avro eight degrees of freedom analogue computer*. Reg Boor, January 2016. Two-page typed article sent to the author (SHL) on 4th January 2016. The career path of Reg Boor, who rose to be Head of Applied Aerodynamics and then Head of Aeroelastics within the company, offers an interesting reflection on post-war Britain. In an e-mail to SHL on 10th March 2016 Reg writes: "I enjoyed my career at Avro and have always thanked my "lucky stars" for such a career. I left high school at the age of 16 feeling I should earn money for my mother but after a 6 month hiatus I went back to school and got an application form from the careers master for a "Professional Engineer Apprenticeship" at Avro, which I obtained. This was 1946 and we were given a day off each week to study for HNC [Higher National Certificate, equivalent to today's GCSE A Levels]. But, of course, by 1969 I was surrounded in the office by 1st Class honours and Ph.D. men so when the Open University started I studied with them immediately for a degree in Maths".

Fig. 6.3 Avro's large analogue computer at Chadderton in 1956

6.3 Personal Anecdotes of Computing at Chadderton

Returning to the Ferranti Mark I* at Chadderton, the activities of Avro's Computer Department may be introduced via the memories of four of the original participants.

Peter Teagle.

Peter Teagle, who became Head of the Design Office at Woodford, remembers[8]: "I started in the Stress Office at Chadderton in 1955 after a year as a Technician under Training so I suppose I was on the ground floor as a user of the facilities of the Mark I*. It was a Stressman, Mr. Arrowsmith, who was put in charge of the Computer Department originally. We had to prepare our calculations into a tabular system and then hand it over to the computer staff to run it. It was a somewhat intermittent form of working and slide rules were still predominant for most work. A tabular operating system was devised for Stressmen and others to use, which

[8]Peter Teagle, e-mail message sent to SHL via Len Whalley on 16th December 2016.

made life simpler for them. This was the TIP system—see[9] and Chap. 15. Suddenly one Stressman could calculate the bending strength of the wing in a week, whereas previously it needed six people! The computer really started to make a difference with the design of the Avro 730 Supersonic Bomber". (The Avro 730 was abruptly cancelled in 1957, as a consequence of the 1957 Defence White Paper).

Sheila Cooper and Ben Cooper.

Memories of the initial group of Mark I* programmers are provided by Sheila Cooper (neé Gray) who joined Avro at Chadderton in August 1954, having graduated that summer from Royal Holloway College University of London with a first-class degree in mathematics.[10] At Avro Sheila was the first woman to be recruited by the company as a programmer. She recalls that she was also the first woman to be appointed to a professional grade—though her salary of £400 pa was £100 less than that of a fresh male graduate who joined at the same time with a second-class degree. Sheila became the only woman in a team of six programmers, led by Peter Morton the Chief Programmer in the Design Office Computer Group. Other members of the programming team were Ben Cooper, Alan Wragg, Alan Fielding, John (surname forgotten) and Ron Lane. All were recent graduates, fired with enthusiasm. Sheila said that it was an exciting time to be working. Her colleagues were intellectually stimulating and "we all knew that we were on the verge of something new".

One of the group, Ben Cooper, who had earlier asked to transfer from the Aerodynamics Department, acted as Sheila's mentor in the early days. Ben said that his former colleagues in the Aerodynamics Department thought him crazy to get involved in writing programs because computers were just 'a flash in a pan'. Ben Cooper left Avro early in 1956 and joined the Computer Department of Ferranti Ltd. Ben and Sheila later married.

Back in the autumn of 1954, Sheila recalls that it took her about six months to learn Mark I* machine-code programming. One of the manuals available was the

---

[9]This was most probably the same as the tabular system in use at Armstrong Siddeley Motors, Ansty, and at Bristol Siddeley Engines Ltd.—see: *Tabular Interpretive Programme*. Bristol Siddeley Engines Ltd., Filton, Bristol. A summary of the language is given in Chap. 15. The Tabular Interpretive Programme, TIP, was devised by the Mathematical Services group of Bristol Siddeley Engines Ltd. in 1957. TIP was made available for the DEUCE Marks I, II and IIA, and the Ferranti Mark I* computers.

[10]*Sheila Cooper's memories of working with a Ferranti Mark I* computer at Avro Chadderton.* Transcript produced by Simon Lavington in April 2017, based on an audio interview of Sheila by Buxton's local historian Vivienne Doyle recorded on 14th April 2016 and supplemented by telephone conversations and e-mails with Sheila in the spring of 2017 and with her daughter Judie Adnett.

132-page *Digital Computer Programmer's Handbook*, written in May 1955 by Peter Morton.[11]

Sheila has vivid memories of running her very first program some time before 1956, under the watchful eye of Ben Cooper. They booked an hour's time on the computer and sat down at the console. Sheila put in her program tape and the program ran for some minutes. Then, as expected, it began to punch out results onto paper tape. Believing the results to be correct, Sheila went to tear off her precious output tape. Much to her surprise and annoyance, Ben leaned over and seized the tape, 'accidentally' dropping it into his mug of tea which, contravening all regulations, was standing on the computer's console. Sheila was horrified because the tape was now soggy and unusable! Ben was unperturbed. He explained that he'd been observing the running of the program and knew for certain it contained errors. A soggy tape was a stark reminder of the value of carefully checking a program before using up valuable computing time.

One of the tasks that Sheila was asked to do at Avro was to write a program that analysed Wind Tunnel results. The objective was to calculate stresses on the air frame at various wind speeds and angles of incidence, having been given the necessary formulae by Avro's Stressing Department. At that time, the only major UK high-performance Wind Tunnel facility was at RAE Farnborough. When it was Avro's turn to use the facility, Sheila and a male colleague had to travel down to Farnborough and stay at a nearby hotel. Sheila remembers two gender-specific incidents about this assignment. Firstly, Avro's Computing Manager Jimmy Arrowsmith was at first very doubtful that it would be appropriate for Sheila and her colleague to stay at the same hotel—even though they had booked separate rooms! Secondly, the Farnborough staff assumed that Sheila was the data-preparation girl who would assist by punching paper tapes. They were taken aback when she announced that she would be writing the actual programs.

Sheila remembers some technical oddities about the Wind Tunnel assignment. There was no automatic way of transferring Tunnel data to a remote computer so Wind Tunnel results had to be printed out by Farnborough's equipment and then typed in on a Teleprinter to produce 5-track paper tape for input to Avro's Ferranti Mark I* computer—very tedious. Sheila's program then produced six coefficients with values between zero and 1, representing six degrees of freedom in the aircraft model, for various sets of speed, inclination, etc. When she asked how many decimal places of accuracy were needed, the aerodynamicists said that four would be fine. No problem, said Sheila, since her program worked to about 12 decimal places and she could round the results to 4 places just before printing the answers.

[11] *Digital computer programmer's handbook*. Compiled by J. P. Morton, Design Office Computer Group, Report no. COMP/M/1008, A. V. Roe & Co. Ltd., Manchester. May 1955. Typed foolscap manual, 132 p. Bound in a soft black treasury-tag-back binder, approximately 1-in. thick with embossed 'Avro' logo on front. To quote from the Preface: "Chap. 6 contains accounts of the various routines in the Sub-routine Library, and will be added to as the library grows". Chapter 6 in the first edition of the handbook consisted of 80 p, describing 29 subroutines. A few of these are dated, the dates lying in the range Nov. 1953–May 1954.

As a test, Sheila's computed results were compared with a previous set of results calculated laboriously by hand. Discrepancies were noted and the traditional aerodynamicists were quick to blame the 'new-fangled computer' for being in error. Sheila decided to examine in detail the methods that had been used during the manual calculations and was able to demonstrate the cumulative effect of progressive rounding errors in the manual calculations. After that, the computer was trusted!

Sheila remembers one potentially frightening incident at Avros. She was alone in the computer room developing a program late at night. Suddenly she smelled burning. A quick look established that the computer itself was not on fire but, on entering the adjacent engineer's room, she saw smoke and flames! An engineer had left a soldering iron switched on and this had caused a cardboard box containing the straw packing surrounding a spare Williams/Kilburn storage tube to catch light. Fortunately the fire was quickly dealt with.

John O'Donnell.

Another former user of the Ferranti Mark I* at Avro was John O'Donnell. He graduated in Physics from Manchester University in 1953 and joined Chadderton in 1956. There he worked on stress calculations for the Armstrong Whitworth AW 660 Argosy, a military transport aircraft. John applied beam theory to the 24 ribs in the tapered wing and did some novel strain energy calculations. An Avro mathematician coded these up for the Mark I* and appropriate parameters were applied. John also remembers that the whole set of stress results for all the ribs was produced within a week.[12] John left Avro in 1959.

Joan Smith.

Joan Smith was yet another mathematician who joined Avro Chadderton in the mid-1950s. She remembers[13] that: "Two teams of mathematicians were engaged to write the necessary programs (software) for the computer: the one for stress, the other for aerodynamics. Ferranti provided some basic software, including subroutines which could form part of a program written by a mathematician. Everything was in machine code.

Although there were computer assistants to punch their programs and data, all programmers were well-versed in punching their own coding and then reading the perforated tape. Double-punching was the best way to detect [punching] errors. For this, the two punched tapes, allegedly identical, were examined under an Anglepoise lamp for possible discrepancies. A duraluminium hand punch made by the engineers at Avro, together with opaque sticky tape judiciously applied to both sides of the paper tape, were used to rectify [punching] errors.

[12] John O'Donnell. 'phone conversation with SHL on 17th December 2015.

[13] Smith, Joan. M., *Early Computing in the Aircraft Industry: Avro's at Chadderton.*

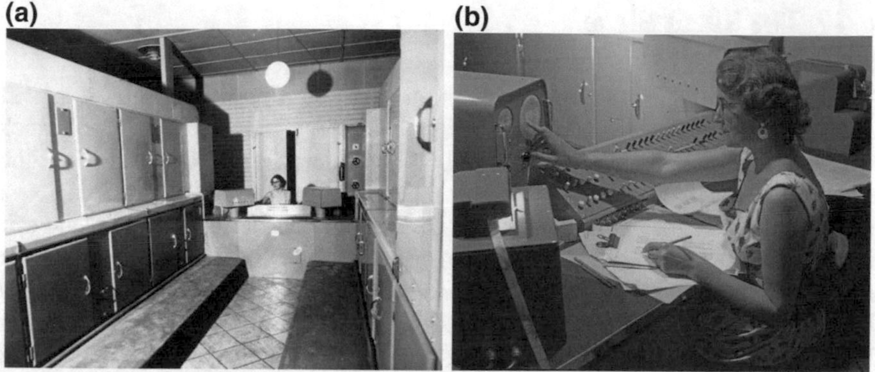

Fig. 6.4 **a** and **b** Avro's Ferranti Mark I* computer in action at Chadderton

"Programmers would develop their own programs on the computer, fault-finding in sessions of perhaps half an hour. When they were developed and added to the program library, it was possible for a computer assistant to take over production runs on the computer with new data applied to the program. Even so, many programmers preferred to run their own, especially if a run would not be straightforward" ".

Joan Smith remembers that: "the mathematicians at Chadderton were not the only Avro employees to use the computer". Besides Woodford staff (see below) "computer time was also taken by Armstrong Siddeley's staff—[Armstrong Siddeley was part of the Hawker Siddeley Aircraft Group, as was Avro]. Programs were run for the design of aero engines: one was the Mamba with contra-rotating propellers, installed in the Fairey Gannet used for coastal work; another was the Viper which went into the Jet Provost trainer. Armstrong Siddeley's would buy time on the Avro's Mark I* until it took delivery of its own machine in 1957". In Chap. 7 we tell the story of the computer installed at Armstrong Siddeley's Ansty, near Coventry, premises towards the end of 1957.

"Research on machine-tool control was carried out in the Computer Department at Chadderton, with a view to making metal models for the wind tunnel if I remember rightly. This was in 1957. The first output tapes were prepared on the computer and run through the computer-controlled machine. Unfortunately the joints were only finger tight, with the result that oil was sprayed under pressure all over the lab!".[14]

Both Peter Teagle and John O'Donnell confirm that that the Ferranti Mark I* ran on a 24-h basis at Chadderton, eventually being used for a range of tasks including payroll calculations. Peter remembers that, in due course, "all day Thursday was used to work out the Avro wages, paid on the Friday".

[14]Smith, Joan. M., *Early Computing in the Aircraft Industry: Avro's at Chadderton*.

6.4 The Silent Slave

By 1964 the digital computing environment at Chadderton had matured. A snap-shot of the organisation in March 1964 is provided by an illustrated article that appeared in an issue of the company's house magazine.[15] Under the headline *Silent slave helps to smash design barriers*, it was reported that "acting as a maid-of-all-work for the design office, the computer department rapidly and efficiently turns out the answers to problems arising from research and investigation into aerodynamics, stress analysis, performance and aero-elasticity". In 1964 the Head of the Computer Department was James (Jimmy) Arrowsmith. The Chief Programmer was Peter Morton, with a programming staff that included Graham Carrol and Peter Cox. The chief maintenance engineer was Gerry Worstencroft, whose team included Ken Shearsmith. A team of data preparation clerks and computer operators was headed by Sheila Whitehead. The 1964 article states that the Mark I* had been "installed in August 1955" but other evidence suggests that this should more correctly refer to the completion of acceptance tests in May 1956. The article then states that the machine was "initially available from 7:30 a.m. to 9:30 p.m. with all-night running on Fridays" but that by 1964 it was available "almost 24 h per day except Sundays".

There is an interesting detail in Fig. 6.5 that has so far escaped surviving technical explanations. Mounted close to the upper edge of the console's set of switches can be seen an extra control panel. This is an add-on, not observable in the surviving photos of other Mark I* installations and not present when the Avro machine was first operational. What was the purpose of this extra panel? Was it part of some special equipment connected to the computer by Avro's engineers or was it part of some experimental input/output equipment attached by Ferranti engineers?

A small shred of evidence pointing towards the first theory has come to light within Reg Boor's description of Chadderton's large analogue computer.[16] There was often a need to provide this analogue computer with an input function based on a particular mathematical equation. Boor writes: "Once, needing a continuous sine wave, I had the sine wave "drawn" on the cylindrical memory drum of the Mark 1* with a lead over the factory roof to the analogue computer!" More generally, Len

---

[15]Hawker Siddeley., News, Aviation Edition.

[16]*The Avro eight degrees of freedom analogue computer*. Reg Boor, January 2016. Two-page typed article sent to the author (SHL) on 4th January 2016. The career path of Reg Boor, who rose to be Head of Applied Aerodynamics and then Head of Aeroelastics within the company, offers an interesting reflection on post-war Britain. In an e-mail to SHL on 10th March 2016 Reg writes: "I enjoyed my career at Avro and have always thanked my "lucky stars" for such a career. I left high school at the age of 16 feeling I should earn money for my mother but after a 6 month hiatus I went back to school and got an application form from the careers master for a "Professional Engineer Apprenticeship" at Avro, which I obtained. This was 1946 and we were given a day off each week to study for HNC [Higher National Certificate, equivalent to today's GCSE A Levels]. But, of course, by 1969 I was surrounded in the office by 1st Class honours and Ph.D. men so when the Open University started I studied with them immediately for a degree in Maths".

3

SILENT SLAVE HELPS TO SMASH DESIGN BARRIERS

Programmers Graham Carroll (left) and Peter Cox try out a new programme on the computer.

Ken Shearsmith (left) and Gerry Worstencroft watch signal waveforms on an oscilloscope connected to the computer.

Chadderton computer department is programmed for progress

Fig. 6.5 Article in the *Hawker Siddeley News* for March 1964. Graham Carrol and Peter Cox are shown in the left-hand photo; Ken Shearsmith and Gerry Worstencroft are shown in the right-hand image. The left-hand photo also shows an extra control panel, positioned just above the normal operator's switches, whose purpose has not yet come to light

Hewitt has looked at the photo in Fig. 6.5 and said: "After further examination I believe whatever the box is it was added by Avro's Electronics Department under Jimmy Arrowsmith. It is certainly an Avro add-on".[17]

Evidence hinting at the second explanation for the extra control panel comes from Ferranti personnel, who describe various attempts to equip Mark I* installations with magnetic tape storage. For example[18] describes a tape system based on 35 mm film stock with a magnetic coating, which was developed by Ferranti for the Mark I* computer at Shell Amsterdam (see Chap. 8). This operated successfully for a short while but was found to have a long-term, time-dependent, flaw so was returned to Ferranti's Moston factory. Various experiments with half-inch magnetic tape systems have also been described, for example,[19] though it is not clear which experiments, if any, were conducted on the Avro computer. Len Hewitt has

[17]Len Hewitt, e-mails to SHL dated 18th May 2016. See Footnote 1.

[18]Johnson, M.H.(Harry). 2002. *My Work with Computers from the Ferranti Mark I to the ICT 1900 (1952–1966).* 84-page typed manuscript, sent to SHL in 2002. Archived at Manchester, see NAHC/SHL/FA1. Of particular relevance is Appendix E: *Ferranti MkI* customers and an early venture into magnetic tape.*

[19]Joan Travis (neé Kaye), e-mail to SHL dated 25th October 2015. Joan joined Ferranti in April 1953 and became involved with the last six Mark I*s. She remembers that: "Generally, except for the Avro computer, I acted as a sort of engineers-mate (analogous to a plumbers-mate) for the design engineers when various 'exotic' peripherals (line printers or mag. tape) were being attached to the computers …. My involvement with the Avro computer was to help two programmers from Rome to use the Mark I* before they got their own".

Fig. 6.6 **a** and **b** Two more photos taken from the 1964 Hawker Siddeley News

added[20]: "I think the mag tape on the Avro Mark 1* was a one-off development which did not go further".

6.5 System Software and Applications Programs

As far as is known, all Avro programs were written in Ferranti Mark I* machine code, which is described in full in Chap. 15. The binary representations, and indeed fine details of some of the instructions, differed from those of the Manchester and Toronto Mark I machines so that there was no direct code compatibility between the Mark I and Mark I*. No autocode systems or symbolic languages were used at Avro.

Like most, if not all, of the Ferranti Mark I* installations, the computer came with a modest amount of system software—essentially little more than what we might today call a loader or an assembler. This consisted of a set of *input routines* stored permanently on drum tracks, the whole being known as the *Radix 32 Input and Organisation Scheme*. This provided a primitive operating environment in which communication between the system and the user was carried out via the tape

[20]Len Hewitt, e-mails to SHL dated 18th May 2016. See Footnote 1.

reader and the console's display tubes and handswitches. A fuller description is given in Chap. 15.

In addition to the Radix 32 Scheme, Ferranti Ltd. made available to Avro various mathematical subroutines for trigonometric functions, etc., and support for more complex activities such as matrix manipulation and equation-solving. By the end of 1955 Joan Smith remembers that there were 29 standard subroutines, "referenced by an identifier A1–A29, for performing common tasks such as output formatting, floating-point operations, trig functions, etc. Ferranti supplied some of the routines, for example a routine for the solution of a set of 84 simultaneous equations, written by Mary Lee Berners-Lee. This was used for wing stressing, for example".[21]

Like all other sites, Avro supplemented these with many internally-written subroutines that were more specific to the company's operations. Avro's Computer Department maintained a stiff foolscap folder containing copies of *Programme Bulletins* issued from time to time by the Computer Group at Chadderton. One surviving folder[22] lists the titles of program bulletins numbered 50–142. Not all the bulletins are dated. For those that are, the dates range from 11th August 1955 to 1st November 1957. Some of the programs are evidently of general use, for example matrix inversion, harmonic analysis, solution of first-order non-linear differential equations, evaluation of convolution integrals. However 77 of the 99 programs are specific to Avro's aerospace applications. Some examples are given in Table 6.1.

6.6 The End of the Ferranti Mark I* at Avro

It is not known for exactly how long the Mark I* remained in operation at Chadderton. It is known that Avro took steps to donate the machine to a worthy destination such as a college. What is certain is that, sometime after January 1966, the computer had been moved to Leicester Museum of Technology where, unable to put it on public display immediately, the museum had stored it in a nearby disused Victorian sewage pumping station.

Under the heading *Computer programmed as showpiece,* an article in the January 1966 edition of the Hawker Siddeley News announced that "A home has now been found for the £100,000 piece of equipment. But its electronic brain isn't likely to tick again for the computer is to become an exhibition piece at the New Museum of Technology at Leicester. … It was used for all the structural, aerodynamic and control calculations for the Vulcan Mark 2. It was also used for the

[21]Smith, Joan. M., *Early Computing in the Aircraft Industry: Avro's at Chadderton.*

[22]Stiff foolscap folder, dark cover, containing copies of Programme Bulletins issued from time to time by the Computer Group at A. V. Roe's main factory at Chadderton near Manchester. A surviving example folder, deposited at the Archive of the Worshipful Company of Information Technologists in London, lists the descriptions of programs numbered 50–142. The exact date-range of these descriptions is not certain but includes the period 11th August 1955 to 1st November 1957.

Table 6.1 Some examples of the 77 applications-specific programs developed by Avro mathematicians at Chadderton in the period up to about November 1957

No.	Title of program	Date of documentation
51	Schuerch wing analysis	11/8/55
54	Flutter determinant	–
59	Multhopp subsonic load grading: 15 × 2 points	22/11/55
61	Kuchemann wing loading	14/3/56
70	Weber and Newby chordwise loading	–
72	Supersonic wind tunnel nozzles	14/3/56
93	Fuel sloshing	–
102	Calculation of RI, UT, dL for honeycomb beams	10/12/56
107	Frame stressing/B	27/8/57
112	Flutter derivatives, Richardson's method	–
115	Etkin-Woodward wing loading	–
132	Machine tool control	–
142	Take off and landing performance	–

design of the 748. … The redundant computer was offered free to anyone who could find it a useful home, preferably for teaching purposes. Offers poured in, but installation, transport and maintenance costs proved too high for the colleges and training schools who made the original offers and for several months the giant machine has laid in store at Chadderton. It will not go on display immediately at Leicester for the museum where the computer will finally rest isn't yet completed".[23]

Further evidence comes from the Ferranti company archivist.[24] In a two-page internal memo written in the early 1970s, Avro's Mark I* computer is described as "stored in the Belgrave sewage pumping station in a disused enginehouse, which also accommodates a collection of old fire engines". The archivist was writing after the computer's whereabouts had been re-discovered by Jane Pugh of the Science Museum (London), in the course of her work to find exhibits for the Science Museums' first Computer Gallery which opened in 1976.

The author was present at a meeting in the disused pumping station, attended by Jane Pugh and representatives from Leicester Museum, Ferranti Ltd., the Royal Museum of Scotland and Manchester University, at which the computer's remains were inspected. They were a sorry sight. A large stack of the Mark I*'s main units

[23]Hawker Siddeley., News, Aviation Edition.

[24]*Manchester University computer Mk I. Result of enquiries made regarding the whereabouts of the first Ferranti Mk I computer*. One-page typed internal memo dated July 1965, assumed to have been written by the Ferranti Archivist at that time. To this document is stapled a two-page addendum entitled *Mk I* computer at Leicester Museum*. This is undated but one can infer that it was written by the Ferranti Archivist (C. J. (Charles) Somers?) sometime between 1973 and 1975. These three pages are currently held in the Ferranti Archive at the Museum of Science and Industry in Manchester.

was piled against an internal wall and had been adorned with pigeon droppings. The computer's console had been detached and was being used as a workbench by the fire engine restorers. To quote,[25] "it was agreed that it was impractical to attempt reconstruction of the complete machine, and that significant parts should be removed and distributed, the remainder being scrapped". The report goes on to say that dismantling was undertaken by Ferranti staff on 6th/7th June 1973—though there is other evidence to suggest that the year might have been 1972[26] or 1974.[27]

According to[28] the components salvaged in the 1970s from the Avro Mark I* computer included:

9 memory units each containing two Williams tubes;
21 door panels containing typical EF50 valve circuits;
1 magnetic drum.

These artefacts were distributed between the Leicester Museum, the Science Museum (London), the Royal Scottish Museum, the Department of Computer Science at the University of Manchester and Ferranti Ltd. In the course of time Ferranti is believed to have distributed its share, which was stored at Orme Mill, Oldham, to the Museum of Science and Industry in Manchester and to the Department of Computer Science at Manchester University. In 1982 the University donated a logic door and a memory unit to the Digital Computer Museum in

[25]*Manchester University computer Mk I. Result of enquiries made regarding the whereabouts of the first Ferranti Mk I computer*. One-page typed internal memo dated July 1965, assumed to have been written by the Ferranti Archivist at that time. To this document is stapled a two-page addendum entitled *Mk I* computer at Leicester Museum*. This is undated but one can infer that it was written by the Ferranti Archivist (C. J. (Charles) Somers?) sometime between 1973 and 1975. These three pages are currently held in the Ferranti Archive at the Museum of Science and Industry in Manchester.

[26]Page 4 of the National Archive for the History of Computing's November 1999 Picture Catalogue has the entry: Ferranti Mark I Star, "rediscovered" near Leicester, 1972. (18 photographs). Unfortunately, by 2015 the NAHC was unable to find its Picture Collection.

[27]The author (SHL) believes that the date of the meeting at Leicester was in 1974. Unfortunately, enquiries at the Leicester Museum of Technology and at the National Museums Scotland (the successor to the Royal Museum of Scotland) have not yielded any documentary evidence whatsoever.

[28]*Manchester University computer Mk I. Result of enquiries made regarding the whereabouts of the first Ferranti Mk I computer*. One-page typed internal memo dated July 1965, assumed to have been written by the Ferranti Archivist at that time. To this document is stapled a two-page addendum entitled *Mk I* computer at Leicester Museum*. This is undated but one can infer that it was written by the Ferranti Archivist (C. J. (Charles) Somers?) sometime between 1973 and 1975. These three pages are currently held in the Ferranti Archive at the Museum of Science and Industry in Manchester.

Fig. 6.7 A twin CRT storage unit from the Avro Mark I*. Each of the two tubes stored 32 × 40-bit words. The sense amplifier and pick-up plate of the upper tube have been swung clear to reveal the tube's screen

Boston, from whence it went to the Computer History Museum, Mountain View, California.[29]

In 1986 another logic door and memory unit were presented to the author by Manchester when he moved to the University of Essex. These two artefacts were displayed in a show-case in the Computer Science Department at Essex. The author retired from Essex in 2002 and, fearing that the showcase would get misplaced in a forthcoming re-ordering of laboratory space, the author arranged for the two artefacts to be donated in 2007 to The National Museum of Computing at Bletchley Park[30] where they are now on display.

[29]Gordon Bell, Vice president of Engineering at Digital Equipment Corporation and co-founder with his wife Gwen of the Computer Museum, Boston, Massachusetts, visited Manchester University in 1982. They exchanged historic computer artefacts with the Department of Computer Science. The Computer Museum at Boston was dissolved in 2000, the Boston artefacts being transferred to the Computer History Museum, Mountain View, California.

[30]When the author (SHL) left Manchester to become Professor of Computer Science at the University of Essex in 1986, his Manchester colleagues presented him with a Mark I* logic door and a Williams Tube unit. These units were moved to the Department of Computer Science at Essex and displayed in a suitable case. On the instructions of professor D. B. G. Edwards the logic door had been re-painted before it left Manchester—better than pigeon droppings but unfortunately in a slightly different colour from its original. SHL retired from Essex in 2002. Five years later the Essex Department was planning to move out of the area in which the Mark I* artefacts were

Fig. 6.8 Some Avro Mark I* units in storage at the National Museums Scotland's facility at Granton, Edinburgh. The magnetic drum is in the centre of the photo

At the time of writing (summer 2018), the largest remaining collection of artefacts is believed to be the Royal Scottish Museum's share, which is currently in storage at National Museums Scotland's Granton facility in Edinburgh. There are

displayed and, anxious for their long-term preservation, SHL arranged to donate the artefacts to the National Museum of Computing at Bletchley Park.

no surviving Accession documents at the National Museums Scotland but, according to[31] the Granton collection consists of the following artefacts:

"Bay D containing 2 Williams tube units, 6 door panels, A & M time base unit;
1 magnetic drum;
1 refrigerator unit".

6.7 What came next at Chadderton and Woodford?

In the mid-1960s the centre of Avro's aircraft design activity transferred from Chadderton to Woodford, ten miles away. The digital computing facilities also transferred to Woodford, where the divisions of Final Testing and Assembly had expanded over the years. For example in 1958 a new Mach 3.5 supersonic wind tunnel had opened at Woodford.[32] Also at about this time, Woodford became involved with tests on the Blue Steel stand-off bomb which was designed to be dropped by Avro's Vulcan bomber—see below.

At first Woodford did not have a computer. Joan Smith remembers that, in the late 1950s, "All work on missiles was carried out at the Woodford site. This meant a car-load of mathematicians from Woodford would regularly descend on the Chadderton site to run their programs. This was secret work, and no one was permitted to enter the computer room when Woodford staff had booked the machine. On returning to Woodford they would take all paper tape output; indeed nothing was left in any waste bin".[33]

Joan Smith also remembers that "three wind tunnels were also housed at Woodford: low speed, the cross section of which was large enough to walk into when necessary to mount a finely-crafted wooden model (of a wing, say) on a sting; a medium-speed tunnel, and a high-speed tunnel with a very small working section". From 1958 periodic transfers of digital data took place from Woodford to Chadderton. For example strain gauge measurements from up to 150 points in the wind tunnel facility were digitised, stored locally, and transferred to punched paper tape which was then taken by road to the Mark I* at Chadderton. Somewhat later, the transfer was made by teleprinter link.

---

[31]*Manchester University computer Mk I. Result of enquiries made regarding the whereabouts of the first Ferranti Mk I computer*. One-page typed internal memo dated July 1965, assumed to have been written by the Ferranti Archivist at that time. To this document is stapled a two-page addendum entitled *Mk I* computer at Leicester Museum*. This is undated but one can infer that it was written by the Ferranti Archivist (C. J. (Charles) Somers?) sometime between 1973 and 1975. These three pages are currently held in the Ferranti Archive at the Museum of Science and Industry in Manchester.

[32]*Avro's Mach 3.5 tunnel – a new intermittent blow-down facility*. Flight Global, 11th July 1958, pages 34 & 35. See: https://www.flightglobal.com/pdfarchive

[33]Smith, Joan. M., *Early Computing in the Aircraft Industry: Avro's at Chadderton*.

The growth of Woodford's own IBM equipment from 1957 to the mid-1960s is remembered by the following three former employees.

Ray Oates started at Woodford in August 1956 as an assessment assistant, at which date there were no computers at Woodford. Ray remembers[34] that "We were given data on either Blue Steel or Blue Streak or perhaps both and charted and graphed it. After an office move the department grew and was called Trials Analysis. We then moved to considerably bigger offices which was when the IBM computer arrived".

David Joy, who joined Woodford early in 1957, remembers "it was later that year [1957] that an IBM 604 computer was installed in offices on the second floor overlooking the machine shop which later became the production line for the Blue Steel. The 604 was a valve machine and created an awful lot of heat, so much so that an extraction system had to be installed (not air conditioning!). The early measurement system was overtaken by the installation of some Benson Lehner equipment that made measuring telemetry channels much easier. This produced punched cards/tape which were read by the 604 and an analysis of these was produced on a very large printer. The 604 lasted for around 18 months to 2 years before a much larger and more powerful IBM 650 was installed in an office at the back overlooking the airfield. I remember it being in a U shape and that the floor had to be reinforced. It was still there when I left in 1965".[35]

Gillian Mills, who started at Woodford in 1960, also worked on the analysis of trajectory data for Blue Steel. She remembers[36] that "We used all IBM ancillary equipment and processing was done by the IBM 650 Computer with Magnetic Drum, Vacuum Tubes and 80 Column cards. We used Marchant electromechanical calculators, great big machines that used to bounce across the desk when multiplying. We used Benson Lehner equipment to read the trajectory data from film onto punch cards. This data was processed by the computer and we then plotted the trajectories on very large flat-bed plotters. The data came from both the Australian Woomera (over land) and the Welsh Aberporth (over sea) test sites".

In addition to the IBM 650 used for the analysis of trials data, various projects within the Avro-Whitworth Division of Hawker Siddeley Aviation Ltd. must have required other, more powerful, computing facilities. By 1965 the computers available throughout the Hawker Siddeley Group included Ferranti Pegasus installations at Kingston-upon-Thames (from October 1956), Coventry (from November 1956), Hatfield (from August 1957) and Brough, Yorkshire (from March 1958). From 1958 Avro's designers could also have used the more powerful Ferranti Mercury computer at Manchester University or, from about 1964, the even more powerful Ferranti Atlas computer at Manchester. There is evidence that the Atlas was indeed used.[37]

---

[34]E-mail from Ray Oates to SHL, on 13th January 2016.

[35]E-mail from David Joy to SHL, on 19th January 2016.

[36]E-mails from Gillian Mills to SHL, on 6th and 7th January 2016.

[37]E-mail from Reg Boor to SHL, on 11th January 2016.

In summary, the sequence of computing facilities at Woodford from 1966 onwards is not quite clear. It is believed that ICT (who took over Ferranti's mainframe interests in 1963) may have supplied an installation that superseded the IBM 650. There is some evidence that this was a Ferranti Pegasus,[38] though the surviving Pegasus installation records do not show any deliveries to Woodford.

By 1972 it is thought that an ICT 1900 series computer had been installed at Chadderton for administrative purposes, with a committed landline to Woodford, and that the ICT 1900 had then been replaced by an ICL 2900 system by about 1981. By then it is thought that Woodford lacked any of its own mainframe computing resource, no doubt sharing facilities at Chadderton and at other British Aerospace sites. The Woodford site was finally closed and sold off by BAE Systems in 2011. In 2015 the Avro Heritage Museum was opened at Woodford.

After the 1980s Joan Smith recalls that: "there was continued development, the aircraft industry leading the way with scientific software, much as in the 1980s the defence industry led the way for CALS (Computer-Aided Acquisition and Logistic Support), based on the SGML (Standard Generalised Markup Language) standard for structured information systems, and conformity to which was a requirement of the US Department of Defence. The UK Ministry of Defence followed suite, with companies such as British Aerospace (into which Avro's had merged) and Rolls-Royce needing to conform in order to gain contracts".[39]

References

Hawker Siddeley. 1966. News, Aviation Edition. 3 (1): 2.

Johnson, M.H.(Harry). 2002. *My Work with Computers from the Ferranti Mark I to the ICT 1900 (1952–1966).*

Avro's Mach 3.5 tunnel—a new intermittent blow-down facility. Flight Global, 11th July1958, pp. 34 and 35. See: https://www.flightglobal.com/pdfarchive.

Smith, Joan. M. 2016. *Early Computing in the Aircraft Industry: Avro's at Chadderton.*

[38]E-mail from Reg Boor to SHL, on 11th January 2016.

[39]Smith, Joan. M., *Early Computing in the Aircraft Industry: Avro's at Chadderton.*

Chapter 7
The Mark I* at Armstrong Siddeley, Ansty, Coventry

7.1 Reluctance at Armstrong Siddeley Motors

Armstrong Siddeley Motors Ltd., which came into being in 1919, had become part of the Hawker Siddeley Aircraft Group in 1935. Amongst other things, the Armstrong Siddeley company made aero engines. They had started work on their first gas turbine aero engine in 1939 at Parkside in Coventry, with testing being carried out at a rural site at Ansty, about five miles north-east of the centre of Coventry. In 1959 Armstrong Siddeley was merged with Bristol Aero Engines, the aircraft engine wing of the Bristol Aeroplane Company, to form Bristol Siddeley Engines Ltd. In 1959 Flight Magazine was able to say of Bristol Siddeley: "In terms of manpower, it is believed that (possibly excluding Russia) the new company is the third largest aero-engine concern in the world, next in size to Pratt & Whitney and Rolls-Royce. In terms of research and development and production facilities the Bristol Siddeley organization is thought to be the leading company of its kind".[1] In 1968 Bristol Siddeley became part of Rolls-Royce Ltd.

Armstrong Siddeley Motors Ltd. (ASM) was at first reluctant to invest in digital computing equipment. This reluctance has been eloquently described by David Evans, ASM's Chief Stress Engineer and head of the Stress Department which was part of the Mechanical Research Department. In a paper entitled *Setting up a computer department,*[2] Evans writes:

"Until the end of 1954 we still tended to regard the computer as a toy, and to consider that an engineer's decision was more likely to be delayed than improved

---

[1]Flight magazine, 6th February 1959, p. 176. This is a six-paragraph editorial piece headed *Bristol Sidney Engines.* See: https://www.flightglobal.com/pdfarchive/view/1959/1959%20-%200371.html.

[2]Evans, D. R. 1958. Setting up a Computer Department. *British Communications and Electronics* 5 (12): 918–922. David Evans was Chief Stress Engineer and head of the stress department. Evans has been described by John McNamara, in an e-mail dated 5th May 2017 to SHL, as "a very pleasant gentleman, very sociable".

S. Lavington, *Early Computing in Britain*, History of Computing,
https://doi.org/10.1007/978-3-030-15103-4_7

by its use. Most engineers are much happier with calculations which they can carry out on their own slide rules, and are reluctant to hand over to mathematicians or computers. The more remote the calculation is from their own work the less happy they are. This is a sound and reasonable attitude … However we engaged a keen young mathematician and looked into the alternative virtues of the different machines available.

"The situation was radically changed when, towards the end of 1955, a sister firm in the Hawker Siddeley Group, A. V. Roe at Manchester, became the first company in the Group to buy an electronic computer. They chose a Ferranti Mark I* and, being somewhat concerned about being able to make full use of it, their Chief Designer called a meeting of representatives of most of the Group firms to discuss the possibility of joint use of the machine …. From then on (October 1956) the small ASM team we had started to build up spent a proportion of almost every week at Manchester and this continued until September 1957". As described in the previous chapter, Avro's Ferranti Mark I* had actually been delivered at the very end of 1955 or, more probably, in early 1956 and had passed its acceptance tests by about the end of May.[3]

Evans continues: "At one time we had hoped that we could reduce the amount of very tedious travelling between Coventry and Manchester, by the use of a teleprinter operated from Coventry, with a machine operator [at Manchester] to put the tapes on the machine and teleprint the resulting tape back to us. … The standard of accuracy was insufficient … Also at that stage the great majority of our work was on programme development for which the presence of the programmer was essential".

7.2 Enthusiasm Grows

By 1956 Armstrong Siddeley Motors was re-considering its investment position. In the Minutes of a Board meeting that took place on 31st October 1956[4] it is recorded that: "Mr. W. H. Lindsey stated that he was anxious to obtain approval to buy a Computer, as he felt that the load in future on the Avro Computer would be such that to share was no longer a practical proposition. The Board approved a spend of up to £75,000 on this project. This expenditure was strongly supported by Sir Arnold [Hall, Director, Hawker Siddeley Group Ltd.]".

By early 1957 enthusiasm at Armstrong Siddeley had grown. The Minutes of a Board Meeting held on Thursday 17th January 1957 contained a section headed

[3]Len Hewitt, e-mail exchange with SHL, September 2015. Len joined Ferranti Gem Mill in mid-1955, then went as the Ferranti maintenance engineer on the Avro Mark I* and left Ferranti in mid-1957 to maintain a Pegasus computer at ICI.

[4]The Armstrong Siddeley Board Minutes have been transcribed by Peter Barnes, the Librarian of the Coventry Branch of the Rolls-Royce Heritage Trust. Peter kindly sent the information to SHL in a series of e-mails starting in October 2015.

Aero Engine Development, within which was a report on the P176 turbojet engine. A capital spend of £2,458,792 was proposed for machine tools, test equipment, building works, etc. Amongst the list was:

Provision Electronic General Equipment	£35,000
Provision Electronic Digital Computer	£80,000
Commercial Computer	£35,000

By about this time, ASM was using the Avro Mark I* for about five hours every week and this involved a cost which would make purchase of a medium-sized computer an economic proposition. According to Evans[5] ASM considered buying a Ferranti Pegasus, an Elliott 402/404, an English Electric DEUCE, or a Ferranti Mercury computer—though the latter was soon ruled out as the quoted price rose dramatically. Mercury was in any case much more powerful than the other three machines. David Evans adds[6]:

"We had already realised that the true economies were almost impossible to assess. The value of the work done by the machine lay in doing jobs more quickly than before, and in carrying out work in a manner which would have been impracticable without the machine. More indirectly it raises the efficiency of use of technical engineering staff by relieving them of the labour of arithmetical calculations. We have found with the Ferranti Mark I* that a ratio of 100-to-one compared with desk calculations covers most production-programme rates approximately. The machine is basically much faster than this, but the time includes the extra detail and precision provided and also certain processes and decisions in which a human operator is relatively more efficient".

Evans recalls that: "One of the chief difficulties now was the question of the delivery date. For most of the new machines, 18 months was the minimum time and in view of the general teething troubles it seemed more likely to be two years …. At this time [early 1957] it came to our notice that a Mark I* machine, at a Ferranti factory in Manchester, was completing the development work on magnetic tape on which it was engaged and would be available for sale. Although it was an older design of machine, and was second-hand, it would be available soon. Furthermore, it had a large storage capacity, which is a great convenience for many reasons even if it is not often used fully …. Another point considered was that the more modern machines were not, in our opinion, really more advanced than the Mark I* …. Ferranti made a certain adjustment to the price and agreement was reached".

[5]Evans, D. R., Setting up a Computer Department. *British Communications and Electronics* 5 (12): 918–922.

[6]Evans, D. R., Setting up a Computer Department. *British Communications and Electronics* 5 (12): 918–922.

7.3 Installation and Maintenance

Armstrong Siddelely took delivery of the last Ferranti Mark I* computers to have been built, the difficult-to-sell *Machine Number 9* of Chap. 5, in the autumn of 1957. The machine had passed its Acceptance Tests at Ferranti's Gem Mill factory in July 1957 and the tests were satisfactorily repeated at ASM's Ansty site at the beginning of December.[7]

Armstrong Siddeley's Mark I*, DC9, was installed in a new R&D building at Ansty that had come into use earlier in 1957[8]—see Fig. 7.1. At Ansty the Mark I* was part of the Stress Office section and so the computer was located in that area of the building. The Ferranti commissioning engineer was Owen Ephraim. One of the Armstrong Siddeley electronics engineers who was involved, John McNamara, remembers that Ephraim was due to carry out work on the Ferranti Mark I* installation in Rome before Christmas and was hurrying to complete the Ansty Acceptance Tests and finish the installation in good time. He was successful.[9]

The installation was a matter of considerable local interest. Under the headline *Armstrong Siddeley install £90,000 'Schoolboy's Dream'*, a contemporary issue[10] of the Coventry Evening Telegraph described how "a £90,000 electronic brain just installed at Armstrong Siddeley Motors, Ansty, … can carry out calculations at 100 times the speed of a skilled operator using a mechanical desk calculating machine." The article was accompanied by the photo shown in Fig. 7.4.

At the time of installation, Armstrong Siddeley had appointed their first computer maintenance engineer, Eric Griffiths. He was soon joined by John McNamara and then, in due course, by about three more engineers including Bill Blakemore and Don Mather.[11] To begin with, "neither had any foreknowledge of digital

[7]Evans, D. R., Setting up a Computer Department. *British Communications and Electronics* 5 (12): 918–922.

[8]*Armstrong Siddeley at Ansty*. Anonymous illustrated article in The Aeroplane, 27th June 1958, pp. 893–896.

[9]John McNamara, telephone conversation with SHL, 24th February 2017. John had joined Armstrong Siddeley at Parkside, Coventry, in 1954 after National Service as a radar technician in the RAF. He started in Armstrong Siddeley's Mechanical Research Department as part of a small group solving problems arising from blade vibration, compressor rotating stall, etc.

[10]It is believed that the article did appear in the Evening Telegraph but the preserved piece of original newsprint unfortunately does not contain a head-of-page date. A quick search of back-numbers of this newspaper from November 1957 to February 1958 has been carried out but the article was not noticed. Further research is needed.

[11]John McNamara, telephone conversation with SHL, 24th February 2017. John had joined Armstrong Siddeley at Parkside, Coventry, in 1954 after National Service as a radar technician in the RAF. He started in Armstrong Siddeley's Mechanical Research Department as part of a small group solving problems arising from blade vibration, compressor rotating stall, etc.

Fig. 7.1 Armstrong Siddeley's new (1957) R&D building at Ansty

computers nor indeed had ever seen one. They were given a crash course at Ansty and then left to fend for themselves in looking after the machine".[12]

David Evans recalls that: "Although Ferranti were responsible for the first year's maintenance, it was desirable for us to train our own men as soon as possible, and the initial installation and commissioning process—carried out very efficiently by Ferranti staff who thought they'd seen the last of Mark I*s—was a very valuable experience for our engineer. It later turned out to be very fortunate that [Eric Griffiths] had acquired this background, as it became necessary for him to take over the maintenance of the machine in March 1958". Eric Griffiths went to Bristol in 1959, where the Bristol Siddeley Aero Engine Company had an English Electric DEUCE computer. John McNamara then became the head of Ansty's computer engineering group.

John McNamara remembers[13] that: "Initially, fault diagnosis and repair was mind bending, but having set a policy of not going home until the machine was in full working order the learning curve was greatly improved. Nevertheless, many long hours and late nights were encountered. The maintenance of the Ferranti

[12]Williams, David. 2003. Information Technology at Ansty. Article in The Sphinx. *The Coventry Branch Magazine* (53): 13–18. (This is the magazine of the Rolls-Royce Heritage Trust, Coventry Branch). One or two of the numbers in this paper have subsequently been corrected by John McNamara, in an e-mail to SHL dated 6th March 2017. The corrections only involve minor matters.

[13]McNamara, John. 2012. *The Ferranti Computers at Armstrong Siddeley Motors Limited, Coventry.* 6-page typed paper, July 2012. Circulated privately.

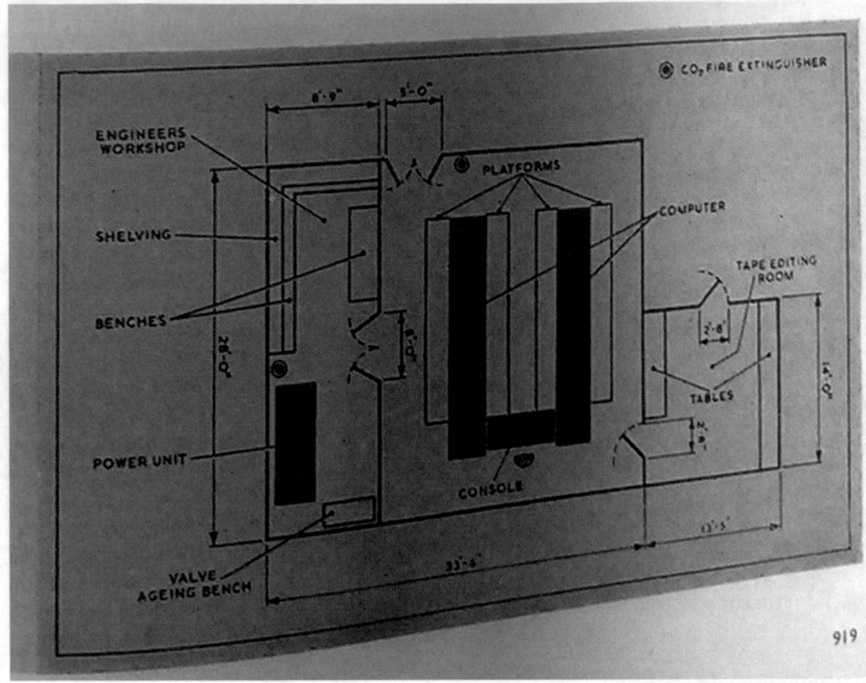

Fig. 7.2 The initial floor plan of the Ferranti Mark I* installation at Armstrong Siddeley's Ansty site. The darker units represent the computer and its power-control unit. Not shown are the motor-generator set and the air conditioning equipment

Computer was not just the understanding of the computer logic and its electronic circuitry, but its power supplies (300, 200, −150 V DC) and cooling system … Additionally, there was a considerable amount of peripheral equipment [manufactured by Creed & Company of Croydon] in the form of paper tape readers, punches and teleprinters that were used for program preparation and as part of the input/output facilities of the computer".

There was a three-hour daily preventive maintenance period. In mid-1958 David Evans observed that "three hours may seem long but there is no doubt about its necessity and value at present … However, with the machine settling down and with greater operational experience, we hope in time to reduce the three-hour period without impairing efficiency …. Our experience has, I think, been similar to other groups in that the mechanical parts have given a surprisingly high proportion of the troubles. The refrigeration was the most serious problem … the [paper tape] punch also gave a fair amount of trouble … All types of fault are reducing steadily as the machine settles down. Our engineers have now [mid-1958] acquired considerable confidence in handling the machine and, although we do not hesitate to call in Ferranti in case of difficulty, we are in general capable of looking after the machine

Fig. 7.3 Two maintenance engineers at work. *Left-to-right:* Don Mather studies a logic diagram and adjusts an oscilloscope whilst Eric Griffiths monitors a logic signal

ourselves. The drum as yet we know very little about and we are—as instructed—just leaving it alone".

Evans made this general comparison about errors: "Calculators on [electro-mechanical] desk machines make frequent errors, so much so that if a really high degree of accuracy and reliability is required, many checks and cross-checks are needed. Electronic machines [i.e. digital computers] have various checks built-in and applied in the programmes, but these checks are much less necessary than they are for desk-machine work".

7.4 The Programmers at Armstrong Siddeley

The Chief Programmer at Ansty was David Atherley (see Fig. 7.4), who ran an initial group of four or five young programmers straight from university that included Alan Sercombe and John Coleman and probably Gordon Bonsall. Alan's recruitment story is typical. Nearly 60 years after the events, Alan recalled his introduction to the Ansty computer as follows.[14]

"I graduated in 1955 with an honours degree in Mathematics from Queen Mary College then, after a delay of some months, commenced two years National Service in December 1955. After training I spent 18 months in Singapore, returning home for demob at the end of November 1957. I spent my demob leave on job interviews. The most interesting offers were both as a computer programmer, one with Rolls Royce and the other Armstrong Siddeley. The salaries were 12 pounds 5 shillings and 15 pounds 5 shillings per week respectively so I sent myself to Coventry!

"I started in mid-January 1958 and the computer was fully operational at that time. It was located next to the Stress Office and the whole computer section reported to the Chief Stress Engineer [David Evans]. It was staffed by a Chief Programmer [David Atherley] and I made up a complement of 4 or 5 programmers. I think I was the last programmer to join the team. Also there were 4 maintenance engineers employed by the company not Ferranti. Eric Griffiths [the leading engineer] was a heavy smoker and the evidence of the late nights worked to keep the machine operational was the cigarette butts on the floor of the computer room in front of the cabinets they had worked on.

"The computer had 13 Williams Tubes of main memory, 5 hole paper tape input and output, a magnetic drum and a character printer (Teleprinter). Subsequently the Ferranti maintenance engineers installed 3 additional tubes of memory bringing it up to the maximum of 16 tubes. This was done to make it compatible with the Mark I* installed in the Royal Dutch Shell Group [in Amsterdam, see Chap. 8].

"As they weren't ready to start my programmer training I was given a desk in the Stress Office to learn about calculating the stress profile of 'vaned impellers' of aero engines. The calculations were then being done on a manual Comptometer and took a week. My first job was to replicate this with a computer program. I was also given a programming manual which I believe was written by another user [probably Avro], not Ferranti. Part of my training was to write and test subroutines to perform division and other mathematical functions including the trig functions. When I started there was a suite of such programs already installed which may have been imported from either Ferranti or another user or maybe written by David Atherley. My development was just a training exercise.

---

[14]A series of e-mails from Alan Sercombe to the author (SHL), May 2016 to March 2017. One or two of Alan's numbers in his e-mails have subsequently been corrected by John McNamara, in an e-mail to the author (SHL) dated 6th March 2017. The corrections only involve minor matters and have been agreed by Alan. Alan now lives in Melbourne.

Fig. 7.4 David Atherley, the Chief Programmer at Ansty, at the Mark I*'s console

"The Programming Room between the Stress Office and the Computer Room was along one side of the Drawing Office. As the Programming Room only held four, I was found a desk in front of the first row of draughtsmen. [The huge Drawing Office is shown in Fig. 7.5]. After moving there I found developing the subroutines fairly straightforward but a complete program was more complicated. Handling input and output; transferring control within a program; absolute addresses; modifying a program, etc.

"After finishing the program for the impeller stresses it took five minutes on the computer as opposed to one week manually. Even so, on one occasion when after several iterations the computer was not providing an acceptable stress profile David Evans, definitely old school, sent it back for manual calculation—which was acceptable. Smiles all round in the Stress Office, but not from me. It took me a couple of weeks to point out the mistakes made by the manual operator, to get final agreement that the computer was correct".

Fig. 7.5 The drawing office at ASM Ansty

Discrepancies between manually-calculated and computer-calculated results, usually due to inappropriate rounding in the former, are also mentioned in Chap. 6.

Alan Sercombe continues: "The only other program that I remember writing was to optimise the fuel mixture for the Black Knight rocket then being tested at Woomera [in South Australia]. It was some time in 1959 that it was announced that the Aero Engine Division of Armstrong Siddeley Motors was to be acquired by Bristol Aero Engines, with the likelihood that a move to Bristol some time would be on the cards. About the same time an advertisement for programmers for Standard Triumph's LEO installation in Coventry caught my eye. [Standard Triumph was a long-established motor car manufacturer based in Coventry]. I applied and was one of six who were successful. After four weeks LEO training, as the only one with previous programming experience, I was appointed senior programmer. That put an end to my time with the Mark 1*."

7.5 Integrating Computing Within the Company's Activities

In 1958 it was not clear how a new computer should be integrated into an aerospace company's existing working practices. In particular, should existing engineering staff be encouraged to write programs or should specialist mathematical programmers write the code? The programmers were generally new recruits so there was bound to be a culture gap between them and the more traditional aero engineers. To quote Evans[15]:

"There are two broad schools of thought concerning the handling of a large computer: one, that it should be run entirely by a specialist group of programmers, and the other, that the central group should carry out training and general work, with specialists in all departments trained to do their own programming. We favour the first approach for many reasons. It takes a long time for a programmer to become reasonably efficient on a machine such as the Mark I*. He may 'learn programming' in three months or less, but cannot usually be regarded as a trained programmer in under a year .... Departmental specialists would in many cases find it difficult to reach this standard on a part-time basis, with consequent inefficient use of both the machine time and, even more important to us, the instruction and explanatory time of the senior programmers".

By the end of 1958 Armstrong Siddeley Motors were employing nine programmers, all mathematics graduates. It is believed that these programmers did their own data-preparation. "Many organisations employ young women as machine operators and tape punchers. We have only one, with an 'A' level School Certificate, but once again the proportion may change if more production work is done".[16]

7.6 Applications

What was the range of tasks performed by the Ferranti Mark I* at Ansty? By mid-1958 Flight magazine reported that the computer "is working on engine performance calculations, predictions of compressor and turbine characteristics, whirling speeds and disc stressing".[17] A contemporary internal company document survives, which is headed: *A Design Method for Centrifugal Compressors, Using the Ferranti Mk.1 Electronic Digital Computer.*[18] Later on, the uses included "the analysis of the vast quantity of data from tests on combustion chambers and pumps … At the start all programs were written in machine code so that only programmers with an intimate knowledge of the Ferranti instruction repertoire could use the

[15]Evans, D. R., Setting up a Computer Department. *British Communications and Electronics* 5 (12): 918–922.

[16]Evans, D. R., Setting up a Computer Department. *British Communications and Electronics* 5 (12): 918–922.

[17]Flight magazine, 6th June 1958. This is a brief editorial piece headed *Computers in service.* See: https://www.flightglobal.com/pdfarchive/view/1958/1958%20-%200772.html.

[18]See document PA1716/5/9/577, which is held in a collection headed *Armstrong Siddeley Motors Project Design Office* at the Coventry History Centre. Apart from the header sheet's title of *A Design Method for Centrifugal Compressors, Using the Ferranti Mk.1 Electronic Digital Computer* the rest of the document consists of a set of drawings. [This document might refer to a rocket application, rather than to a gas turbine application].

Fig. 7.6 The two people using the paper tape editing equipment are John Coleman *(left)*, a programmer and Anne Blackburn a machine operator

machine. Later on, user languages were developed, thus enabling various engineering departments at Ansty and the Bristol Engine Division to use it".[19] One such user language was TIP (Tabular Interpretive Programming), see [20] and Chap. 15. A special user was the High Altitude Test Plant at Bristol, where data was processed at Ansty and returned while testing was in progress. A special Post Office tie-line was rented for the data-transmission between Ansty and Bristol.[21]

Programmers at Ansty also used Intercode, a kind of modest Autocode developed at Fort Halstead—(see Chap. 11). Intercode was "an interpretive simplified coding scheme, designed for 'small' computations only … it also serves as an introduction to proper programming".[22] It was used at Ansty for small tasks, "bringing certain simpler types of problem, which would not normally have justified programming, into the range of the machine".[23] More information about Intercode is given in Chap. 15.

---

[19]Williams, David., Information Technology at Ansty. Article in The Sphinx. *The Coventry Branch Magazine* (53): 13–18.

[20]*Tabular Interpretive Programme*. Bristol Siddeley Engines Ltd., Filton, Bristol. June 1961. 8-page printed manual, approximately A4 size, bound in buff card. Tabular Interpretive Programme, TIP, was devised by the Mathematical Services group of Bristol Siddeley Engines Ltd. in 1957. TIP3, to which this manual refers, was completed in August 1960. "To date, TIP has been made available for the DEUCE Marks I, II and IIA, and the Ferranti Mark I* computers".

[21]McNamara, John., *The Ferranti Computers at Armstrong Siddeley Motors Limited, Coventry.*

[22]Berry, F.J. 1959. Intercode, a Simplified Coding Scheme for AMOS. *Computer Journal* 2 (2): 55–58.

[23]Evans, D. R., Setting up a Computer Department. *British Communications and Electronics* 5 (12): 918–922.

From time to time work was done for the Royal Dutch/Shell Laboratory, Amsterdam, which had taken delivery of its own Ferranti Mark I* in mid-June 1954—see Chap. 8. Alan Sercombe recalls that "Shell wanted to do a 5 quarter forecast of their world wide activities over a weekend but could not do it within the timescale on their machine only". John McNamara remembers that "there was a requirement to keep the machine running for two and a half hours without error. This was a hard target to meet but we managed it on a regular basis. The Shell job was secret and no Armstrong-Siddeley staff were allowed in the computer room during the runs". The exact nature of Shell's computing jobs has not been recorded.

7.7 Acquiring a Second Mark I*

The computing load on the Ferranti Mark I* at Ansty grew until, by 1959, the machine was reaching saturation. John McNamara remembers that quite by chance, an advert in Wireless World was spotted, offering a second-hand Mark I* for £4,000 and Eric Griffiths phoned the Manchester number, which turned out to be that of a scrap dealer. The dealer did have a Mark 1* but there would be difficulties in arranging to view it. Only three people would be permitted to do so and they had to be cleared by MI5. Given this clearance the Armstrong Siddeley party was met at Cheltenham railway station whence they were taken to GCHQ. The party consisted of Dennis Boston (Head of Mathematical Services, Bristol Engine Division), Eric Griffiths and John McNamara. They were allowed to view the computer under a GCHQ police escort.[24]

The background to the GCHQ computer, which was known as DC3 by Ferranti and Cleopatra at Cheltenham, and its original shipment in the autumn of 1953 is given in Chap. 10.

John remembers[25] several visits to Cheltenham in 1959, when the GCHQ machine was carefully taken apart and the smaller pieces taken back to Ansty in a 30cwt van. Finally two low-loader lorries were hired to take the heavier pieces. The cost of re-installation at Ansty, including equipping a second machine room complete with a Faraday Cage, etc., was £8000. Armstrong Siddeley therefore acquired their second Mark I* for a total of £12,000—a real bargain.

No clear images of the second Mark I* at Ansty have so far been located, though the computer is just visible in the photograph of Fig. 7.7.

The period 1960–63 was a busy one for computational activity at Ansty and at Bristol. The computer engineers at Ansty oversaw all tape editing equipment for both sites and were also in charge of data transmission and instrumentation

[24]McNamara, John., *The Ferranti Computers at Armstrong Siddeley Motors Limited, Coventry.*

[25]John McNamara, telephone conversation with SHL, 24th February 2017. John had joined Armstrong Siddeley at Parkside, Coventry, in 1954 after National Service as a radar technician in the RAF. He started in Armstrong Siddeley's Mechanical Research Department as part of a small group solving problems arising from blade vibration, compressor rotating stall, etc.

Fig. 7.7 The two Ferranti Mark1* computers in operation at Ansty in February 1961. The nearest machine is DC9 with Ann Blackburn (?) at the console. The computer in the background is DC3, the former GCHQ computer, with an unidentified man at the console

transmission for the two sites. Eric Griffiths and John McNamara found themselves travelling frequently between Ansty and Bristol in the period 1960–63. During this time the two Ferranti Mark I*s remained fully operational whilst at Bristol the DEUCE computer, which had been delivered in June 1956 and fully operational by 1957,[26] was kept busy until the end of its useful life. In fact, in due course there were two English Electric DEUCE computers within the Bristol Aeroplane Company's embrace: one at Bristol Aircraft Ltd (delivered June 1956) and the other at Bristol Aero Engines Ltd. (delivered in February 1957). It is believed that both were replaced in about 1964 by an English Electric KDF9 computer, with Eric Griffiths heading up the engineering aspects of the installation.

[26]*The Brain comes to Filton.* Bristol Review (the Bristol Aeroplane Company's house magazine), Vol. IV, No 4, Autumn 1956, pp. 12–13.

Fig. 7.8 The Computer Engineer's Office at Ansty in February 1961, with the Programmer's and Operator's Office in the background. In the left foreground is John Blakemore, a programmer. In the centre foreground is John McNamara, who is setting up the pantograph on an electrolytic-tank model of a cooled turbine blade. The three men towards the centre of the photograph are *(left-to-right)* Alan Lewis who is seen testing a thermionic valve at a valve-ageing unit; Bill Blakemore seated at the drawing board; Don Mather standing in front of an oscilloscope. In the background, seated at his desk and on the 'phone, is Eric Griffiths

7.8 What Came Next?

In 1964 the two Mark I*s at Ansty were both scrapped. By then the main focus of Bristol Siddeley's computing activity had moved to Bristol and the KDF9, where computing was organised by Mathematical Services under Dennis Boston. Ansty was connected to Bristol by landline, with day-to-day use being administered by the Rocket Research group.[27]

The computing history at Ansty from 1964 until the 1980s has been described by David Williams.[28] Throughout the period, Ansty's scientific computing facilities were determined by the evolving scene at Bristol. A second KDF9 was acquired by Bristol at the Patchway site and in 1970 a Digital Equipment PDP10 computer was

[27]Williams, David., Information Technology at Ansty. Article in The Sphinx. *The Coventry Branch Magazine* (53): 13–18.

[28]Williams, David., Information Technology at Ansty. Article in The Sphinx. *The Coventry Branch Magazine* (53): 13–18.

Fig. 7.9 Storage racks for spare thermionic tubes (valves) for the Ferranti Mark I* computers at Ansty in 1962

linked with the two KDF9 s. Terminals were connected to the PDP10 and on-line computing from Ansty became available from 1971. In 1974 a multiplexer and a modem together with a PDP8E workstation were installed at Ansty. In 1974/5 Bristol installed a Control Data Cyber 74 to replace both KDF9 s. The number of terminals at both Ansty and Bristol were duly increased. In the early 1980s the Control Systems department at Ansty obtained its own PDP11 computer which had a 300 megabyte disc.

In 1983 the company decided to standardise on IBM mainframe computers. An IBM 360/165 was installed at Bristol, with Ansty having connections via high-speed lines. In 1986 a 6-station CADD4X was installed at Ansty for computer aided design work. Meanwhile, for commercial data-processing and administrative purposes an ICL 1901 was installed at Ansty in the late 1970s. This was soon augmented by an ICL 1902.

References

Berry, F.J. 1959. Intercode, a Simplified Coding Scheme for AMOS. *Computer Journal* 2 (2): 55–58.

Evans, D.R. 1958. Setting up a Computer Department. *British Communications and Electronics* 5 (12): 918–922.

McNamara, John. 2012. *The Ferranti Computers at Armstrong Siddeley Motors Limited, Coventry.*

Williams, David. 2003. Information Technology at Ansty. Article in The Sphinx. *The Coventry Branch Magazine* 53: 13–18.

Chapter 8
The Ferranti Mark I* Installation in Amsterdam

8.1 KSLA, the Shell Company's Laboratories in Amsterdam

A Ferranti Mark I* computer was delivered to Shell's Amsterdam Laboratory in 1954. Before describing the machine and its uses, it is helpful to introduce the company and to outline the gradual adoption of digital computers in Holland in the immediate post-war years.

Shell, the multi-national oil and chemicals company, was formed in 1907 through the amalgamation of the Royal Dutch Petroleum Co. of the Netherlands and the "Shell" Transport and Trading Co. of the United Kingdom—(the quotation marks were part of the registered name). Until its unification in 2005 the firm operated as a dual-listed company in which the British and Dutch enterprises maintained their legal existence but operated as a single-unit partnership. The parent company of the Shell group is now called Royal Dutch Shell plc. In 1914 Shell established its first Laboratory in Amsterdam, the organisation becoming known as KSLA (Koninklijke Shell-Laboratorium, Amsterdam).

In due course KSLA became the name of a large multi-purpose Shell development, located west of Buiksloterweg in Amsterdam-North. It covered an isolated, fenced, part of an island, accessible only by a company ferry from the city or across the gated bridge of Tolhuisweg. As part of the KSLA complex, a new Laboratory

S. Lavington, *Early Computing in Britain*, History of Computing,
https://doi.org/10.1007/978-3-030-15103-4_8

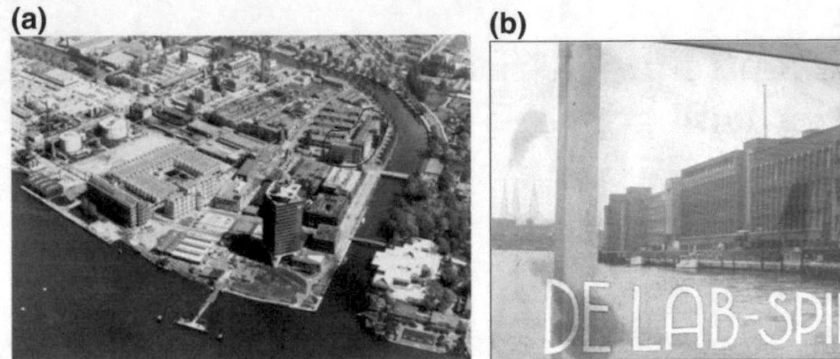

Fig. 8.1 **a** and **b** The left-hand aerial view, taken in 1970, shows the 'new' Laboratory (ringed in red) in relation to the overall KSLA site in Amsterdam-Nord. The right-hand photo, dated June 1954, comes from Shell's monthly magazine *Lab Spiegel (Lab Mirror)* for June 1954 and shows the Laboratory's front face from the Buikslotter Canal. Shell no longer occupies this site and a secondary school now stands where the 'new' Laboratory used to be. Today the modern buildings of KSLA are situated at an adjacent section of the island

building was started in 1939 but only completed in 1953[1] due to the disruptions of the Second World War. The new Laboratory, containing 3000 m^2 of useful space, was formally opened by the Mayor of Amsterdam on 3rd November 1953. Into this new building Shell's first digital computer, a Ferranti Mark I*, was installed in 1954 (see below). Newspaper articles of the time[2] describe the new laboratory as "not only the most modern, but also the largest of its kind in Europe". Photographs of Shell's KSLA site, with the new laboratory indicated, are shown in Fig. 8.1.

8.2 Shell's Search for a Digital Computer

Research within the Shell Corporation necessarily involved mathematical modelling and, until the early 1950s, the use of human *computers* equipped with electro-mechanical desk calculating machines. Analogue computers were also used, including ones designed within the company.[3] These were employed for simulation,

---

[1](a) De Tijd (a daily newspaper), 3rd November 1953.; (b) Algemeen Handelsblad (a trade magazine), 2nd November 1953.; (c) De Tijd (a daily newspaper), 4th February 1955.; With thanks to Onno Zweers for finding and translating these articles from the Dutch. There is a useful search system for Dutch publications here: http://www.delpher.nl.

[2](a) De Tijd (a daily newspaper), 3rd November 1953.; (b) Algemeen Handelsblad (a trade magazine), 2nd November 1953.; (c) De Tijd (a daily newspaper), 4th February 1955.; With thanks to Onno Zweers for finding and translating these articles from the Dutch. There is a useful search system for Dutch publications here: http://www.delpher.nl.

[3]Alberts, Gerard, and Bas, van Vlijmen. 2017. *Computerpioniers: het begin van het computertijdperk in Nederland*. Amsterdam: Amsterdam University Press. ISBN 978-94-6298-378-6.

analysis and control of chemical processes. Shell's head office in The Hague used punched card equipment for administration, mathematical direction being given by B. L. van der Waerden who also worked one day a week at the Mathematical Centre (*Mathematisch Centrum*) in Amsterdam.

The Mathematical Centre, now known as CWI, was founded in February 1946 as a national research institute. Adriaan van Wijngaarden was appointed head of the Centre's computing department in January 1947. In 1946 he had been one of a dozen young University of Delft scholars sent to the UK on a mission to study the latest post-war research in science and technology. Van Wijngaarden quickly decided that developments in mathematical machines was to be his focus. He spent the better part of a year visiting relevant establishments in the UK and the USA. Particularly, he developed close links with the National Physical Laboratory (NPL).[4] Wishing to encourage digital computing in Holland, the Mathematical Centre decided to build its own computer and also to bid for a UNESCO-sponsored International Computing Centre (ICC) to be established in Amsterdam. Regrettably, UNESCO decided in favour of Rome (see Chap. 9).

In leading the Mathematical Centre's computing activities, Van Wijngaarden soon came to be regarded as the founding father of Dutch computer science.[5] There is no doubt that Shell's management would have consulted Van Wijngaarden when, in 1951/52, KSLA began to consider acquiring its own automatic digital computer.[6] In view of subsequent events, it may be significant that Van Wijngaarden had attended the July 1951 Inaugural Conference of the first Ferranti Mark I computer at Manchester University.[7]

As described in Chap. 5, a team from Shell visited Ferranti's Moston factory in mid-1952. In November 1952 Shell asked Ferranti for a definite quotation and shortly afterwards Shell decided to purchase a Ferranti Mark I* computer. To prepare for this, in the summer of either 1952 or 1953 (probably the former) four employees from the Theoretical Department of KSLA were sent on a programming

[4]*Unsung heroes in Dutch computing history: Adriaan van Wijngaarden*. See: http://www-set.win.tue.nl/UnsungHeroes/heroes/vwijngaarden.html; This is a website created in 2007 by Technische Universiteit Eindhoven, Faculteit Wiskunde en Informatica, for the IEEE Computer Society's Web Programming Competition (CHC61).

[5]ERCIM News 106, July 2016. The European Research Consortium for Informatics and Mathematics, Paris.

[6]Rooijendijk, Cordula. 2010. *Alles moest nog worden uitgevonden (Everything had yet to be Invented):* de geschiedenis van de computer in Nederland. 2nd ed. Published by Olympus Pockets. (First edition published by Atlas Contact, Uitgeverij in 2007). ISBN 9789046744048.

[7]Manchester University Computer: Inaugural Conference 9th–12th July 1951. Attendance list on pp. 38–40 of the Proceedings. Amongst the delegates were B. J. Loopstra and A. van Wijngaarden of the Mathematisch Centrum, Amsterdam, and G. Rodino of INAC, Rome.

Fig. 8.2 Adriaan van Wijngaarden, head of computing at the Mathematics Centre (*Mathematisch Centrum*) in Amsterdam in 1951. On his desk is a Friden electro-mechanical desk calculator

course run by Maurice Wilkes at Cambridge University. The fortunate four were J. Krajenbrink, J. Kruizinga, H. Lauwerier and R. Lunbeck.[8]

The decision to purchase a Ferranti computer was enthusiastically promoted by Rudi Lunbeck who, after obtaining his doctorate in Physics and Mathematics, had come to work at KSLA's Theoretical Division in 1952. He took charge of preparations for the use of the machine. Writing in 2003, Lunbeck remembered what happened prior to the decision to purchase.[9] "It started with a search for available machines, with an important criterion being that ours had to be suitable for immediate use. It soon turned out that very few machines were considered, because very few were in regular production. In America, that was the case with IBM. In Europe, Ferranti had started to advertise their Mark 1*. The choice fell on the latter". To put this in context, the first external delivery of an IBM 701 computer was to the University of California's Los Alamos site in March 1953. As far as is known, no IBM 701 machines were exported outside America.

---

[8]Alberts, Gerard, and Bas, van Vlijmen., *Computerpioniers: het begin van het computertijdperk in Nederland.*

[9]Nouwen, Pieter. 2003. Het mirakel van Amsterdam (The Miracle of Amsterdam). *Shell's Venster Magazine* 19–23. Text (in Dutch) by Pieter Nouwen.

A year or so before the computer's arrived at the KSLA, Lunbeck and Hans Lauwerier spent some time at Ferranti becoming familiar with the Mark I* machine and writing code for some of the standard mathematical functions. "It was all brand new to us", said Lunbeck.[10]

8.3 The MIRACLE Arrives

A Ferranti Mark I* computer (number DC6) was assembled and tested at the Moston factory during 1953/4, before being dis-assembled for shipment to Holland. It reached KSLA in the late spring of 1954. Re-commissioning started in mid-June and the machine had passed its acceptance tests in late November 1954.[11] Preliminary programming work by Shell staff had actually begun at the start of 1953, after a decision to purchase a Ferranti Mark I* had been taken towards the end of 1952. Before the arrival of the hardware in Amsterdam, Shell mathematicians had spent some time in England learning about computer programming, as described above. Alida (Lidy) Zweers-de Ronde, who had started work in September 1951 in KSLA's Theoretical Division, remembers that: "In February [1953] we learned how to make programs for it, which I remember because it was the time of the flood. It was hard to concentrate on the work then, with everyone focusing on food".[12]

KSLA's Mark I* computer was named MIRACLE by Shell. MIRACLE stood for *Mokum's Industrial Research Automatic Calculator for Laboratory and Engineering*, Mokum being a Yiddish word for 'city' and the local name by which Amsterdam is sometimes known. The staff soon came up with an alternative

---

[10]Nouwen, Pieter., Het mirakel van Amsterdam (The Miracle of Amsterdam).

[11]Martin Wingstedt, telephone conversation with the author, 29th Sept. 2015. Martin joined Ferranti Ltd. on 5th November 1951. At first he worked at Manchester University on the Mark I computer and then he went to Moston. He travelled to Holland with the Shell Amsterdam machine and, according to his passport, landed on 15th June 1954. It is believed that Mike Moore was another Ferranti commissioning engineer with the Amsterdam machine.

[12](a) Lidy Zweers (neé De Ronde) was interviewed by Erno Eskens in the spring of 2001 and the Dutch transcript was published on 11th May. An English translation, created on 19th October 2015 by her son Onno Zweers, is here: http://onnoz.home.xs4all.nl/miracle/extra/texts/Interview%20with%20Lidy%20de%20Ronde,%20Miracle%20programmer.txt. The Dutch transcription is here: https://onnoz.home.xs4all.nl/miracle/extra/texts/transcriptie%20interview%20Lidy.txt.; (b) Onno Zweers, e-mail exchanges with the author, August 2015. Onno's mother, Lidy de Ronde, joined Shell Labs Amsterdam in 1950. She has left documents, photographs and an audio interview of her time with the Ferranti Mark I*.

expansion: *May It Replace All Chaotic Laboratory Experiments.*[13] MIRACLE was housed in KSLA's new Laboratory, shown in Fig. 8.1.

At the new Laboratory's opening ceremony in November 1953 the Director, H. W. Slotboom, gave the background to the developments. Summarising Slotboom's speech, it was reported[14] that the new building "puts scientific research here in Amsterdam at the forefront of the entire scientific world. To achieve this, the plans dating from 1939 had to be completely amended. When, after World War II, the Shell people were able to compare their lab with the sister institutions in America, they immediately found a major backlog in the equipment. As a result, in Amsterdam they were behind in the quality and pace of scientific research. The preparatory work that had to be done to get rid of this backlog took years. In the new building, in which a staff of 300 scientists is stationed, the most modern devices are now set up …. All this equipment serves the research into specialized catalytic converters in the field of oil processing; these will make oil processing ever more reliable and allow better management of product quality. How far this research extends can be demonstrated by the fact that one even needs an electronic computer, which makes calculations which so far have been impossible to execute".

The Shell Laboratory's Mark I* was not the first general-purpose automatic digital computer to have come into operation in Holland. That honour should go to the Dutch Post, Telephone and Telegraph Company, PTT, which built the PTERA (PTT Electronische Reken-Automaat) computer which first ran a program in September 1953. This used valves and relays and had a 1 K word magnetic drum store.[15] Close behind was ARRA II, a machine designed and built at the Mathematical Centre in Amsterdam, which first ran a program in December 1953. Its predecessor, ARRA I, was an electromechanical project first demonstrated in 1952.

The Ferranti Mark I* installation in Amsterdam was considerably more powerful than either of the above early research machines. As described subsequently by two Dutch computer historians[16]: "*De MIRACLE was wellicht de eerste serieus*

[13](a) Lidy Zweers (neé De Ronde) was interviewed by Erno Eskens in the spring of 2001 and the Dutch transcript was published on 11th May. An English translation, created on 9th October 2015 by her son Onno Zweers, is here: http://onnoz.home.xs4all.nl/miracle/extra/texts/Interview%20with%20Lidy%20de%20Ronde,%20Miracle%20programmer.txt. The Dutch transcription is here: https://onnoz.home.xs4all.nl/miracle/extra/texts/transcriptie%20interview%20Lidy.txt. (b) Onno Zweers, e-mail exchanges with the author, August 2015. Onno's mother, Lidy de Ronde, joined Shell Labs Amsterdam in 1950. She has left documents, photographs and an audio interview of her time with the Ferranti Mark I*.

[14](a) De Tijd (a daily newspaper), 3rd November 1953.; (b) Algemeen Handelsblad (a trade magazine), 2nd November 1953.; (c) De Tijd (a daily newspaper), 4th February 1955.; With thanks to Onno Zweers for finding and translating these articles from the Dutch. There is a useful search system for Dutch publications here: http://www.delpher.nl.

[15]*PTERA: Unsung heroes in Dutch computing history.* This short piece was produced in 2007 by Faculteit Wiskunde en Informatica, Technische Universiteit Eindhoven. See also Footnote 3 above.

[16]Alberts, Gerard, and Bas, van Vlijmen., *Computerpioniers: het begin van het computertijdperk in Nederland.*

werkende computer in Nederland, zeker de krachtigste rekenautomaat die toen beschikbaar was". (The MIRACLE may well have been the first serious computer in the Netherlands, surely the most powerful computer that was available at the time).

More interestingly, Shell's MIRACLE is believed to have been "the first computer in the world to be bought as a commercial venture rather than with direct or indirect government funding".[17] A 2003 retrospective article in Shell's house magazine states that "The KSLA is one of the world's leading Research Institutes. For example, we are probably the first industrial company to have had an electronic computer. This was made possible because we have the engineers, mathematicians and the technical professionals who are capable of installing, operating and maintaining such a machine".[18] It is said that Shell paid one million Dutch guilders for the computer,[19] roughly equivalent to £3.5 million in 2017. Another source[20] states that the hardware itself cost 850,000 guilders, to which was added a further 850,000 guilders to cover shipment and site-preparation. Whatever the precise figure, the sterling total was presumably close to Ferranti's quoted price of £93,800 plus £15,000 for delivery and installation and maintenance for one year, for the Ferranti Mark I* computer which was installed in Rome in 1955 (see Chap. 9). In contemporary terms, the costs were certainly considerable.

Explaining the acquisition of a Ferranti Mark I*, H. W. Slotboom the Director of KSLA, said in a 1954 article celebrating MIRACLE's arrival that "it is a prerequisite for Shell to be at the forefront of the development of science and technology".[21] The article goes on to suggest that "all possible tools must be employed that can benefit the oil company's development. It was foreseen that certain areas of research would remain impossible, simply because the necessary computing tasks would be impracticable. It is therefore obvious that the installation of this modern calculator greatly increases the possibilities for practicing basic and applied science. In addition, with the help of this machine, important data for the production process can be obtained in a flash. After all, the supplied crude oil from different sources varies greatly in composition. The data obtained from analysis of the oils can be processed by the machine in an instant so that the refining process can be adjusted accordingly. These are just a few examples that could be supplemented with many others". The article ends thus: "In our opinion, it is a gratifying fact that a large industrial concern like the Royal Shell values research so much that it is willing to

---

[17]Johnson, M.H.(Harry). 2002. *My Work with Computers from the Ferranti Mark I to the ICT 1900 (1952–1966).* 84-page typed manuscript, sent to SHL in 2002. Archived at Manchester, see NAHC/SHL/FA1. Of particular relevance is Appendix E: Ferranti MkI* customers and an early venture into magnetic tape.

[18]Nouwen, Pieter., Het mirakel van Amsterdam (The Miracle of Amsterdam).

[19]Rooijendijk, Cordula., *Alles moest nog worden uitgevonden (Everything had yet to be Invented):* de geschiedenis van de computer in Nederland.

[20]Article in Algemeen Handelsblad for 2nd November 1953.

[21]Article in Algemeen Handelsblad for 2nd November 1953.

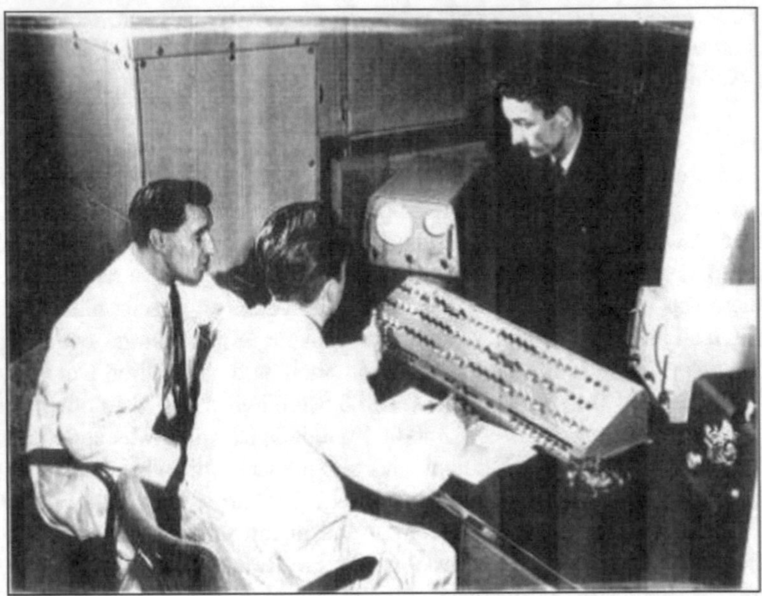

Fig. 8.3 Engineers during the 1954 installation of the Ferranti Mark I* computer at Shell's Amsterdam Laboratories. The computer was called MIRACLE. The person standing is Bruce Brown, the Ferranti engineer in charge of commissioning. The two seated Ferranti engineers are *(left-to-right)* Mike Moore and Martin Wingstedt

pay the huge costs for such a machine. After all, if in our country we want to keep up, the research is a *conditio sine qua non*".

In 1954 the Acceptance Tests for MIRACLE's installation were very carefully drawn up[22] and the document became the pattern for all subsequent Mark I* deliveries. This is possibly a reflection of a perceived need to avoid some of the initial teething troubles with previous Mark I and Mark I* installations. The Amsterdam machine was firstly required to pass several stringent functional tests of particular sub-units, for example the B tube, the drum, etc. Then there was a suite of Operational Tests of the complete system. All programs included in the Operational Tests "shall be provided by the National Physical Laboratory (NPL) and sent to Ferranti and to Shell well in advance". All the tests were to be completed within a period of ten consecutive working days. Certain allowances were made for a test or tests to fail, in which case re-runs were allowed within the spirit of 'at least two out of three re-runs must be successful'. The tests were to be performed at Manchester (before delivery) and then at Shell Amsterdam, with official observers from Shell,

[22]Letter from M. E. Sions (Ferranti Ltd.) to D. Hennessey (NRDC) dated 12th January 1954. This is an 11-page foolscap document, catalogued under NRD C9/1: *Computers, Manchester University/Ferranti Development, Vol. 2". Correspondence. 1952–67* at the National Archive for the History of Computing.

Fig. 8.4 MIRACLE looking back towards the operator's console, which is just behind the white concrete pillar

Fig. 8.5 Close-up of logic circuits (upper) and Input/output chassis

Fig. 8.6 At the opening ceremony for MIRACLE in December (?) 1954. Shell Laboratory's Director H. W. Slotboom is shown on the right, with his back towards the operator's console. In the background is MIRACLE's power supply cabinets. The cooling equipment was on the floor below

the Ministry of Supply (MOS) and NPL. NPL was required to write up and certify the Acceptance Test results. At the time the Mathematical Centre (*Mathematisch Centrum*) in Amsterdam had developed close links with NPL.[23] It is also very likely that KSLA had sought the advice of the Centre when considering which computer to purchase.

8.4 MIRACLE Gets to Work and Expands

Initially MIRACLE was used to support research connected with Shell's oil-refining process. The computer took over the manual computing tasks of a group of about ten 'human computers' who, armed with electro-mechanical desk-top calculators, would apply equations to data as directed by the Laboratory's research scientists. Many of these human computers, all of them female, soon progressed

[23] *Unsung heroes in Dutch computing history: Adriaan van Wijngaarden*. See: http://www-set.win.tue.nl/UnsungHeroes/heroes/vwijngaarden.html. This is a website created in 2007 by Technische Universiteit Eindhoven, Faculteit Wiskunde en Informatica, for the IEEE Computer Society's Web Programming Competition (CHC61).

Fig. 8.7 MIRACLE's Power Supply control cabinets, which are to be seen in the background of Fig. 8.6

from being operators to programmers of the Ferranti Mark I*. The story of one of them, Lidy Zweers, is recounted later.

Ruud Lunbeck recalls that "The mass spectrum people soon arrived, probably because their data had to be quickly processed for use by the other Laboratory departments". A little later, Harry Beckers appeared on the scene. After obtaining a Ph.D. in Physics, Becker started at KSLA in 1955. He soon became head of the Physics and Mathematics Research Department. He eventually ended up as Shell's Group Research Coordinator from 1977 until his retirement in 1991. In 2003 he recalled that: "I immediately got a vision of the way in which I could use the computer for theoretical quantum physics problems. Were we impressed? Yes, for sure, and of course it was specifically the speed of computation that impressed us".[24]

Continuing to reminisce, Beckers recalls that almost every night he spent from ten o'clock to three or four o'clock sitting alone in the machine room. In the morning 'normal' work would begin. "During those nights I worked on a rectification absorber, a distillation column from which at different levels different products could be processed. Once at around three o'clock, the machine stopped working. I then kicked the memory section and then it worked again!" Beginning

[24]Nouwen, Pieter., Het mirakel van Amsterdam (The Miracle of Amsterdam).

Fig. 8.8 Shell operators at MIRACLES's console. Lidy Zweers is in the centre

with MIRACLE and later on other computers, Beckers wrote six different engineering programs, first for refineries and then for the Exploration and Production Department, for which he provided theoretical models.

"The strongest case [for using the computer] that I remember occurred during the Suez crisis of 1956. This resulted in the closure of the Suez Canal, making it necessary for Shell to divert its crude oil transport round the Cape of Good Hope. The calculations for one new, optimal, transport pattern were performed in a single night. By hand it could never have been done as accurately and would have taken days".

The variety of Shell's applications for MIRACLE widened over the years to include activities such as linear programming, operations research and logistics. Amongst those working in this area at KSLA were Jacques Benders and Guus Zoutendijk. The application of linear programming to the routing of ships during the Suez crisis (see above) was their work.[25] It is known that by 1958 Shell was attempting a 5 quarter forecast of their worldwide activities over a weekend. As was

[25]Gerard Alberts, e-mail dated 19th September 2017 to the author.

hinted at in Chap. 7, Shell was not able to complete this task within the desired timescale on their own machine and so ran part of the problem on Armstrong Siddeley's Ferranti Mark I* at Ansty (near Coventry). Both Benders and Zoutendijk went on to hold senior academic posts in Dutch universities.

MIRACLE evidently grew in effectiveness. An American employee at the US Embassy in London prepared a detailed survey of West European computer installations in 1957.[26] Of Shell's Ferranti Mark I* at Amsterdam he commented: "their utilisation record is one of the best the writer has ever encountered".

During its life MIRACLE was enhanced by the addition of various extra input/output equipment—some of it novel. It seems that the Amsterdam site might have been used on occasions by Ferranti to try out new developments, presumably with the agreement of Shell. Joan Travis, neé Kaye, was a Ferranti programmer who, she says, "acted as a sort of engineers-mate (analogous to a plumbers-mate) for the design engineers when various 'exotic' peripherals (line printers or mag. tape) were being attached to the computers".[27] Here are her Amsterdam anecdotes.

"Looking at my passport, it would appear that I had four separate visits to Holland. My first two visits to Shell in Amsterdam were for eight days from 13-10-1955 to 21-10-1955 and for thirty eight days from 21-1-1958 to 28-2-1958. I only know that I started the last two visits on 4-11-1958 and 10-12-1958, but not when I returned. I presumably returned before Christmas 1958 on the fourth visit. On one visit, possibly the third, we could only work overnight. I think the first visit was possibly to do with attaching a line printer (Bull perhaps). The second visit could have been trying to attach 35 mm. mag. tape units. The third and fourth visits could have been attaching ½ in. mag. tape. Of course the design engineers would have gone over before I did. I never had contact with the Shell programmers, only the engineers.

"We were working one weekend and the Dutch engineers decided to give us a treat. The normal tea, which was served from the tea trolley, was herbal very weak and without milk. The engineers tried to make us what they thought was real 'English' tea; this involved them boiling up tea leaves in a container for a long time! When we had to work overnight we would keep our packed food in the drum compartment to keep it cool".

Another comment on the first project to equip MIRACLE with a magnetic tape comes from Harry Johnson.[28] He remembers that: "Along with other work, Shell wanted to use their computer to investigate the possibility of using linear programming techniques in scheduling oil deliveries world wide. For this purpose they

---

[26]Hoffman, A.J. 1957. *New Computers in France and the Netherlands*, 1. This document is archived as Box 216, Backman papers, National Museum of American History, mathematical branch, Smithsonian Institution, Washington DC. The quotation is cited as given on page 164 of: *The digital flood: the diffusion of information technology across the US, Europe and Asia*. J. W. Cordata. Oxford, 2012. ISBN: 978-0-19-992155-3.

[27]Joan Travis, neé Kaye, e-mail to the author on 25th October 2015.

[28]Johnson, M.H.(Harry)., *My Work with Computers from the Ferranti Mark I to the ICT 1900 (1952–1966).*

felt a need to have a backing store more substantial than the standard drum and discussed the problem with the Ferranti design team. A number of American companies had magnetic tape stores under development but none had a product yet available.

"The trickiest aspect of such a store seemed to be the control of the spooling mechanism for thin and easily damaged tape moving at high speed. The Ferranti designers conceived an approach that would eliminate this requirement. They would use standard 35 mm film stock with a magnetic coating. … I cannot now remember the details of the mechanism but the tape was to be threaded through a mixture of driving rollers and read/write heads and then simply allowed to spew into the trough which was referred to as a swizzle box. … The unit was developed and added to the Amsterdam machine and appeared at first to work very reliably. As the Shell programmers became more and more confident in its performance the tape unit was used on bigger and bigger jobs; but it gradually became clear that it had developed some chronic problem".

It transpired that the problem lay in the swizzle box and the project was eventually abandoned. A little later Shell International Petroleum Co. Ltd. ordered a Ferranti Mercury computer for their London establishment. This machine, which was delivered in January 1959, was nearly 20 times faster and had much more bulk storage capability than the Ferranti Mark I*. The initial applications for Shell's Mercury were described as 'linear programming and sales analysis'.[29]

Meanwhile tape decks, manufactured by the ElectroData Division of Burroughs Corporation and using ¾-in. tape, were installed on Shell's Mark I* computer in about 1959. In January 1960 it was reported[30] that: "MIRACLE is now in its fifth year of operation, the working-load being about 5000 h per year. Besides the addition of an on-line printer (150 lines/min) and a medium-speed punch (27 characters per second) in previous years, the storage facilities have recently been extended by the installation of magnetic tape equipment. Information on each of the two Electrodata tape decks (model 546) can be transferred in blocks of 128 twenty-bit words to the fast store via a buffer store and vice versa".

By the 1960s, the oil industry had found many opportunities for using digital computers, in activities covering the geological interpretation of seismic data and reservoir modelling, pipeline operations, shipment optimisation, the chemistry of refining and fluid catalytic cracking, and the data-processing aspects of sales and marketing. MIRACLE remained in use until 1961, even though it had become

[29]Delivery lists for all the major early British computers are given on the Computer Conservation Society's Our Computer Heritage website, together with brief comments on applications—see: http://www.ourcomputerheritage.org/. The information on Shell's Mercury is taken from Swann, B.B. 1975. *The Ferranti Computer Department: A History*. This 1975 document was initially circulated privately and marked 'confidential'. A copy is held at the NAHC, catalogue number FER/C30.

[30]The US Office of Naval Research's Digital Computer Newsletter. Vol. 12 No 1 January 1960, p. 19.

outdated by then. It was also totally unsuitable for administrative work, for which Shell's office in The Hague had purchased an IBM 650 in 1958. This was replaced in 1961 by an IBM 7070 and then later by an IBM 7094.[31] Shell's IBM 7070 at The Hague was officially opened in November 1961,[32] at which time it was reckoned to be the largest computer in Holland.

More details of MIRACLE's end, and what followed, are given later. Before that, it is interesting to recount some personal experiences of life with MIRACLE at KSLA.

8.5 Lidy's Story

Returning to the early days of MIRACLE, here are some memories of an installation that would have been described, if the word had been invented in 1954, as Holland's Amazing Supercomputer. Lidy Zweers (neé De Ronde) started working at Shell's Amsterdam Laboratory in 1951, when she was 22 years old. She was an early user of MIRACLE, so how did she view this new Supercomputer?

Lidy Zweers was interviewed in 2001 and an English translation of her comments is available.[33] In the interview Lidy described her initial job-title at KSLA as *Calculator* which, at the time, signified a mathematical clerk equipped with an electro-mechanical desk calculator. When the Ferranti Mark I* arrived in 1954 she remembers that: "We (the Calculators) were all supposed to work with it .... Two technicians tried to keep the machine up and running, because every now and then there was a malfunction. But the calculations that we used to do manually—mainly statistical calculations to gain better understanding of the experiments that were taken in the Laboratory—they were sometimes very time consuming. At Shell they thought it would be great if a computer could do these much faster.

"Initially, three people from our department [probably including Rudi Lunbeck and Hans Lauwerier] went on a course in England. They returned wildly enthusiastic. We were to learn next, and meanwhile more Calculators were hired. I was in charge of the Calculators, who were all women who clearly had mathematical

[31]Alberts, Gerard, and Bas, van Vlijmen., *Computerpioniers: het begin van het computertijdperk in Nederland.*

[32]Het Vrije Volk (daily newspaper), 11th November 1961. Nieuwe computer voor Shell (New computer for Shell).

[33](a) Lidy Zweers (neé De Ronde) was interviewed by Erno Eskens in the spring of 2001 and the Dutch transcript was published on 11th May. An English translation, created on 19th October 2015 by her son Onno Zweers, is here: http://onnoz.home.xs4all.nl/miracle/extra/texts/Interview%20with%20Lidy%20de%20Ronde,%20Miracle%20programmer.txt. The Dutch transcription is here: https://onnoz.home.xs4all.nl/miracle/extra/texts/transcriptie%20interview%20Lidy.txt.; (b) Onno Zweers, e-mail exchanges with the author, August 2015. Onno's mother, Lidy de Ronde, joined Shell Labs Amsterdam in 1950. She has left documents, photographs and an audio interview of her time with the Ferranti Mark I*.

Fig. 8.9 Some of the team of ten women *computers* at Shell's KSLA laboratory before the arrival of MIRACLE. A Marchant electro-mechanical calculator can be seen on the desk in the background

aptitude. Almost all the girls who were there started working with the machine" (Fig. 8.9).

Lidy described her own mathematical qualifications as giving her "authority to teach at secondary schools—which I didn't want. So I was happy to work at Shell". When asked whether she was afraid of the computer, Lidy replied: "Ooooh, I loved it. I thought it was amazing. I was excited. Not really scared, but I thought, oh boy, if I make a mistake, what will happen? But of course nothing ever happened". Lidy was aware that the computer room, which had no windows, was surrounded by a Faraday Cage to shield the CRT storage tubes from external electro-magnetic radiation. "That idea I found scarier. I thought: we're sitting in a cage and who knows what could happen? But the computer itself didn't frighten me".

When asked whether there were job losses when the computer arrived at Shell, Lidy replied: "No, no. But our department consisted of nine or ten people and that wasn't enough to keep the machine going all the time, because it could work so fast and it was a lot of work to create a program, punch it to tape and test it, and only then could we do production runs. So that took a while … The calculation was so fast. If I needed two weeks for a least squares calculation and the computer does it within ten minutes, then it's a large gain. But the preparations were relatively unbelievable. Many people were needed to do that. So no staff were fired … It took a year or two to get things running smoothly. I worked with the computer for two years and when I left, it finally began to run well, I believe" (Fig. 8.10).

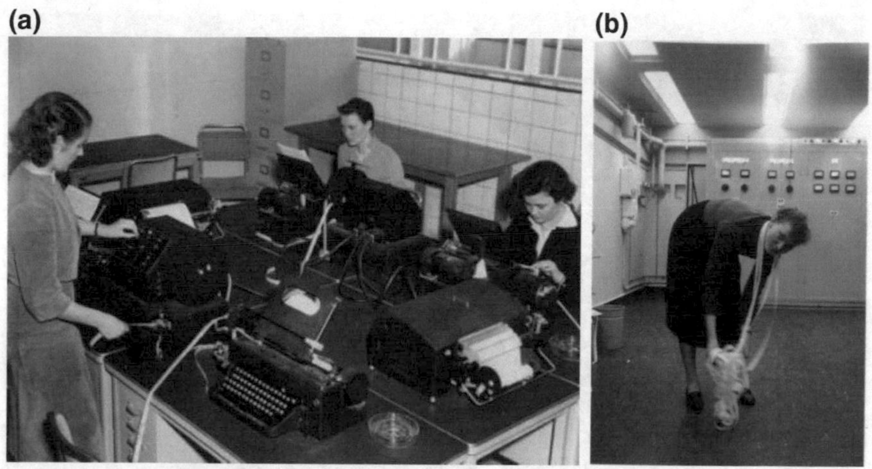

Fig. 8.10 **a** and **b** *(Left)*. Off-line punching, printing, copying and editing equipment for 5-track paper tape. *(Right)*. Untangling paper tape at KSLA. The computer's input equipment had no provision for spooling tape after the reading process. Thus, paper tape was spewed out into a bin, to be tipped out and untangled later

"In due course, a few people from all departments in the Laboratory had to learn to program. They were taught by one of the academics. He gave a course of fourteen days, but only theory, and they couldn't practice on the machine. He didn't have time for that and he couldn't do that himself. We had never seen him at the machine. There was some gossiping about that, as you may understand! When that didn't work out so well a few times, I asked if I could give that course and was allowed to. I was so proud and excited. I knew exactly how I would approach it. Each course was a fortnight. From each department two people were attending so we had a classroom with 20–25 people. In the morning I taught the theory, and in the afternoon I demonstrated how to operate the computer. I sat them at the console and then they had to operate it, from the very first day. At first they couldn't operate it by themselves, but I thought: within a week one should be able to run a small program to do a very basic calculation: 3 + 2 = 5, including punching it, feeding it to the computer and print the result. And quite a few were able to do that after a week. Immensely proud and wildly enthusiastic! And that is encouraging, isn't it? That's the fun!"

Lidy was naturally not the only one involved in providing instruction. For example, Dr. Lunbeck wrote a programming manual for the MIRACLE in 1958.[34]

When asked about computer applications, Lidy remembered that they certainly increased. She then went on to describe the kind of statistical calculations that she was asked to do. "In the Laboratory, tests were done with the oil. Both the raw and purified oils that were produced in the lab were examined. All sorts of properties

[34]Gerard Alberts, e-mail exchanges with SHL, May 2017.

Fig. 8.11 Prince Bernhard *(left)* being Shown MIRACLE by Ruud Lunbeck. The visit took place on 19th April 1956

were expressed in numbers, I'm not sure which. … And for the variations you then get—like the variations of fuel you find at a gas station—they came up with lists of calculations. Then they had to determine what would statistically be the best. There are several statistical calculations to that end which I knew back then. Not any more! One of the calculations which was the most common was the least squares calculation. I made a program for least squares and it worked. To my amazement, it worked great. But within a certain range of numbers, of course, because you could have extremely small or big numbers. But this was not the most common range that I had taken.

"Some operations on the computer produced sound. And there was a smart technician, who had noticed that you could produce a melody. You could create a program focused only on the sound. … I think the National Anthem [Wilhelmus] was amongst the first melodies produced". It was said that when Queen Juliana and Prince Bernhard came to visit in 1956, MIRACLE played Mozart.[35]

When asked about standard library subroutines supplied by Ferranti for common mathematical functions, Lidy replied that initially there were none. "There was nothing, nothing at all. We had to build the most basic actions [for example the

[35]Rooijendijk, Cordula., *Alles moest nog worden uitgevonden (Everything had yet to be Invented):* de geschiedenis van de computer in Nederland.

square root subroutine] ourselves. Everything. And it's fun, for a while. But if you have a lot of work to do, then you think, well, that's that same calculation again, they should put it on a separate tape in the machine somewhere, at an address, so that you can retrieve it again. And that is indeed what happened. By the time that I left, the first standard programs came in".

Asked whether there were visitors to Shell's computer installation, Lidy replied "Yes, people from the business community and the university were interested. Every now and then there were visitors. They were shown around the computer room. Then they ran demonstration programs and so forth, which an outsider could follow … We didn't lease computing time to universities because universities were also early to adopt electronic computers. But our computer at Shell was really the first in Netherlands".

Like most computer installations, MIRACLE would perform tricks for visitors. Here are the impressions of one journalist, who attended a demonstration shortly after MIRACLE came into regular service in 1955[36]: "It was able to say in four languages (French, German, English and Dutch) on a cathode ray tube (a kind of television screen) on which day of the week a certain date, possibly hundreds of years ago, fell. And when somebody wanted to be funny and tried to make MIRACLE say what day 29 February 1955 would fall (a date that does not exist), MIRACLE said on its electron tube, green flashing and very clear: "nonsense"… MIRACLE then managed, after it was fed with another punched tape, to hum the English song *Clementine* on its hooter, as well as the Dutch *In a Blue Checkered Smock-frock* (In een blauw geruite kiel) which some MIRACLE visitors found more creepy than fun …".

To return to Lidy's memoires, she left Shell in about 1956 to get married. "They [Shell] did not employ married women. That was a 'social problem'. We did not know then that it was a problem. Most married women just left. Today, it is very common for women to have a job, but back then it wasn't. They actually made me an exceptional offer. Not that I experienced it like that, because I was done with the job. I had seen enough of it. I thought: the numbers come out of my nose! My boss said: 'would you like to keep working for a while?' Then I said, 'thanks but no thank you'. It had been fun because it was very special with the computer and especially the teaching I found very nice … My immediate boss was very easy and friendly".

8.6 The End of MIRACLE and What Followed

The Amsterdam Mark I* computer was finally taken out of service on the 24th of December 1961—though there is some evidence to suggest that the hardware may not have been physically removed until a little later. In[37] it is suggested that 1965

[36]Article in Het Vrije Volk, 4th February 1955.

[37]Nouwen, Pieter., Het mirakel van Amsterdam (The Miracle of Amsterdam).

marks the eventual departure of MIRACLE's remains. However, 24th December 1961 seems much more likely as the actual switch-off date.[38] The scene has been dramatically described in a novel[39] by Gerrit Krol, who is believed to have worked at one time as a programmer for Shell. Here is a crude English translation of Gerrit's much more melodious—though not necessarily technically accurate—Dutch valediction:

24 December. "Today we buried Miracle, the day before Christmas. It was really moving. There he stood, the beast. His successor is ten times as small and one hundred times faster, of plastic and silent. Miracle is made of iron and enamel, buzzes when he works and when you toggle a switch he whistles. Then he whistles the tune that matches the program. Then you know it's your program and it works. Every morning he starts with the recognition song 15 min the same tune.

"Mozart once composed a number of measures that, when played in whatever order, form a melody. This is of course put on paper tape and every time there are visitors, HBS students from Krommenie, or the management of a Chinese fertilizer factory, the paper roll with the label Mozart will be inserted. Guests can operate a row of buttons, each in turn, and each time the machine produces another melody, although they all somewhat seem to resemble each other, barrel-organ like whining thingies that stop after half a minute as suddenly as they started.

"Indeed: melodies, satisfied, indifferent, as if they always existed. We started to wrap the front [of the computer] in paper. Red white, blue and green and black crepe paper.

"One moment, M. entered with his arms full of paper, which he shared among those present before starting himself. Boxes with paper ribbon, yellow and pink, all of which had to be used. Garlands were made, curtains. The line printer became a black catafalque with candles on it, which were later removed so as not to hurt anyone's feelings.

"Scissors were handed round, adhesive tape, tubes of glue, layers drained on the ground. People shouted and pointed to the latest funny things, and—*mirabile dictum*—in the midst of this noise, seated on a chair moved in front of the machine, Miss Pool was turning the knobs, computing as if her life depended on it.

"Sybrandy was standing with his army shoes on top of her sheets of paper, because that was the way it had to be done. Miss Pool continued smiling, she could understand, but at a certain moment M. saw it and he decided it was time to end it. He stood in a corner and there he pulled a 2000 V lever.

"Like a spinning top, like a siren, the machine died down, the tube monitors stood grey and Miss Pool raised her hands to heaven, pushed them to her chest and looked back beggingly. M. pushed the lever back, but the machine did not start

---

[38]Alberts, Gerard, and Bas, van Vlijmen., *Computerpioniers: het begin van het computertijdperk in Nederland.*

[39]Krol, Gerrit. 1967. *Het gemillimeterde hoofd (The Close-Cropped Head).* Published in Amsterdam: Em. Querido. Thanks to Onno Zweers for assisting with the English translation of this section. Gerrit Krol (1934–2013) spent time working as a programmer of MIRACLE.

again. Miss Pool pushed a lot of buttons for a while, and then rolled her paper ribbon on a small wheel and stooped away. Thus, Miracle was brought to its end".

After being switched off, what happened to MIRACLE's hardware? A 2000 study of surviving artefacts of early Dutch computers[40] found no mention of MIRACLE, though in 2000 the Abbott Room at Shell's Research & Technology Centre in Amsterdam had a collection of MIRACLE photographs. The lack of any surviving MIRACLE hardware has recently been confirmed.[41]

MIRACLE was replaced at KSLA in 1962 by an IBM 1401 system. By then, computation activity had been spreading out to other Shell locations. The headquarters of BPM, Bataafsche Petroleum Maatschappij, the Dutch branch of Shell, acquired an IBM 650 in 1958 at The Hague, as mentioned previously. This was replaced in 1961 by an IBM 7070. At Pernis, Shell's Rotterdam oil refineries, a pair of Bull Gamma 3 electronic calculators with the Extension Tambour (drum) enhancement were installed in 1959. Meanwhile, at KSLA and at the Delft test plant, Shell had developed several analogue computers.[42]

Within Shell's UK establishments a Ferranti Pegasus was delivered to Shell's Stanlow Refinery, Cheshire, in February 1961 and another Pegasus was delivered to Shell Research Ltd., at Thornton, Cheshire, at the same time. A much more powerful Ferranti Mercury computer had meanwhile been delivered to Shell International Petroleum Ltd. in London in January 1959.[43]

Within the UK, Shell's computer installations may be compared with installations at other oil companies over the period 1954 to 1961—see Table 8.1. This data is based on delivery of all computers manufactured in the UK[44] within the period. With the exception of IBM installations at the Atomic Weapons Research Establishment, Aldermaston, in 1957 and 1959, it is believed that no computers of foreign manufacture were imported into Britain before 1960.

---

[40]Alberts, Gerard. 2000. *Collectie Nederland oude computers. Werkgroep Verzamelbeleid Computerhistorie*. (Dutch collections of old computers. Workgroup for Collection Policy, Computer History). Produced by Schiedam, Holland: Dynamiek Productions.

[41]Dooijes, E H.(Edo). 2017. Director of the Computermuseum. Universiteit van Amsterdam, e-mail to the author dated 26th March 2017.

[42]Alberts, Gerard, and Bas, van Vlijmen., *Computerpioniers: het begin van het computertijdperk in Nederland*.

[43]Delivery lists for all the major early British computers are given on the Computer Conservation Society's Our Computer Heritage website, together with brief comments on applications—see: http://www.ourcomputerheritage.org/. The information on Shell's Mercury is taken from Swann, B.B., *The Ferranti Computer Department: A History*. This 1975 document was initially circulated privately and marked 'confidential'. A copy is held at the NAHC, catalogue number FER/C30.

[44]Delivery lists for all the major early British computers are given on the Computer Conservation Society's Our Computer Heritage website, together with brief comments on applications—see: http://www.ourcomputerheritage.org/. The information on Shell's Mercury is taken from Swann, B.B., *The Ferranti Computer Department: A History*. This 1975 document was initially circulated privately and marked 'confidential'. A copy is held at the NAHC, catalogue number FER/C30.

Table 8.1 Computer installations at oil companies operating in the UK, during the period 1955–1961

Year	Computer	Company/location	Stated application(s)
1955	BTM HEC2 M	Esso's Fawley refinery	Application not stated
1957	Two ICT 1201s	Shell's Shellhaven refinery	Oil stock accounting; order stock control, payroll, costs, statistics
1959	Ferranti Mercury	Shell International, London	Linear programming. Sales analysis
1961	Ferranti Mercury	BP	Linear programming. Sales analysis
1961	Ferranti Pegasus	Shell refinery, Stanlow	Application not stated
1961	Ferranti Pegasus	Shell research, Thornton	Application not stated

This Table does not include Shell's installations in Holland from 1954 onwards, which are described in the text

The tentative conclusion from Table 8.1, together with this chapter's account of Shell's Dutch installations, is that Shell led the European oil industry in respect of the early deployment of digital computers.

References

Alberts, Gerard. 2000. *Collectie Nederland oude computers. Werkgroep Verzamelbeleid Computerhistorie*. (Dutch collections of old computers. Workgroup for Collection Policy, Computer History). Produced by Schiedam, Holland: Dynamiek Productions.

Alberts, Gerard, and Bas, van Vlijmen. 2017. *Computerpioniers: het begin van het computertijdperk in Nederland*. Amsterdam: Amsterdam University Press. ISBN 978-94-6298-378-6.

Dooijes, E H.(Edo). 2017. Director of the Computermuseum. Universiteit van Amsterdam.

Hoffman, A.J. 1957. *New Computers in France and the Netherlands*, 1.

Johnson, M.H.(Harry). 2002. *My Work with Computers from the Ferranti Mark I to the ICT 1900 (1952–1966)*.

Krol, Gerrit. 1967. *Het gemillimeterde hoofd (The Close-Cropped Head)*. Amsterdam: Em. Querido.

Nouwen, Pieter. 2003. Het mirakel van Amsterdam (The Miracle of Amsterdam). *Shell's Venster Magazine* 19–23. Text (in Dutch) by Pieter Nouwen.

Rooijendijk, Cordula. 2010. *Alles moest nog worden uitgevonden (Everything had yet to be Invented):* de geschiedenis van de computer in Nederland. 2nd ed. Published by Olympus Pockets. (First edition published by Atlas Contact, Uitgeverij in 2007). ISBN 9789046744048.

Swann, B.B. 1975. *The Ferranti Computer Department: A History*.

Chapter 9
The Ferranti Mark I* Installation in Rome

9.1 The Italian Scenario

In the late 1940s Professor Mauro Picone (1885–1977), the Director of the Instituto Nazionale per le Applicazioni Calcolo (INAC) in Rome, was the principal Italian mathematician interested in promoting Italy's development of automatic computing. It was a difficult time. Picone's contacts with the USA were impeded by post-conflict problems and, instead, he initiated links and exchange visits with Louis Couffignal in France.[1] Picone's visits to Paris may have led UNESCO, based in Paris, to choose INAC and Rome as the seat of the tentative International Computation Centre in 1951. This turned out to be an empty initiative which languished until 1961 when it was transformed into the Intergovernmental Bureau for Informatics. This is not to be confused with the United Nations International Computing Centre (ICC) which was established in 1971 with headquarters in Geneva.

Picone's prospects changed in 1950 when, during the Congress of the International Mathematical Union held in Cambridge, Massachusetts, Picone was at last allowed to visit many of the emerging American digital computing centres.[2]

Thus inspired, Picone's ambition was to build an Italian computer. During 1950/51, the Olivetti engineer Michele Canepa had spent about 18 months working in Howard Aiken's laboratory at Harvard on the Harvard Mark IV. With Canapa's enthusiastic support, Picone tried to initiate a project to build an Italian Harvard Mark IV with help from the Olivetti company. Olivetti, well-known as a manu-

[1]Celli, A., M. Mattaliano, and R. Spitaleri. 2012. *Istituto per le Applicazioni del Calcolo, 1927–2012: 85 years in Computational and Applied Mathematics*. Proceeding of MASCOT 12 & ISGG 12, The Joint 12th Meeting on Applied Scientific Computing and Tools and 12th ISGG Meeting on Numerical Grid Generation, Las Palmas de Gran Canaria, 61–70, Oct 2012. Published by IMACS, the International Association for Mathematics and Computers in Simulation.
[2]Celli, A., M. Mattaliano, and R. Spitaleri., Istituto per le Applicazioni del Calcolo, 1927–2012: 85 years in Computational and Applied Mathematics.

S. Lavington, *Early Computing in Britain*, History of Computing,
https://doi.org/10.1007/978-3-030-15103-4_9

facturer of typewriters, had launched its first electric desk calculator in 1948. In the event, Olivetti was reluctant to provide the necessary resources and the project to build an Italian computer had effectively collapsed by the end of 1951.[3] We shall see that Picone's ambition to obtain a computer was realised in a different manner in 1954—but that his installation was not quite the first in Italy.

Probably the first electronic digital computer to come into operation in Italy was a Computer Research Corporation's CRC102A, the production version of CADAC, a computer designed for the USAF. Luigi Dadda, a graduate engineer from Milan Polytechnic, had spent some months in California working on the CRC 102A and was keen for the Polytechnic to purchase one with funds from the Marshall Plan (the *European Recovery Plan*). A CRC102A computer was installed in Milan in the autumn of 1954. It had 1K words of magnetic drum storage and an instruction time of about ten operations per second. The cost was about $120,000. The machine was unusual because it came without any kind of manufacturer's maintenance agreement or post-sales support.[4]

It is generally accepted that 1954 was the year in which "four different Italian computer projects were begun at roughly the same time".[5,6] Besides delivery of the CRC102A to Milan, a Ferranti Mark I* computer was ordered for Rome (see below). 1954 saw the initiation of the CEP (Calcolatrice Elettronica Pisana) project at Pisa, which came to fruition in 1961, and Olivetti's ELEA 9003 which came to fruition in Pisa in 1958. Let us now concentrate on the Ferranti Mark I*, the most powerful of the early Italian machines.

9.2 INAC and FINAC in Rome

Hard on the heels of the Milan Polytechnic came the Instituto Nazionale per le Applicazioni Calcolo (INAC) in Rome. INAC was a computation laboratory originally founded in Naples in 1927 and transferred to Rome in 1932 to become a division of Italy's Consiglio Nazionale delle Ricerche (CNR). In 1975 INAC's name was changed to L'Istituto per le applicazioni del calcolo "Mauro Picone", abbreviated IAC. More details of the early history of INAC will be found in Celli et al.[7]

---

[3]Celli, A., M. Mattaliano, and R. Spitaleri., Istituto per le Applicazioni del Calcolo, 1927–2012: 85 years in Computational and Applied Mathematics.

[4]De Marco, Giuseppe, Giovanni Mainetto, Serena Pisani, and Pasquale Savino. 1999. The Early Computers of Italy. *IEEE Annals of the History of Computing* 21 (4): 28–36.

[5]De Marco, Giuseppe, Giovanni Mainetto, Serena Pisani, and Pasquale Savino., The Early Computers of Italy.

[6]Cordata, J.W. 2012. *The Digital Flood: The Diffusion of Information Technology Across the US, Europe and Asia*. Oxford. ISBN 978-0-19-992155-3. See also: Biographical notes of Luigi Dadda, https://it.wikipedia.org/wiki/Luigi_Dadda.

[7]Celli, A., M. Mattaliano, and R. Spitaleri., Istituto per le Applicazioni del Calcolo, 1927–2012: 85 years in Computational and Applied Mathematics.

After deciding not to construct their own computer Professor Mauro Picone, the Institute's Director, turned his attention to Ferranti Ltd.,[8] whose Mark I* was considerably more powerful than Milan's CRC102A. Bernard Swann and John Bennett from Ferranti visited INAC in early February 1954, through the good offices of Dr. Ignaxio Mottola, Ferranti's agent in Italy. In a letter to Picone dated 9th February,[9] Ferranti quoted for the supply of "One Manchester Universal Computer including power supplies, tape reader input, punched tape and teleprinter output and tape editing equipment (consisting of two keyboard perforators each with a tape reader, a keyboard transmitter and a re-perforator, one teleprinter and one control unit)". The quoted price was £93,800. Delivery and installation and maintenance for one year was quoted as £15,000. "We expect that a machine will be ready for tests [at the Moston factory] in about 6–8 weeks' time ... In accordance with Professor Picone's request we are reserving until the end of this month, machine number DC4, which is at present unallocated".

The quotation letter included various technical details, as follows. The computer would have 12 CRT storage tubes (each with a capacity of 1280 binary digits); the drum would have 256 tracks, each storing the same amount of information as two CRT units; the input paper tape reader would transfer at 200 characters/s; the power supply came with its own isolating motor/alternator set. There was a cautious note that the external cooling system (compressor and fan-cooled condenser) could only deal with ambient temperatures up to 90 °F (32 °C).

Picone had evidently asked about the availability of additional backing storage such as an extra drum and a magnetic tape deck. Ferranti advised that they could supply a two-reel magnetic tape unit with a CRT storage tube as buffering, at an additional cost of £15,000. The eventual addition of a magnetic tape system in mentioned below.

Picone was minded to accept Ferranti's quotation. The order for a Mark I* was received by Ferranti in March 1954 and the contract was finally approved on 28th July.[10] The funding came from Azienda Rilievo Alienazione Residuati (ARAR). ARAR was an Italian government entity created in October 1945 for "... the recovery, custody and disposition of material remnants of war, left behind by the Allies or abandoned by the Germans in Italy or otherwise acquired". The ARAR ordinance permitting this expenditure was ARAR—MPA 203-122 Administration—Law 21/3/1953.[11] The background to the negotiations is described in more detail in Chap. 5.

---

[8]A collection of 14 papers related to FINAC and/or to the Instituto Nazionale per le Applicazioni Calcolo (INAC) in Rome, 1952–1962, collected and edited by Paolo Ercoli, Rome. See: http://www.jnorman.com/cgi-bin/hss/39086. Most of the papers in this collection were originally distributed as *Pubblicazioni INAC* and sent to several universities world-wide.

[9]Four-page letter dated 9th February 1954 from B. B. Swann (Ferranti Ltd.) to Professor Picone (INAC).

[10]Pollard, B.W. 1956. Rome, December 14th 1955: The Inauguration of a Ferranti Computer. *Ferranti Journal* 9–11.

[11]CNR memo (raccomandata a Mano) dated 24th May 1974 to Prof Ilio Galligani, Director, IAC, on the subject of Calcolatori Elettronici. CNR reference N47538.

INAC's Ferranti Mark I* computer was known as FINAC. In 1951, three years prior to FINAC's arrival, Mauro Picone had sent two of his students, Dino Dainelli and Enzo Aparo, to the United States "to study the art of computer programming". After the installation of FINAC Dainelli and Aparo, together with Corrado Böhm, became key members of the INAC software team.[12] There is evidence that INAC programmers also obtained some practical experience on the Ferranti Mark I* installed at A. V. Roe's Chadderton factory—see Chap. 6—before the delivery of their own computer. The Ferranti programmer Joan Travis, neé Kaye, remembers that one of her tasks was "to help two programmers from Rome to use the Avro Mark 1* before they got their own. No love was lost between the two of them. I suspect the help involved was to keep them apart! One was Corrado Böhm".[13]

FINAC passed its factory acceptance tests at Moston in October 1954. It was then stripped down, packed, and shipped out of Liverpool in the *SS Pinto*. The ship reached Naples on 6th January 1955. There was said to have been some initial misunderstanding about getting the machine through customs because Ferranti "did not realise that Italian Customs officers required to be paid".[14] Part of the problem was that Italian law required a tax to be paid for each thermionic valve (tube) and a Customs stamp to be glued on each valve. This was possible for a radio but impossible for a computer with thousands of valves. Mauro Picone negotiated an agreement whereby the Customs stamps were "paid for and virtually received and virtually glued".[15]

The FINAC computer was installed on the fourth floor of the Consiglio Nazionale delle Ricerche (CNR) building in the Piazzale delle Scienze 7 (now Piazzale Aldo Moro 7), to the east of Rome's central Station (Termini) and close to the University of Rome's main campus. INAC was a formal part of the University of Rome after World War Two until approximately 1955. Special cranes were constructed for the computer's installation in the grand CNR headquarters,[16] as dictated by the building's position which is shown in Fig. 9.1.

---

[12]A collection of 14 papers related to FINAC and/or to the Instituto Nazionale per le Applicazioni Calcolo (INAC) in Rome, 1952–1962, collected and edited by Paolo Ercoli, Rome. See: http://www.jnorman.com/cgi-bin/hss/39086. Most of the papers in this collection were originally distributed as *Pubblicazioni INAC* and sent to several universities world-wide.

[13]Joan Travis, e-mail to the author dated 25th October 2015.

[14](a) Ellson, Allan. 2015. *Ferranti MkI* Commissioning and Installation: Some Recollections.* 14th Oct 2015.; (b) Alan Ellson, letter to the author dated 16th January 2000 and telephone conversation with the author on 20th September 2015 and e-mail to David Link on 26th July 2017. Alan joined Ferranti in May 1951 and was involved in commissioning the Mark I at the University of Manchester. He then transferred to Moston and was involved in the installation of FINAC. He left Ferranti in the summer of 1956.

[15]Andrea Celli, e-mail dated 29th September 2017 to the author. Celli is actively researching the history of INAC.

[16]Pollard, B.W., Rome, December 14th 1955: The Inauguration of a Ferranti Computer.

FINAC completed its final acceptance tests in Rome on 22nd June 1955. The INAC and Ferranti personnel present at the tests are listed.[17] In addition to the normal Ferranti tests, INAC had specified a special program involving 62 linear equations, which was successfully run at the first attempt in a time of 3.75 h. "A quarter of an hour after this final test had been started, the public utility supply to the building was interrupted and the computer closed down automatically. However, when the supply was restored, and the computer switched on again, it was found possible to re-start the computation at the point where it had been interrupted by the supply failure".[18] Enzo Aparo remembers the 62 linear equations as taking 5.5 h to run[19] but this is possibly total elapsed times including the period during which the computer's electricity supply had dropped out. By way of approval, Aparo commented that "in fact, during the Second World War, Dainelli and I spent about eight hours a day for a complete week in order to solve a system of 21 linear algebraic equations".

Ferranti's one-year guaranteed maintenance of FINAC commenced, by agreement, on 25th June and the machine came into regular operation. Allan Ellson remembers that "It was June, the students were sleeping on the roof below us and the cooling system was working flat out. The Italian staff criticised us because we stopped for tea breaks but we countered that, for them, every 10th day was a Saint's Day and no work was done. When it came to the Acceptance Tests we had the Mayor of Rome and the Chief of Police present. Before the computer left home someone [probably D G Prinz] had laboriously programmed the machine to play the Italian National Anthem—one note at a time. The moment eventually came for us to run this and the tune began to grind out. At this point one of the Italian engineers present poked me in the back and whispered "That's the Fascist National Anthem!"—but it was too late to stop it!".[20]

FINAC was officially inaugurated on December 14th 1955 by the Italian Head of State, President Giovanni Gronchi, but without an anthem being played! Sir Vincent de Ferranti made a short speech and Sebastian de Ferranti was amongst several other senior company personnel present. In addition to various demonstration programs the computer played the March from Aida (see also below). One

[17]Test Report (Verbale di Collaudo), June 1955, IAC archived document reference 17210–2. B.586/1. This report lists the following people as being present during the running of the FINAC Acceptance Tests: Dr. Enzo Aparo, Dr. Corrado Borhu, Dott. Dino Daizelli, Ing. Guilio Rodino and Ing. Giorgio Sacerdoit representing INAC, and Sig. Ing. Keith Lonsdale, Ing. Robert E Hodgkinson and Dr. Ignaxio Mottola representing Ferranti Ltd.

[18]Pollard, B.W., Rome, December 14th 1955: The Inauguration of a Ferranti Computer.

[19]De Marco, Giuseppe, Giovanni Mainetto, Serena Pisani, and Pasquale Savino., The Early Computers of Italy.

[20](a) Ellson, Allan., *Ferranti MkI* Commissioning and Installation: Some Recollections.*; (b) Alan Ellson, letter to the author dated 16th January 2000 and telephone conversation with the author on 20th September 2015 and e-mail to David Link on 26th July 2017. Alan joined Ferranti in May 1951 and was involved in commissioning the Mark I at the University of Manchester. He then transferred to Moston and was involved in the installation of FINAC. He left Ferranti in the summer of 1956.

Fig. 9.1 A modern view of the Consiglio Nazionale delle Ricerche (CNR) building in central Rome. A Ferranti Mark I* called FINAC was installed in 1955 on the fourth floor, just below the CNR inscription

of the maintenance engineers has said that "Our requests to put a Ferranti badge on the computer while newspapers and TV were around, was turned down for some unknown reason".[21]

Returning to the story of the wrong National Anthem being played in June, was the offending tune the Fascist Party anthem (*All'armi*) or merely the old monarchist National Anthem (*Marcia Reale*) which was current from 1861 to 1946? The INAC historian Andrea Celli asked Roberto Vacca, who was present at the 14th December Inauguration, about the *Fascist National Anthem* issue and he is sure that no anthem was played in December, neither the fascist anthem, nor the monarchic anthem, nor the republican anthem. Celli adds[22]: "A magazine a few days before inauguration wrote that Italian and UK Anthems will be played. After December 14th, several newspapers published a chronicle of the ceremony, all wrote of Aida being played, none wrote of any anthem". A 1956 sound clip of FINAC playing the March from

[21]Owen Ephraim, letter and e-mails to the author, October and November 1999. Owen joined Ferranti in 1953. His first job was to join the Mark I maintenance team at Manchester University. He then moved to Moston as commissioning engineer on Mark I*s. He took a Mark I* to Rome and remained there for a period as a maintenance engineer. He eventually left the Rome site to work on Pegasus.

[22]Andrea Celli, e-mail dated 1st October 2017 to the author.

Fig. 9.2 Professor Mauro Picone speaking at FINAC's official opening ceremony in December 1955. On his left are Professor Colonnetti, Sir Vincent de Ferranti and Professor Ghizzetti

Fig. 9.3 President Giovanni Gronchi examines a print-out of results. On his left is Professor Picone. The programmer nearest the camera is Dr. Enzo Aparo

Aida may be heard.[23] According to,[24] when played on 14th December 1955: "this was a little difficult to identify but nevertheless drew considerable laughter from the Presidential party".

9.3 FINAC at Work

After FINAC's inauguration two working groups were established at INAC, one devoted to hardware and the other to software. The three members of the hardware team consisted of Paolo Ercoli, Roberto Vacca and Giorgio Sacerdoti. Sacerdoti, who had worked on the installation of FINAC, was the first Italian student to graduate with a thesis on electronic computers.[25] The software group included Dino Dainelli, Enzo Aparo and Corrado Böhm. Many of these people can be seen in the photo in Fig. 9.4. Amongst other contributions to FINAC's software, Corrado Böhm designed a system called INTINT (INTerpretation-INTegration) for handling vectors.[26]

At the request of INAC, D. G. Prinz of Ferranti was involved in the writing of some early programs for FINAC and spent a 6-month period in Rome from 29th September 1955.[27] The Ferranti engineer Owen Ephraim remembers[28] that "the first

---

[23]L'Officina dei Calcoli. This film is Number 34 in a series called Oggi e Domani (Today and Tomorrow), organised by Archivio Audiovisivo del Movimento Operaio e Democratico (the Audiovisual Archive of the Democratic and Labour Movement). The date in the Archive catalogue is given as 1956/1958. See: https://www.youtube.com/watch?v=cBEpEgZsRkY. The film consists of three 10-min sections, respectively featuring Italian traditional umbrella manufacture, the Ferranti Mark I* and the design of stamps for the United Nations. The Mark I* section is at time-interval 4:40–7:45. Good sequences are included of FINAC's data-preparation equipment, programmers at the console, the CRT display during program execution and the computer's hardware. At one point we see what is believed to be a Bull lineprinter with several engineers adjusting it. The Ferranti programmer Joan Travis (nee Kaye) remembers visiting Rome from 23rd May to 9th June 1956 to work on attaching a Bull lineprinter to FINAC. I deduce that the OGGI E DOMANI No. 34 film sequence showing FINAC must have been shot sometime after April 1956—and possibly during June 1956. The film also includes a few seconds' sound of FINAC playing the March from Aida.

[24]Pollard, B.W., Rome, December 14th 1955: The Inauguration of a Ferranti Computer.

[25]A collection of 14 papers related to FINAC and/or to the Instituto Nazionale per le Applicazioni Calcolo (INAC) in Rome, 1952–1962, collected and edited by Paolo Ercoli, Rome. See: http://www.jnorman.com/cgi-bin/hss/39086. Most of the papers in this collection were originally distributed as *Pubblicazioni INAC* and sent to several universities world-wide.

[26]INTINT programmazione indiretta per calcolatrici elettroniche, Manuali per le Applicazioni Tecniche del Calcolo, Vol. 3, Cremonese, Roma (1958).

[27]Pollard, B.W., Rome, December 14th 1955: The Inauguration of a Ferranti Computer.

[28]Owen Ephraim, letter and e-mails to the author, October and November 1999. Owen joined Ferranti in 1953. His first job was to join the Mark I maintenance team at Manchester University. He then moved to Moston as commissioning engineer on Mark I*s. He took a Mark I* to Rome and remained there for a period as a maintenance engineer. He eventually left the Rome site to work on Pegasus.

Fig. 9.4 Members of the Consiglio Nazionale delle Ricerche (CNR) standing in front of FINAC in 1955. *(Left to right)*: Corrado Bohm, Paolo Ercoli, Wolf Gross, Aldo Ghizzetti, Mauro Picone, Giorgio Sacerdoti, Roberto Vacca, Enzo Aparo, Dino Dainelli

calculation I saw was for the pressures etc. arising in a pipe feed to a hydro-electric plant. There was concern because, if quickly operated, the safety valves caused the pipes to explode". Roberto Vacca[29] also remembers that: "Dr. Prinz wrote a program to play music in FINAC and it was used for the March in Aida etc. Once an English flute player (Rainer Schulein) heard FINAC playing pretend Mozart music starting with random 20 bit numbers, based on a program written by Prinz[30] and inspired by a paper of Mozart on how to get tunes from random numbers. Then Schulein on his flute played a parody of the computer parodying Mozart".

Significant upgrades to FINAC were made during its working life—see also below. For example the Ferranti programmer Joan Travis remembers visiting Rome from 23rd May to 9th June 1956 in connection with the attachment of a Bull lineprinter. "I believe the engineer in charge in Rome was Giorgio Sacerdoti, he was very pleasant—he had suffered from polio when he was younger".[31] A

[29]E-mail from Roberto Vacca to Owen Ephraim, 7th November 1999 (and copied by Owen to the author).

[30]Andrea Celli, e-mail dated 29th September 2017 to the author. Celli is actively researching the history of INAC.

[31]Joan Travis, e-mail to the author dated 25th October 2015.

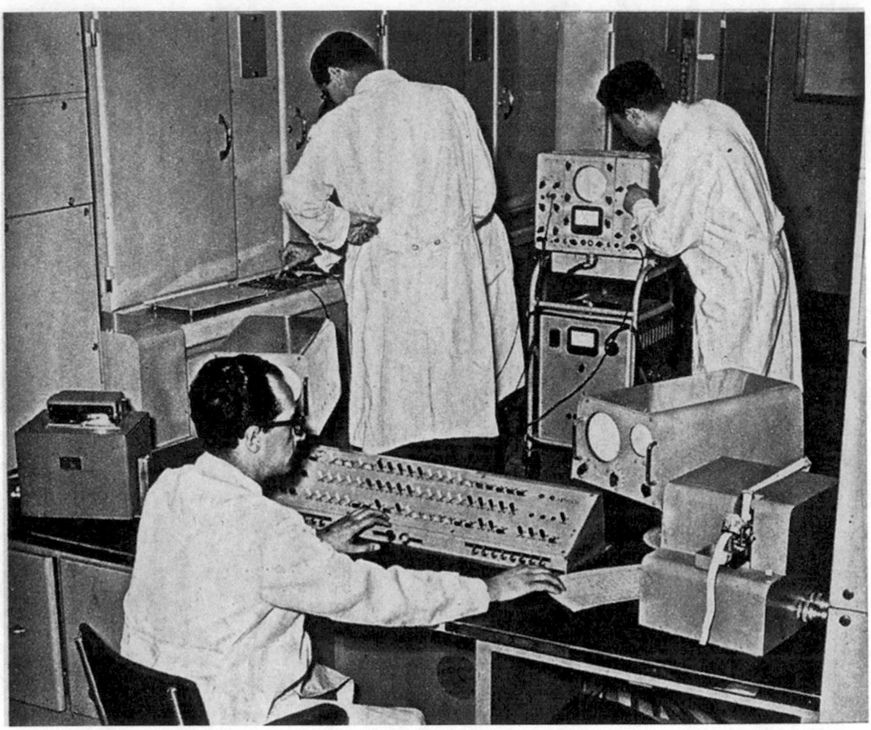

Fig. 9.5 A photo of FINAC which appeared in the January 1956 edition of the magazine *Science and Life (Scienza e Vita, Number 84, p. 15).* The person in the foreground seated at the computer's console is possibly Paolo Ercoli. The person on the right in the background is believed to be Dino Dainelli

ten-minute film sequence of FINAC is available[32] which shows, amongst other things, the Bull lineprinter in operation with engineers making adjustments. It is possible that the film was shot in the summer of 1956.

[32]L'Officina dei Calcoli. This film is Number 34 in a series called Oggi e Domani (Today and Tomorrow), organised by Archivio Audiovisivo del Movimento Operaio e Democratico (the Audiovisual Archive of the Democratic and Labour Movement). The date in the Archive catalogue is given as 1956/1958. See: https://www.youtube.com/watch?v=cBEpEgZsRkY. The film consists of three 10-min sections, respectively featuring Italian traditional umbrella manufacture, the Ferranti Mark I* and the design of stamps for the United Nations. The Mark I* section is at time-interval 4:40–7:45. Good sequences are included of FINAC's data-preparation equipment, programmers at the console, the CRT display during program execution and the computer's hardware. At one point we see what is believed to be a Bull lineprinter with several engineers adjusting it. The Ferranti programmer Joan Travis (nee Kaye) remembers visiting Rome from 23rd May to 9th June 1956 to work on attaching a Bull lineprinter to FINAC. I deduce that the OGGI E DOMANI No. 34 film sequence showing FINAC must have been shot sometime after April 1956 —and possibly during June 1956. The film also includes a few seconds' sound of FINAC playing the March from Aida.

Fig. 9.6 FINAC in operation. The equipment behind and just above the operator's console is an oscilloscope and maintenance instrumentation, mounted on a trolley

Reliability and upgrades.

Glimpses of the first four years of FINAC's operation in Rome may be had from issues of the US Office of Naval Research's *Digital Computer Newsletter,* which seems to have developed a particular interest in FINAC's progress.[33] Here are some examples from issues of the *Newsletter*, which give an idea of the performance of the machine and the units of new equipment that were installed as time went by.

July 1955: "During the first year, FINAC is scheduled for a regular 8-hours-per-day schedule. It will be mainly devoted to partial differential equations, periodogram analysis and inversion of matrices … An extra high speed parallel output will be added about January 1956. It will print 150 lines per minute, 64

[33]The US Office of Naval Research's Digital Computer Newsletter was published regularly during the 1950s and 1960s. In addition to providing updates on the American computing scene, the Newsletter reported from time to time on the European arena. Here is a list of the relevant FINAC references. Vol. 7 no. 3, 1955, p. 11; Vol. 8 no. 1,1956, p. 17ff; Vol. 8 no. 3, 1956, p. 11ff; Vol. 9 no. 1, 1957, p. 11ff; Vol. 10 no. 3, 1958, p. 11; Vol. 11 no. 3, 1959, pp. 8 and 9; Vol. 19 no. 1, 1967, p. 7ff.

Fig. 9.7 Close-up of FINAC's console. This image is in the form of a contemporary postcard, published by INAC. The white rectangle on the wall at the back of the room is a display board for the maintenance engineers' notes and diagrams

characters each row (92 including spaces). Courses on programming are given at the Institute".

January 1956: "Programming is in progress on the solution of m simultaneous algebraic equations in n unknowns by orthogonalization procedures. The machine [FINAC] is used presently eight hours per day, four by the engineers and four by the mathematicians. About 1-1/2 h are used for training purposes. The staff is formed by three engineers and three groups of four programmers each. Courses on programming are being printed".

July 1956: "A parallel high speed printer is being added to the machine. It is expected to be operating in the first days of June .... Good operating time of the machine during this year has been 89% of the total operating time".

January 1957: "During the first year of maintenance (June 1955–June 1956) the average weekly efficiency of the computer was 89.1%". .... [A detailed analysis of faults is then given in the *Newsletter*]. "New overflow instruction. The logic and the electronics for a new "overflow" instruction were recently designed in Rome and the physical circuitry was built in the FINAC. At present the new Instruction is working and at full disposal of INAC mathematicians. The OVERFLOW Instruction (code 01110) is: 'Test the setting of the OW flip-flop; if the flip-flop is set, transfer control to the instruction whose address is specified and reset the OW flip-flop, otherwise use the Instruction next in numerical order in the usual way'. ...

The components involved in the construction of the new circuitry and in the modification of the existing circuits are the following: 36 valves (10 pentodes, 14 double triodes and 12 double diodes), 10 capacitors, 10 crystals and 140 resistors. A whole new chassis is needed".

July 1958: "During the second year of operation (June 1956–June 1957) the total useful time was 2089 h and the average weekly efficiency of the computer was 91.2%. Corresponding values for the first year: 912 h and 89.1%" ....

"New equipment installed and new order codes. On September 13, 1957 a new Creed Model 25 Perforator was installed. The acceptance test of the perforator consisted in punching about 108,000 characters, which it did with no errors. The average punching speed is about 26.5 characters per sec., with activated check contacts. 7 months experience has shown that the punch is very reliable when properly set. Only occasional adjustments have been necessary. Before the end of the summer a second Model 25 Perforator and two Ferranti 35 mm. Magnetic Tape Units will be installed.

"In connection with the installation of the magnetic tape units, new circuits have been added to be used for double precision arithmetic. These two new features are expected to be particularly useful for the solution of large systems of linear equations, especially in cases of ill conditioning. For double precision two orders are used to set or reset a flip-flop. When the flip-flop is in the reset state the computer works as usual, that is, the instructions regarding accumulator and multiplier refer to 40 bits. When the flip-flop is in the set state the same instructions refer to 80 bits. A "jump if overflow occurred" instruction controls both simple and double precision arithmetic. This allows fixed point operation also in case of ill conditioned systems of linear equations, speeding up computing time. For the new circuits 106 valves have been installed and 37 removed".

July 1959: "During the first three years of operation (30 June 1955–30 June 1958) the efficiencies were 89.1, 91.2, and 92.2%".... [Detailed fault-analysis is then given in the *Newsletter*]. "Machine use (3rd year):

Time Useful	1645 h
Not useful	60
Down time	93
Idle time	23
Engineering	844
Total	2665

New Equipment Installed. Another Creed Model 25 Perforator has been provided. These paper tape punches have proved to be very reliable. The second punch has a speed of about 30 characters per second. The double precision facility, for arithmetic on 80 bit word (see *Digital Computer Newsletter*, July 1958) has proved to be extremely useful and reliable. A transistorized flip-flop register has been installed to monitor the address of the last Jump-instruction performed by the machine. This facility is particularly useful in debugging and trouble shooting. A set of transistor counters driving electromechanical counters is being installed for the purpose of analyzing the frequencies of occurrence for the various types of instructions. The connection between valve circuits and transistor circuits is done through more or less conventional crystal diode circuits.

9.4 The Wider Picture

By 1960 digital computing activity in Italy had achieved considerable momentum. An Auerbach Report[34] states that: "As of June 1960, nearly 150 computers were installed or on order (most of them from the United States) with about 85 in actual service. …. The most active computer development work currently being done is at the University of Pisa and the Olivetti Laboratories near Milan". Olivetti's first prototype digital computer had appeared in 1957. As was mentioned above, the CEP (Calcolatrice Elettronica Pisa) computer project was started at Pisa in 1954 and first worked in 1961. A CEP simulator was implemented on FINAC.[35]

Olivetti began a collaboration with INAC in June 1961, when Professor Picone met Mario Tchou, one of Olivetti's leading engineers.[36] It was agreed that Olivetti would build a scientific computer, to be named CINAC (but also known as Elea 9004) with design input from INAC staff. The plan was that CINAC would replace the Ferranti Mark I*. Tragically, Tchou was killed in a car accident in November 1961. His place was eventually taken by Giorgiou Sacerdoti. Meanwhile, Olivetti got into financial difficulties and in 1964 its electronics division was sold to the American company General Electric. The CINAC project was revived but only a prototype was completed. This CINAC prototype was acquired by INAC towards the end of 1965 and began useful work in the following year. At this point FINAC, the Ferranti Mark I* computer, was gradually taken out of service (see also below).

In January 1967 it was reported[37] that: "INAC has acquired a new computer. Acceptance tests were successfully completed in February 1966. The computer, called CINAC for short, is the result of a four year joint effort by the Italian National Research Council and the electronic division of the Olivetti Company (later Olivetti General Electric). The system consists of: a central processing unit, a main memory unit and up to eight Synchronisers, each capable of dealing with up

[34]*European Information Technology*. Report 10/48/TR, 15th January 1961, by Auerbach Electronics Corporation under contract to the Office of Naval Research. See: http://www.dtic.mil/dtic/tr/fulltext/u2/250143.pdf, p. 60 onwards is relevant to Italy.

[35]Celli, A., M. Mattaliano, and R. Spitaleri., Istituto per le Applicazioni del Calcolo, 1927–2012: 85 years in Computational and Applied Mathematics.

[36]Celli, A., M. Mattaliano, and R. Spitaleri., Istituto per le Applicazioni del Calcolo, 1927–2012: 85 years in Computational and Applied Mathematics.

[37]The US Office of Naval Research's Digital Computer Newsletter was published regularly during the 1950s and 1960s. In addition to providing updates on the American computing scene, the Newsletter reported from time to time on the European arena. Here is a list of the relevant FINAC references. Vol. 7 no. 3, 1955, p. 11; Vol. 8 no. 1,1956, p. 17ff; Vol. 8 no. 3, 1956, p. 11ff; Vol. 9 no. 1, 1957, p. 11ff; Vol. 10 no. 3, 1958, p. 11; Vol. 11 no. 3, 1959, p. 8 and 9; Vol. 19 no. 1, 1967, p. 7ff.

to 16 peripheral control units, each controlling communication with a peripheral unit, with external lines, or with another computer. The main memory can have up to 65,536 directly addressable words (24 or 48 bits each) and is automatically time-shared by the central processing unit and the synchronisers".

CINAC's initial configuration had 24,576 words of main memory with a cycle time of 2.7 μs. "A special custom-built console is the reproduction of the console and man-machine interface of a previous computer [the Ferranti Mark I*] so that old programs can be used directly through a simulator".[38] This console may be part of the artefact displayed in the Museum of Computing Machinery at Pisa—see Figs. 9.8 and 9.9 below. The notion of equipping the console of a follow-on computer with the look and feel of an earlier computer was also used by Fort Halstead, when specifying the COSMOS replacement computer for Fort Halstead's Ferranti Mark I* in the mid-1960s—(see Chap. 11).

9.5 How Did It All End?

With CINAC fully operational, it is believed that FINAC was finally switched off some time in 1966. So ended the long and active life of Rome's first computer. FINAC had been used by "hundreds of individuals and by such organisations as the Air Ministry and the Treasury Ministry, various institutes located in Rome, Milan and Turin largely for engineering, scientific and mathematical applications".[39] One of the most complex tasks that FINAC undertook was to verify the stability of the Vajont dam, completed in 1959 and said at the time to be the tallest dam in the world. The calculations involved solving a system of about 400 linear equations, taking FINAC several days.[40]

The remains of the original FINAC hardware gradually shrunk and finally vanished. Here is the story, according to Andrea Celi.[41] After being dismantled, a few selected FINAC components were displayed in a showcase cabinet to honour Rome's first computer. Other parts were given to various research centres or to schools. Some accessories such as teleprinters and paper tape editing equipment were stored in the basement under the CNR congress hall. When INAC moved in

[38]The US Office of Naval Research's Digital Computer Newsletter was published regularly during the 1950s and 1960s. In addition to providing updates on the American computing scene, the Newsletter reported from time to time on the European arena. Here is a list of the relevant FINAC references. Vol. 7 no. 3, 1955, p. 11; Vol. 8 no. 1,1956, p. 17ff; Vol. 8 no. 3, 1956, p. 11ff; Vol. 9 no. 1, 1957, p. 11ff; Vol. 10 no. 3, 1958, p. 11; Vol. 11 no. 3, 1959, p. 8 and 9; Vol. 19 no. 1, 1967, p. 7ff.

[39]Cordata, J.W., *The Digital Flood: The Diffusion of Information Technology Across the US, Europe and Asia.*

[40]De Marco, Giuseppe, Giovanni Mainetto, Serena Pisani, and Pasquale Savino., The Early Computers of Italy.

[41]Andrea Celli, e-mail dated 19th November 2017 to the author.

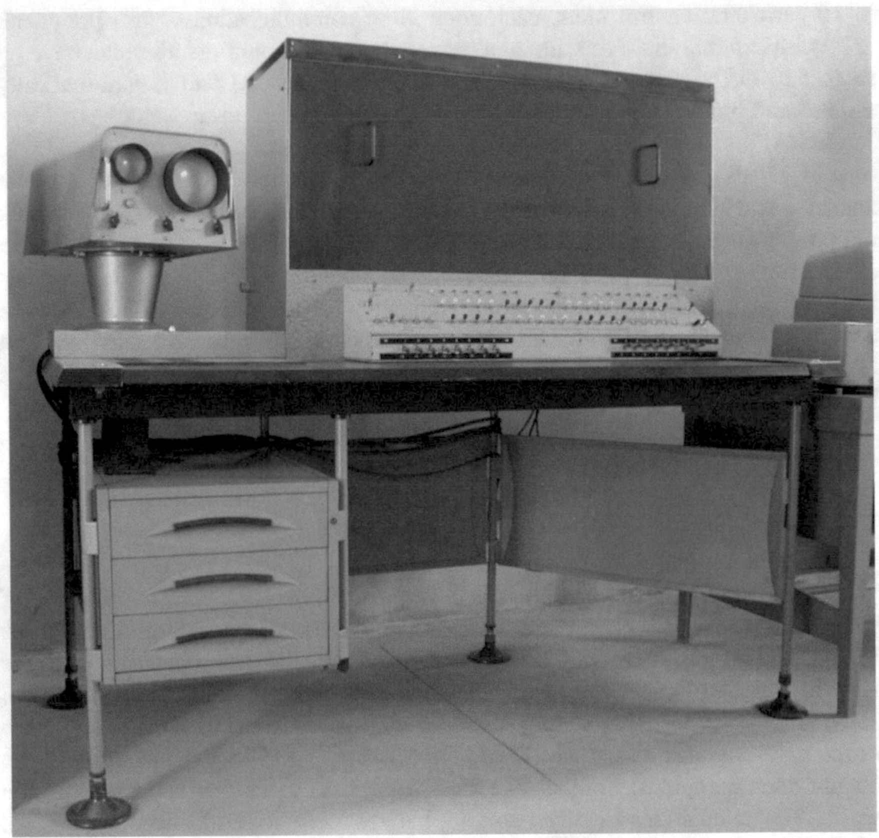

Fig. 9.8 The simulated FINAC switches on the simulated console at Pisa

1971 from the grand CNR headquarters to a separate building, the showcase cabinet and the FINAC archive were stored in the basement of the new building. Then in 1974 the CINAC computer ceased operation and CNR decided to find room to store CINAC hardware. At this point the INAC Director was asked if he wanted save FINAC's equipment. He decided to destroy much of the remaining FINAC parts so as to have room to store CINAC. In 1982, INAC (now IAC) reorganised the basement in order to enlarge a garage. The FINAC showcase cabinet and a large part of the FINAC archive were destroyed in favour of the garage. It therefore seems most unlikely that any fragment of FINAC has survived today.[42] CINAC, on the other hand, ended up in the Museum of Computing Instruments at Pisa.

[42]Franchi, Stefano, and Francesco Bianchini. 2011. *The Search for a Theory of Cognition: Early Mechanisms and New Ideas.* Published by Rodopi. ISBN: 978-90-420-3427. See Chap. 4, *The periphery of the rising empire: the case of Italy, 1945–1958*, by Claudio Pogliano.

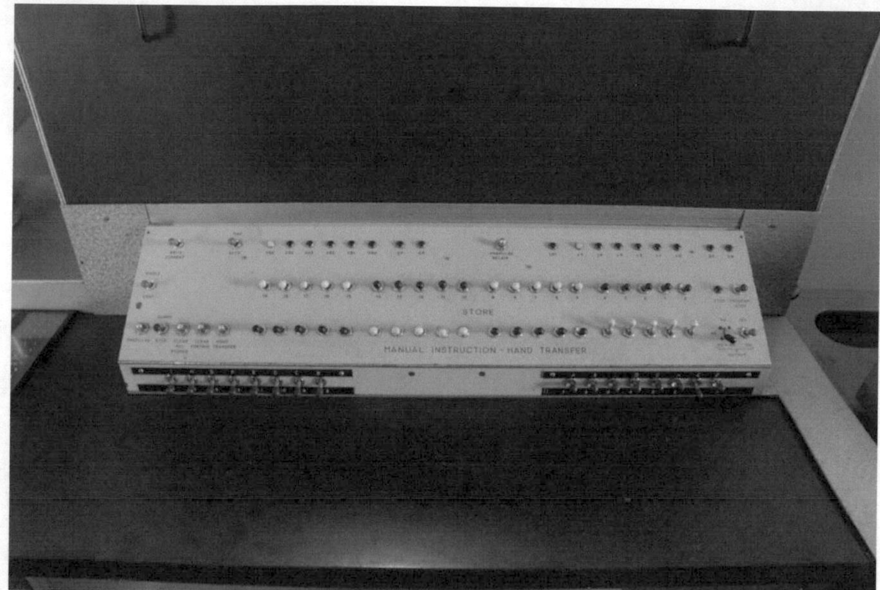

Fig. 9.9 A close-up of the operator's console at the Museum of Computing Machinery at Pisa

References

Celli, A., M. Mattaliano, and R. Spitaleri. 2012. Istituto per le Applicazioni del Calcolo, 1927–2012: 85 years in Computational and Applied Mathematics. In *Proceeding of MASCOT 12 & ISGG 12, The Joint 12th Meeting on Applied Scientific Computing and Tools and 12th ISGG Meeting on Numerical Grid Generation, Las Palmas de Gran Canaria*, 61–70, Oct 2012. Published by IMACS, the International Association for Mathematics and Computers in Simulation.

Cordata, J.W. 2012. *The Digital Flood: The Diffusion of Information Technology Across the US, Europe and Asia*. Oxford. ISBN 978-0-19-992155-3.

De Marco, Giuseppe, Giovanni Mainetto, Serena Pisani, and Pasquale Savino. 1999. The Early Computers of Italy. *IEEE Annals of the History of Computing* 21 (4): 28–36.

Ellson, Allan. 2015. *Ferranti MkI* Commissioning and Installation: Some Recollections*. 14th Oct 2015.

Franchi, Stefano, and Francesco Bianchini. 2011. *The Search for a Theory of Cognition: Early Mechanisms and New Ideas*. Published by Rodopi. ISBN: 978-90-420-3427.

Pollard, B.W. 1956. Rome, December 14th 1955: The Inauguration of a Ferranti Computer. *Ferranti Journal* 9–11.

Chapter 10
GCHQ Cheltenham's Mark I*

At the time of writing (2018) the relevant authorities are still coy about releasing information on post-war codebreaking, though since about 2012 the Americans have been more forthcoming than the British. It has not yet proved possible to tell the full story of the Ferranti Mark I* computer that was delivered to GCHQ's Cheltenham establishment in 1953. We can, however, sketch the overall Anglo-American cryptanalytical context in which this computer was chosen and describe the interactions between GCHQ and Ferranti Ltd. that led to its delivery and use.

10.1 US and UK SIGINT at the Start of the Cold War

The British Government's Code and Cipher School (GC&CS), and the equivalent cryptanalytical organisations in the American Army and Navy, had achieved brilliant successes during World War II at decoding German and Japanese ciphers. They had done this by a combination of hand analysis, the use of IBM electro-mechanical punched card equipment and the construction of special-purpose *Rapid Analytical Machines*. In the latter category were the Turing-Welchman Bombes and Colossus—and many other less well-known specially-designed devices. Some of these used novel techniques—for example the machine designed by Vannevar Bush based on photo-electric comparisons between two texts.[1]

In the immediate post-war years the British greatly reduced the staff at Bletchley Park and re-formed the GC&CS as the Government Communications Headquarters (GCHQ). This was at first principally located at Eastcote in Middlesex, before

[1]Wheatley, L.H. 1953. *Cryptanalytic Machines in NSA*, 365. ID: A60928. NSA 34. Declassified (with redactions) in 2014. See: https://www.nsa.gov/news-features/declassified-documents/friedman-documents/assets/files/reports-research/FOLDER_107/41743419078275.pdf.

S. Lavington, *Early Computing in Britain*, History of Computing,
https://doi.org/10.1007/978-3-030-15103-4_10

moving in the early 1950s to Cheltenham in Gloucestershire. The Americans initially kept their separate Army and Navy organisations, respectively known by acronyms such as SIS and OP-20-G before being amalgamated into the Armed Forces Security Agency (AFSA) and finally, in 1952, into the National Security Agency (NSA).

There were differences of emphasis in tackling Signals Intelligence, SIGINT, on each side of the Atlantic. During the war, the Americans had increasingly relied upon standard IBM punched card equipment such as Tabulators, Collators, Sorters and Reproducers.[2] In general, the following types of data analysis could be done (on text or numbers) using punched-card equipment:

Counting, printing	(Tabulator)
Merging, selecting	(Collator)
Expanding, reproducing	(Reproducer)
Distributing, sorting	(Sorter)

Typical problems tackled included the initial statistical evaluation of an intercept. This involved: (a) searching the encoded message for monographic, digraphic or other polygraphic frequencies; (b) examining derivatives (e.g. 'replacing each letter by the difference between itself and the preceding letter' of the original encoded message; (c) searching for 'double hits', perhaps indicating that two messages had been enciphered with the same additive.[3] From the start to the end of the war, the total number of IBM punched card machines used by SIS had risen from 13 to about 400; the equivalent figures for OP-20-G were 16 and 200.[4] Sometimes the standard IBM equipment was specially modified for a particular cryptanalytic task, for example when the GEEWHIZZER circuitry was added to an IBM 405 or 407 Tabulator.[5]

Of course, punched card equipment contained no memory (except for a few counters) and only primitive arithmetic and logical functionality. But America's wide adoption of procedures based on punched cards may have made the American SIGINT community more receptive to the advent of the new type of general-purpose computers—once these became a practical proposition.

During the war, Bletchley Park had been slower to make use of punched card equipment. On a visit to Washington in August 1941 Alastair Denniston, the operational head of GC&CS, had reported that: "They [the Americans] make far

[2]Budiansky, Stephen. 2001. Codebreaking with IBM Machines in World War II. *Cryptologia* 25 (4): 241–255.

[3]*Survey of where mechanised operation can benefit the COMINT effort on literal texts.* Document ID: A60837. Author(s) not given. 43 p. Date unclear but possibly 1959. Declassified (with redactions) in 2014. See: https://www.nsa.gov/news-features/declassified-documents/friedman-documents/assets/files/reports-research/FOLDER_107/41743209078242.pdf.

[4]Budiansky, Stephen., Codebreaking with IBM Machines in World War II. *Cryptologia* 25 (4): 241–255.

[5]Wheatley, L.H., *Cryptanalytic Machines in NSA,* 365.

greater use of these machines to avoid use of personal effort, but I am not convinced that these mechanical devices lead to success. Close personal effort makes one intimate with the problems which, when served up mechanically, fails to appeal".[6]

After the war, many GC&CS cryptanalysts were dispersed, typically back to into academia where they remained relatively out of touch with industry. At the same time, their American colleagues had had closer links with American companies and retained these links after the war. Another post-war difference between the two sides of the Atlantic related to funding: it is believed that GCHQ was relatively less well supported than SIS and OP-20-G during the period 1945–55, a decade of extreme economic hardship in the UK.

In conclusion, there were several reasons why there were different rates at which general-purpose stored-program computers were adopted during the early 1950s on each side of the Atlantic. The Americans were the first to take computers seriously, as we describe in Sect. 10.2.

In passing, some urgency had been injected into Allied SIGINT operations after so-called *Black Friday*, 29th October 1948. On that day, the Soviet Union made fundamental changes to its codes and ciphers, effectively destroying all Anglo-American cryptanalytic access to Moscow's high-level communications. This came on top of other ominous developments in 1948/49 such as the Berlin Blockade and the creation of the People's Republic of China. The Korean War started in June 1950. The code-breakers urgently needed to up their game.

10.2 The Possibilities of General-Purpose Computers

The first major, public, lectures on the possibilities for stored-program computers took place in the summer of 1946, at the Moore School of Electrical Engineering, University of Pennsylvania. Among the attendees was Lt. Commander James T. Pendergrass of the US Navy's Communications Supplementary Activities, Washington (CSAW), which was related to OP-20-G. In December 1946 Pendergrass produced a report recommending that CSAW construct a computer similar to the one then being proposed by John von Neumann at the Institute of Advanced Study, University of Princeton.[7] CSAW's computer was eventually called ATLAS. It was delivered to Washington on 8th December 1950 and was operational before Christmas. It was the first general-purpose computer to be

[6]Report, A G Denniston, 31st October 1941. In The National Archive (PRO), Kew, document HW 14/45.

[7]Snyder, Samuel S. 1964. *History of NSA General-Purpose Electronic Digital Computers.* Washington DC: Dept of Defense. Doc ID: 6586784. Typewritten illustrated report of approx. 110 pages. Approved for release in 2004. Contains photographs of ATLAS, ABNER, etc., though the images are unclear. See: https://www.nsa.gov/news-features/declassified-documents/nsa-early-computer-history/assets/files/6586784-history-of-nsa-general-purpose-electronic-digital-computers.pdf.

installed in any SIGINT establishment, though it lacked programmable input/output facilities. An account of ATLAS is given in Sect. 10.3.

Pendergrass's report was also received by the Army Security Agency (ASA), which was related to SIS. During 1948 ASA consulted several organisations, especially the Bureau of Standards Digital Computer Laboratory which was to build its own computer called SEAC. The consultations resulted in an ASA report recommending the construction of a general-purpose computer but based on a four-address instruction format rather than the one-address format chosen by CSAW for ATLAS. ASA's computer was eventually called ABNER. ABNER became operational at ASA in June 1952—see Sect. 10.3.

Before going on to describe ATLAS and ABNER in some detail it is pertinent to wonder whether, in 1946/47, GCHQ was similarly inspired to invest in general-purpose computers? Clearly GCHQ had an understanding of the new type of computer, not least because of the close links with their American SIGINT colleagues who, in turn, had John von Neumann as one of their consultants. The buzz had also reached other UK establishments such as the National Physical Laboratory (NPL)—and indeed NPL had recruited Alan Turing in October 1945 to design an Automatic Computing Machine. In May 1946 the Royal Society had awarded a grant to Professor Max Newman at Manchester University to construct a computing machine. Newman had on his staff Jack Good and David Rees. David Rees had attended the Moore School lectures but no evidence has come to light about whether Rees talked to Pendergrass whilst the two of them were there. Indeed, as discussed in Chap. 14, the signs are that Rees did not personally follow up on anything he'd imbibed at the Moore School.

Turing, Newman, Good and Rees had all worked at Bletchley Park during the war and it is reasonable to assume that GCHQ could have (probably would have?) consulted them on whether it was worth planning for a general-purpose computer to be installed at Eastcote. GCHQ staff at Eastcote certainly remained interested in special-purpose machinery for SIGINT and for its converse, COMSEC (Communications Security). However, it seems that no significant investment in general-purpose computers was made by Eastcote until negotiations with Ferranti Ltd. were set in motion in the autumn of 1951—see below.

Jack Good, who left Manchester University in April 1948 to join GCHQ at Eastcote, probably played a part in persuading GCHQ to act. In January 1950 he circulated a document entitled *The Cryptological Uses of a General Purpose Computer.*[8] This four-page document starts with the sentence: "I have been asked to consider this problem at very short notice …". About a third of Jack Good's text then deals with the general differences between the new type of stored-program computer and the variety of special-purpose Rapid Analytical Machines on which GCHQ was relying at that time. Good notes that: "For some problems there is little

[8]Good, I.J. 1950. *The Cryptological Uses of a General Purpose Computer.* According to the official GCHQ Historian Tony Comer, this GCHQ document was a 'think-piece', indicating a discussion paper intended only for internal circulation. A redacted version of the document was released by GCHQ in November 2018.

doubt that special-purpose machines would be more efficient than a general-purpose computer. But it should be realised that if in an emergency a demand arises for a special purpose machine, there will be a considerable transitional period during which the general-purpose computer (if it existed) would fill the gap". In some cases, "it should be possible to operate the general-purpose computer in combination with a Colossus". We'll come to Jack Good's particular recommendations in Sect. 10.6.

Sometime after Good's report, a GCHQ internal Survey on the subject was undertaken in 1950. Then all apparently went quiet for a while. Then in July 1951 six GCHQ staff went to Manchester University to attend the Inaugural Conference of the Ferranti Mark I computer. The staff, in the order of their assumed influence within GCHQ, were: Dr. G. W. (Gerry) Morgan, Dr. G. (Geoff) Timms, Dr. I. J. (Jack) Good, S. D. (Toby) Harper, Alan Bruce and Miss V. Allinson. By the time of the Manchester Conference, GCHQ had already placed contracts with Ferranti Ltd. for certain special-purpose equipment, as described in Sect. 10.4. It is very probable that the Conference offered an opportunity for GCHQ staff to ponder the possibility of Ferranti providing them with a copy of the Manchester computer.

At any rate, in a follow-up memo dated 7th November 1951,[9] Dr. G. W. (Gerry) Morgan summarised the position as follows:

"Two of the recommendations of the [1950?] Survey were that we [GCHQ] purchase a Computer and send a mathematician to AFSA to work there for some months on the programming of cryptanalytic tasks for ATLAS and ABNER. The latter are general mathematical computers and more beside, and it was very clearly recognised that GCHQ could not have its own equivalent for, probably, three years at the least. It could be hoped, however, that when it did come it would be better than ATLAS and ABNER and partly because of them. It is very clearly recognised also that quite 90% of the value of a computer does not become apparent until it is in use, and that we must begin to accumulate programming experience.

"I am strongly convinced that the recommendations are sound and, while the implementation of the second must await an invitation from AFSA, there is no reason for delay in carrying out the first. This can be done in one way only, by purchasing a replica of the Manchester Computer from Ferranti, who had hoped to produce a line of six identical machines. The first is already installed in the University; the second and third are being assembled, one for the firm itself [in the event, this actually went to Toronto—see Chap. 3] and the other for the Ministry of Supply [probably, at this stage, intended for Fort Halstead—see Chap. 11]. We could secure the fourth. The cost of a machine has been quoted to us as £60,000.

"There are quite a number of points of security that need attention. The first is the desirability of suppressing completely the fact that GCHQ has purchased a computer. If the very existence of our computer can be concealed so much the better …."

---

[9]Morgan, G.W. 1951. *GCHQ & Ferranti.* GCHQ Minute Sheet X/565/1802.

Fig. 10.1 I. J. (Jack) Good, who worked at Bletchley Park during the war. He joined Professor Max Newman's Mathematics Department at Manchester University in 1945 and then re-joined GCHQ in April 1948

Amongst the devices recommended by Morgan for security was that the ordering of the computer should be done through the Ministry of Supply (MOS) "which we already use as a cover address in connection with Elliotts" [see Sect. 10.4]. Another recommendation by Morgan was that "I make proper contact with Sir Ben Lockspeiser (DSIR), Sir David Brunt (Chairman, DSIR), and Brigadier Hinds (MOS), for the triple purpose of establishing at any rate non-hostile relations, of getting their assistance to covering our tracks and possibly of getting some positive aid from them".

Gerry Morgan then goes on to summarise nine existing GCHQ projects with Ferranti Ltd., which are listed in Sect. 10.4. His Minute then ends with this comment on personnel: "We have so far dealt with Ferranti as GCHQ, have indoctrinated several of their staff, and have had numerous meetings. They know Harper, Bruce, Ridlington, Timms and myself as GCHQ. It is important to review the whole relationship before we go any further and decide on the best and most secure policy".

In the light of the known technological innovations produced by Ferranti Ltd. and Elliott Brothers Ltd. in the period 1948–1951, two things may be deduced from Gerry Morgan's document:

(a) GCHQ had not been idle in developing special-purpose cryptanalytic equipment in the period 1948–1951;
(b) None of this equipment had any real connection with general-purpose stored-program computers.

In this respect, GCHQ was playing catch-up with the Americans. A technical analysis of the ATLAS and ABNER projects will suffice to demonstrate the size of the gap. We will then return to GCHQ's links with Ferranti in Sect. 10.4.

10.3 Comparing ATLAS and ABNER with the Ferranti Mark I*

As mentioned above, GCHQ hoped to have some experience of using the pioneering ATLAS and ABNER computers before arranging that its own first computer, a Ferranti Mark I*, was delivered to Cheltenham in 1953. It is thus of interest to compare the three machines. The basic design for each one was decided, in differing circumstances, in 1948/49. The three trajectories from concept to finished product were very different. Some background history is useful to emphasise the cultural differences between Ferranti Ltd. and the equivalent American enterprises that designed and built ATLAS and ABNER. This also serves to emphasise the early involvement of the American security services with the design and production of pioneering digital computers that eventually reached the open market.

10.3.1 The ERA ATLAS Project

Although ATLAS was a commercially-produced computer, its conception and construction were intimately—some were later to say too intimately—connected with CSAW. ATLAS was designed and built by Engineering Research Associates (ERA), a company founded in January 1946 by two former CSAW staff, William Norris and Howard Engstrom. Norris and Engstrom hired forty of their code-breaking colleagues and ERA took over a redundant factory in St. Paul, Minnesota, sharing the site with the Naval Computing Machine Laboratory. The ERA premises became a Navy Reserve base and armed guards were posted.

At first the ERA company was occupied with developing new code-breaking machines under contract for the US Navy. These were special-purpose devices and made much use of magnetic drum storage. ERA's first machine, GOLDBERG, was completed in 1947. It used a primitive drum and, thereafter, the company became expert in this form of storage.

In 1947, as a direct consequence of the December 1946 Pendergrass Report, the Navy awarded ERA a contract, known as *Task 13*, to develop a stored-program computer. ATLAS was the result. By the summer of 1948 the system design was essentially complete and prototyping was under way. The register-level architecture was based on von Neumann's plans for his IAS computer—which first performed useful computation in the summer of 1951. ATLAS's circuitry was influenced by

MIT's plans for WHIRLWIND. WHIRLWIND was working reliably enough to do useful work by March 1951.

Although the original *Task 13* proposal had specified a high-speed random-access internal memory based on the RCA Selectron tubes, these proved unavailable. Instead, Atlas employed a magnetic drum of a type similar to one already used by ERA for other special-purpose equipment built for the Navy.

The first ATLAS was delivered to CSAW in December 1950 and became fully operational before Christmas. It was said to be very reliable. An ERA publicity brochure produced in 1951 stated that the first ATLAS "operated for its first 500 h with only 16 h of unscheduled maintenance. We believe this to be an unusual record for a new electronic computer". ATLAS cost $950,000,[10] equivalent at that time to £339K and therefore over thrice as costly as the Ferranti Mark I*.

ERA then worked on a successor, ATLAS II, which had a longer word-length (36 not 24 bits), a two-address instruction format, 1K words of additional electrostatic (probably Williams/Kilburn) storage and program-controlled input-output instructions. ATLAS II was installed in October 1953.

Meanwhile, ERA had started to sell ATLAS I commercially as the ERA 1101. However, the company experienced financial difficulties and was bought in 1952 by Remington Rand. A year or so before that, Remington Rand had purchased the Eckert–Mauchly Computer Corporation, whose first commercial product was UNIVAC I. The ERA and UNIVAC Divisions were run separately within Remington Rand but ERA's successor computers were given the *UNIVAC* label because the word UNIVAC had become well-known on the open market. It was on 31st March 1951 that the first UNIVAC computer was delivered to the US Census Bureau—thereby becoming the first truly commercial computer to have been delivered in America.

A slightly-modified ATLAS II became the UNIVAC 1103, while a more heavily modified version with ferrite core memory and floating-point hardware became the UNIVAC 1103A. We shall see in Sect. 10.8 that GCHQ Cheltenham eventually purchased a UNIVAC 1103A as the successor to their Ferranti Mark I*.

10.3.2 The ASA ABNER Project

As a direct result of the December 1946 Pendergrass Report, the Army Security Agency (ASA) also initiated a project to acquire a general-purpose computer. In 1948, after consulting all the existing American computer research groups, ASA decided on a four-address instruction format architecture similar to the one being proposed for the EDVAC and RAYDAC machines. The resulting ASA computer, called ABNER, was named after the main character in the satirical American comic strip *Li'l Abner*.

[10]Snyder, Samuel S., *History of NSA General-Purpose Electronic Digital Computers*.

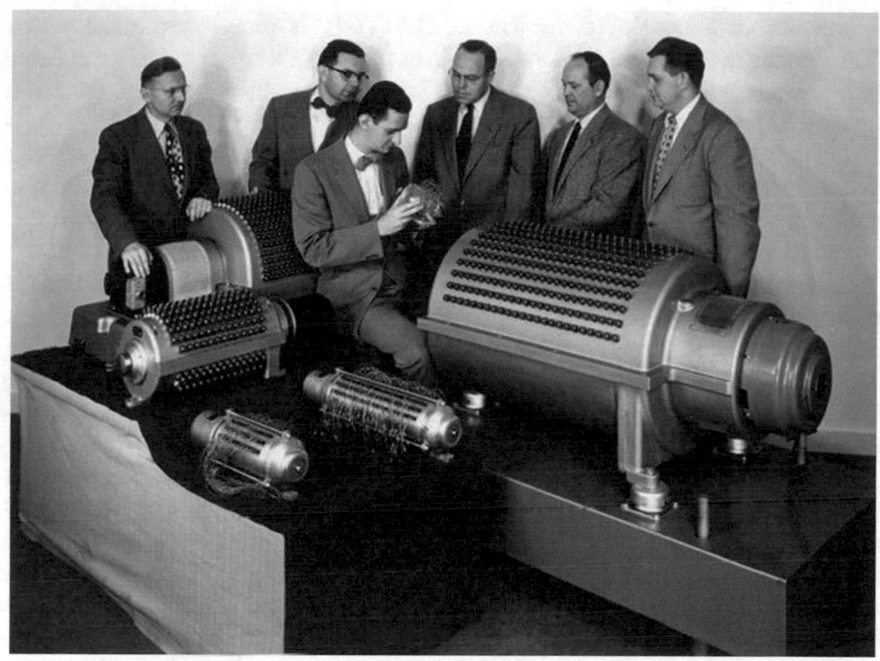

Fig. 10.2 A group of engineers from Engineering Research Associates (ERA) in about 1953, displaying the variety of magnetic drum storage devices that the company had designed

ASA had great difficulty finding a company prepared to design and build ABNER. In the end, a contract was given to Dr. Samuel Lubkin, a former member of the Moore School's ENIAC team, to design a machine that would be built by ASA's own engineers with certain sub-units being sub-contracted to outside companies. Thus, ABNER's mercury delay lines were sub-contracted to the electronic component manufacturer Technitrol Inc. (founded in 1947) and ABNER's magnetic tape units were provided by Raytheon Manufacturing Co.

By July 1949 Lubkin, who held a post at the National Bureau of Standards (NBS), had become too busy with NBS's own computer SEAC, which first worked in May 1950. When NBS support evaporated, ASA took over all responsibility for ABNER. Whilst this may have slowed down ABNER's progress, it did have the consequence of fostering a closer relationship between ASA's in-house engineers and mathematicians—a fruitful relationship that led to the specification of several special cryptanalytical instructions for ABNER and a rich repertoire of Input/output facilities.[11]

By September 1951 ABNER was compete and commissioning commenced. The checking of all ABNER's facilities proved quite complex but had been finished by April 1952; the system became operational soon after. The computer cost

[11]Snyder, Samuel S., *History of NSA General-Purpose Electronic Digital Computers.*

approximately $600,000 (equivalent to £214,000). It proved less reliable than ATLAS and extremely difficult to maintain but ABNER's cryptanalytic instructions (see below) allowed it carry out certain ASA data manipulations "more efficiently than certain other computers having inherently higher speed circuitry".[12]

In April 1955, a second ABNER was delivered to NSA, having been constructed under contract by Technitrol Engineering Corporation. ABNER II differed from its predecessor in several respects, particularly the inclusion of 128 words of rapid-access memory as well as 1 K additional words of main memory.

It has not proved possible to describe all of ABNER's special instructions in detail but they generally gave support for three classes of cryptanalytic task: character transformations, data-stream manipulations, and paired stream comparisons. The following example of an ABNER *Paired Stream Comparison,* taken from,[13] will suffice to give the flavour. "Nicknamed *Swish* to depict its action, this instruction accomplished the logical equivalent of a complete, high-speed comparator. In effect, it (a) passed two streams of five-bit characters from memory through an analytic unit, (b) compared groups of characters (group size between 1 and 63 characters) for coincidence, (c) counted the number of group coincidences, (d) stored the coincidence count in a memory location specified in the same instruction, and (e) restored one of the streams to memory at a specified offset from its original place in memory (offset between zero and eight characters)".

10.3.3 Technical Comparisons of the Three Computers

Table 10.1 lists the main features of a standard Ferranti Mark I*, ATLAS I and ABNER I. The basic design for each of these three machines dates from 1948/49 but the finished products are seen to be strikingly different.

When comparing computers from a programmer's view, the Ferranti Mark I*'s index registers and the random-access properties of its primary memory (both based on Williams/Kilburn cathode ray tubes) would have been welcomed. ATLAS's lack of programmed input/output would have been regretted but the superior reliability of this machine would have been enjoyed. For the trickier cryptanalytical tasks, ABNER's special instructions and its rich variety of input/output equipment would have appeared interesting and, it is assumed, very useful.

---

[12]Snyder, Samuel S., *History of NSA General-Purpose Electronic Digital Computers.*

[13]Snyder, Samuel S., *History of NSA General-Purpose Electronic Digital Computers.*

Table 10.1 Technical comparison between Ferranti Mark I*, ATLAS and ABNER

	Ferranti Mark I*	ATLAS I	ABNER I
Word length, bits	40 bits	24 bits	45 bits
Instruction length, bits	20 bits. *Note* 1	24 bits	45 bits
Primary memory type and size, words	CRT Electrostatic RAM, 256 words. *Note* 2	Magnetic drum, 16,384 words. *Note* 3	Mercury delay line, 1024 words. *Note* 4
Secondary memory type and size, words	Magnetic drum, 16,384 words. *Note* 5	None	(Magnetic tape)
Input/output facilities	Paper tape. *Note* 6	Paper tape. *Note* 7	Paper tape; punched cards; magnetic tape. An off-line media converting unit
Special cryptanalytical instructions?	No. *Note* 8	No.	Yes. *Note* 9
Instruction format	1-address	1-address	4-address
No. of instructions in the instr-set	32	42	31
Serial or parallel architecture?	Serial, 10 μs digit period	Parallel, 2.5 μs digit period	Serial, 1 μs digit-period

Notes

1. Addressing was to the 20-bit half-word level
2. A basic Mark I* had 8 Williams/Kilburn tubes, giving 256 words. This was extended in most of the Mark I* installations up to 16 tubes (512 words). The access-time (time to retrieve a 40-bit word from primary memory) was 400 μs
3. ATLAS's drum memory had 200 fixed tracks. The access-time (time to retrieve a word from memory) took between 32 and 17,000 μs depending upon the position of a word on a track. To help a programmer optimise access-times for a particular routine, a 'Skip' feature was added to ATLAS in May 1951 that allowing the Program Counter to be advanced by intervals greater than one (actually 9, 17, 33, or 65)
4. ABNER's delay-line memory was divided into 8-word sections, each section having a total delay of 384 μs. Thus, the access-time (time to retrieve a word from memory) took between 48 and 384 microseconds, depending upon the position of the particular word of interest
5. Information was transferred between the Mark I*'s drum and CRT primary store in terms of tracks. There were 256 tracks, each of 64 words. Track transfers took 36 ms for read and 63 ms for write
6. A Bull lineprinter and its CRT buffer was added to most Mark I* installations, but not to GCHQ's machine
7. ATLAS Input/output was not under program control
8. GCHQ's particular Mark I* probably included the sideways ADD (population count); it also had memory-mapped external serial highways to connect with special GCHQ apparatus—see text
9. ABNER had 19 conventional and 12 special analytic instructions. Of the specials, the *Swish* instruction was particularly useful—see text

10.4 GCHQ's Links with UK Industry, Especially Ferranti Ltd

Writing on 7th November 1951,[14] Dr. G. W. Morgan noted that GCHQ's contracts with Ferranti Ltd. already included the following tasks:

(i) "The original magnetic drum—(no development);
(ii) Photo-electric tape readers: no development yet but some contemplated;
(iii) High-speed tape punches: development under discussion;
(iv) Magnetic drum for OEDIPUS: development of switching circuits already placed;
(v) Magnetic drums for COLOROBS: drums and some circuits already ordered, more development under discussion;
(vi) Cathode ray tube stores ordered for delivery this year.

"It is anticipated to include:

(vii) Cathode ray tube store developments;
(viii) Magnetic tape development;
(ix) Hollerith card cum magnetic drum and/or tape development;
(x) Purchase of a Computer.

Of these, (iv) is tied up with Elliotts and (ix) at least with BTM".

All except (x) of the above tasks were concerned with special-purpose SIGINT equipment, for which the Ferranti contacts had been on-going for at least a year. The OEDIPUS and COLOROB *Rapid Analytical Machines* are mentioned again in Sects. 10.6 and 10.7. The Ferranti 300 characters/sec. optical paper tape reader (task (ii)) was to gain a reputation for usefulness on both sides of the Atlantic.

Ferranti's drum developments probably lagged behind the ERA equivalents in America. Nevertheless, to quote J. H. Cane[15] "Technical staff from GCHQ spent a period with Ferranti and Manchester University, the main interest being the magnetic drum. A 10-in. drum was purchased together with its control, and a few input/output circuits, and used for development, particularly the form of storage named *Revolver*. This used the drum track in the continuous write mode, the read head placed immediately before the write head, the loop completed via input/output circuits enabling individual bits to be changed".

Revolver was just one of several special hardware projects at GCHQ in the early 1950s[16]; more are mentioned in Sect. 10.6.

---

[14]Morgan, G.W., *GCHQ & Ferranti.* GCHQ Minute Sheet X/565/1802.

[15]H. J. (John) Cane, letter to the author dated 21st November 2004. Quoted in Footnote 16. John Cane is a former engineer whose career at GCHQ lasted from the middle of the war until the early 1980s.

[16]Lavington, Simon. 2006. In the Footsteps of Colossus: A Description of Oedipus. *IEEE Annals of the History of Computing* 28 (2): 44–55.

GCHQ's links with the Borehamwood Research Laboratories of Elliott Brothers (London) Ltd. go back to 1949 when, in conjunction with the Admiralty, GCHQ placed an initial contract CP/12439/49 for what became the Elliott 153 computer.[17] This SIGINT project concerned the high-speed plotting ('pin-pointing') of DF fixes, originally via analogue means but, from March 1950 onwards, via a digital computer. The resulting equipment, the Elliott 153 general-purpose computer, was installed at Irton Moor, near Scarborough, in August 1954 where it was well-connected to receiving ('listening') stations at home and abroad via the government's Defence Teleprinter Network (DTN). Because the 153 was entirely devoted to DF plotting, its stored-program capability is not thought to have had any influence on Cheltenham's 1951 decision to purchase a Ferranti Mark I*.

In summary, GCHQ's links with Ferranti Ltd. were well-established when, late in 1951[18] a Ferranti general-purpose computer was ordered. At that point, Ferranti already had a second Mark I computer in production but the company was considering certain enhancements that led, eventually, to the Mark I* version. The discussions were initiated by Dr. J. F. (John) Bennett of Ferranti Ltd. who, on 21st March 1951, circulated his ideas for modifications to staff at Manchester University and Ferranti Moston—(see Chap. 4). Since GCHQ had suddenly become the highest-priority customer for a Mark I*, GCHQ contributed its own requirements to the debate.

GCHQ's first introduction to the Mark I* proposals came on 7th/8th November 1951, when Alan Bruce of GCGQ visited B. W. (Brian) Pollard, Ferranti's Moston Computer Department Manager, to discuss a number of on-going contracts.[19] Bruce was told that the Mark I's original instruction set had been reduced from 53 to 32 orders with a consequential simplification of the input routines, that new *Shift* and *Standardise* instructions had been included and that a Bull lineprinter and its buffer was being included. All this is explained more fully in Chap. 15.

A formal written specification of the Ferranti Mark I*'s instruction set was received by GCHQ on 3rd December 1951.[20]

During January to March 1952 GCHQ proposed a series of special modifications to the Mark I*'s instruction set.[21] Briefly, these initially included:

(a) the addition of four external registers (in effect, wires connected to the computer) of which three were for outputting data to other equipment and one for

[17] Lavington, Simon. 2011. *Moving Targets: Elliott-Automation and the Dawn of the Computer Age in Britain, 1947–67*. Springer. ISBN 978-1-84882-932-9.

[18] Tony Comer, GCHQ Departmental Historian, e-mail dated 24th November 2015 to the author.

[19] *Visit to Messrs Ferranti Ltd.* A. Bruce. GCHQ Minute 4845/7442, 27th November 1951.

[20] *Ferranti Universal High Speed Computer.* GCHQ Minute sheet X-7103/0154, December 1951.

[21] GCHQ's proposals and the resulting comments from Ferranti Ltd. are described in the following four GCHQ Minutes: (a) Minute X/565/1950, Author X20a, 29th January 1952; (b) Minute X/7103/0177, Author G. Timms, 21st February 1952; (c) Minute X/7103/0181, Author G. Timms, 28th February 1952; (d) Minute X/7103/0188, Author G. Timms, 7th March 1952.

input from such equipment. Access was required "from storage and also, if possible, from accumulator". The nature of GCHQ's external equipment was not specified at this stage—but see comments below;

(b) the number of B lines (index registers) to be increased from 8 to 16;

(c) the following additional instructions to be included: sideways ADD (population count) as had been included in the original Mark I's instruction set, random number generator (also originally included), logical OR (also originally included), Cyclic shift;

(d) High-speed paper tape input and output to be provided. The desired character rate was not specified.

Requirement (a) is interesting, for it appears to open up possibilities for the standard Mark I*s facilities to be augmented by special equipment designed at GCHQ. On 14th February 1952 Ferranti explained that the external leads (actually, serial highways) could in principle communicate either with the primary store or with the accumulator. For the first case, special instructions would be needed of the form *Transfer contents of external register X to storage location S*, or its inverse *Transfer contents of storage location S to external register Y*. For the second case, each lead would have to be allocated an (unused) primary store address—i.e. the highways would be memory-mapped. Memory mapping would be cheaper and easier for Ferranti to implement.

By 22nd February Ferranti appeared to be offering only memory-mapping. There still seemed to be some confusion about the exact method of referencing the desired "six pairs of external leads, each consisting of one source and one destination" but it was made clear that "the intention is to use some of these for performing orders which Ferranti are unwilling to provide". Then by 6th March it was suggested that spare bit 14 in the Mark I*'s instruction field could be invoked to provide one pair of external leads with the ability to communicate directly with primary memory. No further evidence has come to light to ascertain whether this was ever implemented. Pending declassification of secret GCHQ documents, all that can be said is that the Mark I* as delivered to Cheltenham probably had six pairs of external leads for special input/output from/to external equipment and that programmers saw these leads as memory-mapped addresses.

Returning to the previous list of GCHQ's requirements, Ferranti rejected the request that the number of B lines (index registers) should be increased from 8 to 16. Item (c), the request for additional instructions, was debated at length because of the conflicting demands of necessary hardware modifications. A final listing of the Mark I*'s instruction set as actually delivered to Cheltenham is still classified but a study of[22] implies the following: the disjunction instruction (logical OR), cyclic shift and random number generator were rejected; the sideways ADD (population

[22]GCHQ's proposals and the resulting comments from Ferranti Ltd. are described in the following four GCHQ Minutes: (a) Minute X/565/1950, Author X20a, 29th January 1952; (b) Minute X/7103/0177, Author G. Timms, 21st February 1952; (c) Minute X/7103/0181, Author G. Timms, 28th February 1952; (d) Minute X/7103/0188, Author G. Timms, 7th March 1952.

count) was included. The omission of the hardware random-number generator was no real loss: it was not secure enough for operational use, for example, in producing one-time pads. In any case, GCHQ had developed its own in-house random number generator system called Donald Duck.[23] It is thought that several Donald Duck systems were in operational use.

It is believed that Cheltenham's Mark I* initially had 12 Williams/Kilburn storage tubes. A Bull lineprinter and its buffer store were considered but probably not initially included. Once installed, the computer was named CLEOPATRA by GCHQ.

10.5 Delivery of CLEOPATRA, GCHQ's Ferranti Mark I*

According to,[24] it was late in 1951 that MOS placed an order for what was to become the first Ferranti Mark I*, CLEOPATRA. GCHQ was initially given October 1952 as the delivery-datc. This turned out to be ambitious, bearing in mind that the final documented specification of the Mark I*'s instruction set, together with various other details such as the form of the Acceptance Tests, did not appeared until July 1952.[25] By this time the construction of Cheltenham's Mark I* at Moston had already begun.[26] Meanwhile, Ferranti's engineers were wrestling with the installation of FERUT, the Mark I computer that went to the University of Toronto—(see Chap. 3). From Ferranti's viewpoint, the Moston Manager Eric Grundy reported in 1953 that "the development of Number 3 computer for MOS [in effect for GCHQ] had been a much bigger job than had been anticipated and that a

[23]Correspondence between the author and D. C. (Don) Horwood (ex-GCHQ) between 2001 and 2004, with enclosed technical documents on DONALD DUCK. This correspondence is catalogued in Box K4 at http://www.ourcomputerheritage.org/CatK.pdf and the material is held in the Bodleian Library, Oxford University. The documents include a patent assignment and extracts from the Minutes of a 19th Sept. 1945 meeting of the Cypher policy Board—(papers CAB 21/2522).

[24]Tony Comer, GCHQ Departmental Historian, e-mail dated 24th November 2015 to the author.

[25]*Schedule 1: Specification for computing machines. Final draft*, 18th July 1952. Nine typed foolscap pages, no author indicated. National Archive for the History of Computing, document NRD/C7/5: NRDC, Computer, Ferranti, Miscellaneous. Correspondence, 1950 on.

[26]Internal memo ('Note for File') from the Managing Director (NRDC): *Electronic Computers*. 1st May 1952. National Archive for the History of Computing, document NRD/C9/1: NRDC, Computers, Manchester University/Ferranti, Development, Vol. 2. Correspondence, 1952–67. This is a two-page typed foolscap document, summarising a long discussion that had taken place between Halsbury, Hennessey and Crawley (all of NRDC) and Vivian Bowden (of Ferranti Ltd.) on 29th April 1952.

year had been lost on the development side".[27] Ferranti's Moston Computer Department was clearly under a certain amount of stress.

The detailed installation arrangements between Ferranti and MOS were, and are, still classified. Alan Ellson, who joined Ferranti in May 1951, recalls that "A rather strange couple of chaps from 'Station X' lived with us during the commissioning [of the Cheltenham Mark I* at Moston] but I was surprised that they had been able to absorb enough to install it without help".[28] Harry Johnson[29] remembers that the computer was "very thoroughly commissioned and set to work in the factory then dismantled into the largest transportable pieces and loaded into containers that were delivered by Ferranti drivers to a designated remote site (possibly in Scotland,[30]) a lay-by from which it was collected by MOS drivers for onward delivery. From time to time spares were ordered, so it was assumed that it was working satisfactorily. Speculation was that it had been delivered to GCHQ which was then at Cheltenham, but this was not admitted until about 25 years later".

The precise location at which GCHQ's computer was installed was known as Oakley. By way of background explanation, GCHQ moved its main operations from Eastcote to the Cheltenham area between the spring of 1950 and late 1953.[31] Two sites were selected for development: Benhall Farm to the west of Cheltenham and Oakley Farm to the east. Oakley was the more important of the two. On each site there already existed six standard 12-spur single-storey temporary office blocks offering a total of half a million square feet of space. The main wartime occupants had been the US Army's European Theater of Operations. After the war, Oakley was initially used by the UK's Ministry of Pensions and for teacher training before

---

[27]National Archive for the History of Computing, box NAHC/NRD/C9/1 "Computers, Manchester University/Ferranti, Development, Vol. 2". Correspondence 1952–67. File note of a visit by Lord Halsbury (Director, National Research Development Corporation) to Ferranti Moston on 24th July 1953.

[28]Alan Ellson, letter to the author dated 16th January 2000. Alan joined Ferranti in May 1951, was involved in commissioning the Mark I at the University of Manchester and then transferred to Moston. He left Ferranti in the summer of 1956.

[29]Johnson, M.H.(Harry). 2002. *My Work with Computers from the Ferranti Mark I to the ICT 1900 (1952–1966)*. 84-page typed manuscript, sent to SHL in 2002. Copy held at the National Archive for the History of Computing, Manchester, see NAHC/SHL/FA1. Of particular relevance is Appendix E: *Ferranti MkI* customers and an early venture into magnetic tape.*

[30]Williams, David. 2003. *Information Technology at Ansty*. Article in The Sphinx. *The Coventry Branch Magazine* (53): 13–18. (This is the magazine of the Rolls-Royce Heritage Trust, Coventry Branch). The particular information about the Cheltenham Mark I* comes from John McNamara who was a computer maintenance engineer involved in moving the computer from Cheltenham to Ansty, Coventry. John joined Armstrong Siddeley Motors in 1954. He has written up his experiences in: *The Ferranti computers at Armstrong Siddeley Motors Ltd. Coventry*. 5-page typed manuscript, July 2012. John and the author had telephone and e-mail correspondence in the spring of 2017.

[31]*How GCHQ Came to Cheltenham.* Peter Freeman, GCHQ, 2002. This illustrated 34-page booklet was part of an exhibition in Cheltenham's Art Gallery and Museum to mark the 50th anniversary of GCHQ's arrival in Cheltenham. Copy available in Cheltenham Reference Library: Cheltenham.reference@gloucestershire.gov.uk

Fig. 10.3 **a** and **b** The Oakley site at Cheltenham in 1955 and in 1993. GCHQ staff moved to Oakley from Eastcote, Middlesex, in the period 1950–1953. CLEOPATRA, GCHQ's Ferranti Mark I* computer, was installed at Oakley in September 1953

GCHQ took over and extended the premises. Oakley was finally abandoned at the end of 2012, by which time GCHQ was inhabiting the iconic large *doughnut-shaped* building at Cheltenham.

According to,[32] CLEOPATRA arrived at Oakley in September 1953 and became operational, except for the drum, in February 1954. To quote[33]: "At about this time the drum in OEDIPUS ran over-speed and damaged some of its tracks". GCHQ, in discussion with Ferranti, therefore arranged for modifications to be made to CLEOPTRA's drum so that the same thing could not happen. Drum modifications caused a pause in machine availability but CLEOPATRA was again operational from May to September—but with reliability measured as only 42%. On 30th Sept. 1954 CLEOPATRA was shut down for the installation of two of the memory-mapped output highways. These were connected respectively to an IBM electromatic typewriter and to a gang punch (an electro-mechanical device for punching a given pattern of identical holes in a deck of punched cards). This took until the end of the year. The machine was operational during most of January and

[32]Ellis, J.H. 1955. *A Short Report on CLEOPATRA*. GCHQ Minute X/3829/2445. Two typed pages. 27th Apr 1955.

[33]Ellis, J.H., *A Short Report on CLEOPATRA*.

February 1955 but considerable trouble was experienced due to the electromatic and gang punch outputs and to the drum bearings. "It was found that the drum surface was in very bad condition owing to heads having been rammed in during previous setting up". Finally the drum bearings seized up and the drum was replaced by Ferranti.

To quote,[34] "satisfactory operations of CLEOPATRA were not achieved till well into 1955". Very few details of this computer's life at GCHQ have been released at the time of writing (2018). Some general, and informal, comments are offered in the next section. Regrettably, no photographs of the Ferranti Mark I* at Cheltenham have been made available, though some images of what is thought to be CLEOPATRA's special I/O circuitry under construction have been released, but without comment—see Fig. 10.4.

10.6 The Possible Applications of CLEOPATRA

Returning to Jack Good's Report of January 1950[35] it is interesting to list the cryptanalytical areas for which Good thought a general-purpose computer might be useful. Judging from the Report's redacted list released in 2018, these candidate applications appear to include the following:

Calculation of Fourier transform of cipher and digraph of Fourier transforms;
Working back problems for machines with irregular motion, e.g. Sturgeon;
Looking for isomorphs, which involves computing correlation coefficients, etc.;
Dragging of cribs at high speed (with dictionary);
Transposition ciphers;
Electronic Bombe—a computer plus a single mechanical Enigma machine.

Four or five years later, by which time CLEOPATRA was fully operational, the above list of candidate applications could have changed. It may be assumed that, by 1955, a few GCHQ mathematicians had already had two or three year's exposure to programming ATLAS and ABNER in America. Recently-released National Security Agency (NSA) documents, some of them heavily-redacted, give the names but not the details of many cryptanalytic programs run on the ATLAS and ABNER computers. It is likely that GCHQ considered writing similar applications for CLEOPATRA.

Here are some sample American ATLAS/ABNER programs, taken from an NSA paper declassified in 2017[36]:

---

[34]Tony Comer, GCHQ Departmental Historian, e-mail dated 24th November 2015 to the author.

[35]Good, I.J., *The Cryptological Uses of a General Purpose Computer.*

[36]Snyder, S.S. 1980. *ABNER:* The ASA Computer. Part II: Fabrication, Operation and Impact. Declassified 2017 with redactions. *NSA Technical Journal* 63–85. See: https://www.nsa.gov/news-features/declassified-documents/nsa-early-computer-history/assets/files/6586518-abner-the-asa-computer-part-ii.pdf

Fig. 10.4 **a** and **b** Two photos released without comment by GCHQ, with an indication that the images are of in-house equipment designed to be connected to GCHQ's Ferranti Mark I* computer

LOON, a program designed to detect instances of key re-use by statistical and coincidence tests.
AFSA W-7200 Counts, a program producing a series of statistics to evaluate or assess the randomness of one-time key tapes, each 10,000 characters in length.
HAYSTACKS, a program which examined pages of key in a search for certain cyclic characteristics.
STETHOSCOPE (and SUPERSTETH), a set of initial diagnostic tests which were typically applied to cipher texts of unknown systems.

Several other ATLAS/ABNER programs are quoted by name only, such as: ARC, ORIOLE, TURTLE, KEVIN, RAIN-D, SAIL, SHOWDOWN, FARO, CZAR, LABYRINTH, MAZE, WALNUT, RENO and SUICIDE. Although no further details have come to light, it is clear that the American cryptanalysts must have embraced their early computing facilities with enthusiasm.

Only two (unrelated) CLEOPATRA programs have thus far been officially admitted by GCHQ.[37] These are:

KING. This program was used to test whether a given number is a primitive root of a Mersenne prime.
JUNGLE. This program produced letter checks for PORTEX, a British eight-rotor cipher machine consisting in effect, of two cyclometrically-stepping four-rotor machines in series, with an end crossover. Letter-checks were an added security procedure to ensure that an operator had set up a cipher machine correctly.

Besides cryptanalytic programs, there could have been other applications for CLEOPATRA that came under the heading of conventional data-processing. One such area is intercept traffic analysis. Formerly carried out by electro-mechanical punched-card equipment, some of this traffic analysis activity may have been transferred to the Ferranti Mark I*.

More generally, two of the special serial output highways installed for CLEOPATRA were known to have been connected respectively to an IBM electromatic typewriter and to a gang punch. It is not known what these, or indeed the other special I/O highways, were used for. The few GCHQ source-documents available for examination do not suggest that CLEOPATRA was closely linked to any particular *Rapid Analytical Machine*.

It is believed that CGHQ still made use of special *Rapid Analytical Machines* until at least the end of the 1950s. These "still had some advantages— for example their fast input speeds—for some years after the War".[38] The last of the two retained Colossus II machines did not go out of service until about 1961. One of these machines had been re-built in a more generalised form to meet new challenges.[39] Furthermore, during the 1950s Cheltenham's in-house engineers continued to design and build special-purpose machines such as new Robinsons, a 'Super-Rob', Revolver, Autolycus, Johnson and Colorob and, in collaboration with Elliott Brothers and Ferranti Ltd., Oedipus.[40] CLEOPATRA is not thought to have been connected to any of the above equipment—and indeed the Ferranti Mark I*'s basic digit frequency may have rendered it too slow for any useful partnership.

CLEOPATRA did not last long at Cheltenham. Before describing its successor, we make a small digression to illustrate why CLEOPATRA was considered comparatively slow in cryptanalytical terms.

---

[37]Tony Comer, GCHQ Departmental Historian, e-mail dated 31st July 2018 to the author.

[38]Peter Freeman (former GCHQ staff and latterly, in retirement, GCHQ historian) in a letter to the author dated 20th September 2004. Peter Freeman died in about 2007.

[39]H. J. (John) Cane, letter to the author dated 21st November 2004. Quoted in Footnote 16. John Cane is a former engineer whose career at GCHQ lasted from the middle of the war until the early 1980s.

[40]Lavington, Simon., In the Footsteps of Colossus: A Description of Oedipus.

10.7 Comparative Rates of Computation: Oedipus

Oedpius was a special-purpose GCHQ machine which had a Ferranti-designed magnetic drum, anticipation-pulse Williams/Kilburn CRT storage and a unique Content-Addressable (i.e. Associative) semiconductor *Dictionary* designed in conjunction with Elliott Brothers Ltd. Oedipus is described in detail.[41] The Oedipus story illustrates two facts:

(a) early general-purpose computers were less than satisfactory for many of GCHQ's front-line tasks;
(b) GCHQ possessed considerable in-house hardware design skills.

Oedipus was intended for a particular pattern-matching task. Oedipus's hardware *Dictionary* could perform a match between a five-character interrogand and 4000 stored five-character groups in just 50 microseconds. Contemporary computers of the 1950s, which usually had no Table Lookup instructions, would have had to implement this search via a loop of about five instructions traversed 4000 times. If the fastest contemporary computer had an instruction rate of about 40 Kilo Instructions per second (KIPS)—see Table 10.2—then the Oedipus *Dictionary* search was about 10,000 times faster than a general-purpose computer's search.

Oedipus started to become operational in May 1954, was declared fully operational by mid-1955, and was thus in use at the same time as CLEOPATRA. Oedipus worked satisfactorily until dismantled in February 1962, although it had actually been retired from active service late in 1959.[42] The IBM 704 of Table 10.2 came onto the market in 1955 and was reckoned to be the fastest general-purpose computer of the time. However, for the particular cryptanalytical task for which Oedipus was designed there was no point in GCHQ considering the purchase or rental of a relatively costly IBM 704.

As a footnote on cost, NSA in America could certainly afford an IBM 704. Indeed, during the period under discussion the following IBM computers were amongst those rented by NSA[43]:

IBM 701: delivered to NSA in April 1953, returned December 1955;
IBM 702: delivered to NSA in the spring of 1955, returned in April 1956;
IBM 704: three delivered between Jan 1956 and Sept. 1957;
IBM 705: five delivered between May 1956 and Feb. 1961.

During this period, NSA was also developing its own special Rapid Analytical Machines. It seems likely that, in comparison with NSA, GCHQ's annual budget was not so great. It is not known whether Cheltenham rented any IBM machines

---

[41]Lavington, Simon., In the Footsteps of Colossus: A Description of Oedipus.

[42]Peter Freeman (former GCHQ staff and latterly, in retirement, GCHQ historian) in a letter to the author dated 20th September 2004. Peter Freeman died in about 2007.

[43]Snyder, S.S., *ABNER:* The ASA Computer. Part II: Fabrication, Operation and Impact.

Table 10.2 A rough comparison of contemporary speeds and memory capacities

Computer; year first installed	Kilo instructions per second (KIPS)	Typical online storage (Kbytes)
Ferranti Mark I* 1953	0.8	66 (Note 2)
Oedipus 1954	100 (Note 1)	108
IBM 704 1955	42	54 (Note 3)

Notes

1. The figure of 100 KIPS is based on OEDIPUS's character-manipulating instruction time of 10 μs per 5-bit character. Of course, the Ferranti Mark I* and the IBM 704 processed many more bits per instruction

2. Assumes 12 (out of a max of 16) Williams-Kilburn tubes, each holding 64 20-bit words plus 200 (max 256) tracks of drum, each track holding 128 of 20-bit words

3, Assumes 4K of 36-bit words of ferrite RAM plus one drum holding 8K words

during the late 1950s but, by the 1960s, it is believed that GCHQ had become a large user of high-performance general-purpose computers.

10.8 What Followed CLEOPATRA at Cheltenham?

In an internal memo dated 25th September 1957,[44] it is implied that a joint NSA/GCHQ decision had been made in 1956 to replace CLEOPATRA with a UNIVAC 1103A computer code-named EFFIGY. It was not considered sensible to run the two computers in parallel, so arrangements were made to transfer all programming and engineering staff from CLEOPATRA to EFFIGY once the latter had become operational at Cheltenham. It was estimated that EFFIGY would be delivered at the end of November 1957. A maintenance team from Remington Rand would stay at Cheltenham for six months after delivery. Anticipating an initial EFFIGY error-fixing period of two or three months, followed by an overlap period, GCHQ planned for CLEOPATRA to be effectively shut down in about June 1958. Shortly after this, CLEOPATRA would be disposed of.

Meanwhile, it appears from[45] that Cheltenham had acquired a relatively modest BTM 555 computer (strictly, a programmable calculator with a drum store) and that some of CLEOPATRA's smaller programs were being ported to the 555 (Fig. 10.5).

A follow-up internal memo dated 27th September 1957[46] emphasised the uncertainties attached to EFFIGY's arrival date. No evidence has been released that

[44]*CLEOPATRA: closing date*. GCHQ Minute W/65/8305/1743. Author illegible. 25th September 1957.

[45]*CLEOPATRA: closing date*. GCHQ Minute W/65/8305/1743. Author illegible. 25th September 1957.

[46]Alexander, C.H.O'D. 1957. *CLEOPATRA: Closing Dates*. GCHQ Minute H/0566/IXP3a, 27th Sept. 1957.

Fig. 10.5 A BTM 555 computer (strictly, a programmable calculator with a drum store), of a type installed at GCHQ Cheltenham in about 1957. (This particular photo was taken at another government installation)

gives a definite date for the shut-down of CLEOPATRA but we do know that, by 1959, the disposal of this machine was in the hands of a Manchester scrap metal dealer who had placed an advertisement in a popular electronics magazine called *Wireless World.* CLEOPATRA—just the basic Mark I* hardware and minus any GCHQ special additions—was being offered for sale for £4,000. In Chap. 7 we describe how this computer was bought by Armstrong Siddeley Aero Engines and installed in their Ansty factory (near Coventry) alongside their existing Ferranti Mark I*. It gave good service until 1964, when both computers were scrapped.

References

Alexander, C.H.O'D. 1957. *CLEOPATRA: Closing Dates*. GCHQ Minute H/0566/IXP3a. 27th Sept. 1957.

Budiansky, Stephen. 2001. Codebreaking with IBM Machines in World War II. *Cryptologia* 25 (4): 241–255.

Ellis, J.H. 1955. *A Short Report on CLEOPATRA*. GCHQ Minute X/3829/2445. Two typed pages. 27th Apr 1955.

Good, I.J. 1950. *The Cryptological Uses of a General Purpose Computer.*

Johnson, M.H.(Harry). 2002. *My Work with Computers from the Ferranti Mark I to the ICT 1900 (1952–1966).*

Lavington, Simon. 2006. In the Footsteps of Colossus: A Description of Oedipus. *IEEE Annals of the History of Computing* 28 (2): 44–55.

Lavington, Simon. 2011. *Moving Targets: Elliott-Automation and the Dawn of the Computer Age in Britain, 1947–67*. Berlin: Springer. ISBN 978-1-84882-932-9.

Morgan, G.W. 1951. *GCHQ & Ferranti.* GCHQ Minute Sheet X/565/1802.

Snyder, Samuel S. 1964. *History of NSA General-Purpose Electronic Digital Computers*. Washington DC: Dept of Defense.

Snyder, S.S. 1980. *ABNER:* The ASA Computer. Part II: Fabrication, Operation and Impact. Declassified 2017 with redactions. *NSA Technical Journal* 63–85.

Wheatley, L.H. 1953. *Cryptanalytic Machines in NSA,* 365.

Williams, David. 2003. Information Technology at Ansty. Article in The Sphinx. *The Coventry Branch Magazine* 53: 13–18.

Chapter 11
The Mark I* at the Armaments Research Development Establishment, Fort Halstead

11.1 Armaments Research, Pre-war and Post-war

Fort Halstead, in Kent, was constructed between 1895 and 1897 as part of the outer defences of the London area. By the end of the Second World War the site housed various armament R&D groups, eventually brought together in 1955 to form the Armament Research and Development Establishment (ARDE). It is known that the centre of the UK's early development of the atomic bomb was initially based at Fort Halstead, in a branch called High Explosive Research (HER), before a separate dedicated development site was ready at Aldermaston. There is therefore a direct connection between digital computing activity at Fort Halstead and that at the Atomic Weapons Research Establishment (AWRE) at Aldermaston—described in Chap. 12.

Until 1942 the main work of ARD, the Armaments Research Department, was carried out at Woolwich Arsenal in London. Electro-mechanical desk-top hand calculators were the main aids to computation, supplemented by a Differential Analyser at the Mathematics Laboratory at the University of Cambridge. At the outbreak of war, Professor John Lennard-Jones from Cambridge had been seconded to become Chief Superintendent of Armament Research to the Ministry of Supply. The Mathematics Laboratory, of which Lennard-Jones was the Director, was taken over for ballistics calculations.

In 1942 a committee headed by Dr. H. L. Guy reorganised armament R&D, drafting in scientists and engineers to replace senior military personnel. The Headquarters was transferred from Woolwich Arsenal to Fort Halstead and a new branch for theoretical research was established under Professor Neville Mott. In addition to Brunsviga and Burroughs desk calculators, six new large calculating machines, the NCR National Accounting Machine series 2000 and 3000, were introduced.[1]

[1]Pattison, Gordon. 2015. *Playing Chess with Turing: How Computers Came to Fort Halstead.* DSTL Fellow. Lecture given at Fort Halstead on 11th Dec 2015. The author had many helpful e-mail discussions with Gordon Pattison during the summer of 2017.

S. Lavington, *Early Computing in Britain*, History of Computing,
https://doi.org/10.1007/978-3-030-15103-4_11

In 1947 ARD under its post-war leader Dr. William Penney had been given the responsibility for developing Britain's nuclear weapons programme, for which access to computers was essential. Sometime after 1949 ARD began to use the EDSAC computer at Cambridge and then, from late 1951, the Ferranti Mark I at Manchester University. The two active Fort Halstead personnel were A. E. (Alick) Glennie and K. N. (Ken) Dodd. Glennie transferred to Aldermaston in 1953 and Dodd transferred to RAE Farnborough in 1954, where they both had distinguished careers in computing.

11.2 Arranging the Purchase of a Computer for Fort Halstead

There was some initial uncertainty surrounding the Ministry of Supply's (MOS's) order for the Ferranti Mark I* that was eventually delivered to Fort Halstead in July 1954. The man who was keenest to see a computer installed at Fort Halstead appears to have been Brigadier G. H. Hinds of the Ministry of Supply. He was a senior officer in the Royal Artillery who had been involved in MOS assessments of advanced German weapons systems at the conclusion of the Second World War.[2] In 1950 Brigadier Hinds had become Director of Weapons Research (Defence) in the MOS. In this role he seems to have kept himself abreast of developments in computing machines.[3]

Brigadier Hinds sent Ferranti Ltd. a Letter of Intent to purchase a digital computer in the spring of 1951, based upon his knowledge of the computer then being installed at Manchester University. By June 1951 this letter had been withdrawn by MOS because, so Ferranti was told, the sum necessary to acquire a Ferranti Mark I exceeded the amount that Brigadier Hinds was empowered to authorise. Meanwhile three unrelated things were happening that served to further complicate MOS's position. Firstly, as discussed in Chap. 10, a strategic priority seems to have arisen that GCHQ should get a computer ahead of other potential government customers. Secondly, mathematicians at Ferranti and NRDC had started to suggest alterations that might be made to the Mark I's instruction set, with consequential delays to GCHQ placing their order for a computer. Thirdly, the importance of the UK's atomic weapons programme dictated that a site separated from Fort Halstead should be created just for atomic weapons research. Some, or maybe all, of these factors could have delayed the placing of Fort Halstead's formal contract.

---

[2] *Service of Brigadier G. H. Hinds, Royal Artillery, with the Ministry of Supply in Germany, 1945–1947.* Imperial War Museums catalogue item HU 103674.

[3] Reiners, C.A. 1953. *Survey of Computing Facilities in the UK.* Published by the Ministry of Supply. NAHC document MIS/D3act. The Foreword, written by Brigadier G. H. Hinds, contains statistics on the numbers of Differential Analysers, desk calculators, etc., available at various establishments.

These uncertainties caused Lord Halsbury to observe[4] in May 1952 that the MOS "are still dickering about placing an order for the second machine". It is not clear what Halsbury meant by 'the second machine'. Later on Halsbury adds that "Of the two machines ordered by MOS No. 1 is being dealt with by Evans, and No. 2 by McColl working under Penney [Sir William Penney] at Fort Halstead". By August 1952 Crawley of NRDC was able to say[5]: "MOS has in fact placed a contract for a second machine (Number 5). This contract, No. 6/WT/23833/CB.15b, was placed on 2nd August". In his 1975 retrospective history,[6] Swann remembers the contract-date for both the Fort Halstead *and* Aldermaston computers as being October 1952. For these reasons, one might take it that 'autumn 1952' is an indication of the time at which MOS had definitely made up its mind.

Further evidence comes from a retrospective history[7] which states: "Although The Fort would concentrate on conventional weapons following the setting up of the Aldermaston site which began in 1950, the requirement for computing power still existed and so it was decided to buy a Mark 1 for installation at The Fort. Between 1951 and 1954, facilities to house the machine were assembled in S11 building. By 1954 however, Ferranti had brought out an improved machine known as the Mark 1*, and it would be one of these which would be built at The Fort. In July 1954, the installation of the Mark 1* to be known as AMOS (and one of only nine Mark 1* machines built by Ferranti, others going to computationally demanding sites such as AWRE and GCHQ) was complete". Actually, there were two Mark I machines and seven Mark I* machines built.

11.3 Installation of AMOS; Early Computing Activity

The Fort Halstead Mark I*, which cost approximately £90,000, was known locally as AMOS.[8] The origins of this name are not quite clear, though several myths exist. The most likely explanation is that AMOS was the first of the two additional Ministry of Supply machines (i.e. after the GCHQ Cheltenham one), which were

---

[4]NRDC Internal memo (*Note for File from the Managing Director (NRDC): Electronic Computers*). 1st May 1952. NAHC file NRD C9/1. This is a two-page typed foolscap document, summarising a long discussion that had taken place between Halsbury, Hennessey, Crawley and Bowden on 29th April 1952.

[5]NRDC Internal memo, Crawley to Hennessey, dated 2nd August 1952. In NAHC file NRD/C9/1. The group of documents catalogued as C9 are labelled *Computers, Manchester University/Ferranti, Development, Vol 2, Correspondence* 1952–1967.

[6]Swann, B.B. 1975. *The Ferranti Computer Department: A History*. This 1975 document was initially circulated privately and marked 'confidential'. A copy is held at the NAHC, catalogue number FER/C30.

[7]Pattison, Gordon. 2000. *Early Computers at the Fort.* DERA News, Oct 2000. This paper's source reference for the largest program on AMOS, the battle intelligence system, is: TNAK DEFE 15/1995.

[8]Pattison, Gordon., *Playing Chess with Turing: How Computers Came to Fort Halstead.*

known at Ferranti's Moston factory during manufacture as MOS (A) and MOS (B). Certainly, Allan Ellson recalls Alick Glennie saying in 1954 that MOS "were expecting two machines called AMOS and BMOS but could not be sure they would arrive before XMAS".[9]

James Gawlik, who had joined Fort Halstead in 1950 as a mathematician, became the first Head of the Computing Section following the departure of Ken Dodd. John Berry, who had been recruited in 1951, became the Deputy Head under Gawlik in 1954. Berry had done two years post-graduate mathematics research at Manchester University and was a contemporary of Audrey Bates, Alan Turing's first research student. (Audrey Bates' computing activity is described in Chap. 2). Both Gawlik and Berry took the EDSAC course at Cambridge University and regularly visited Manchester University.[10]

The first AMOS course was organised at Fort Halstead by H. J. Gawlik from 1st to 12th November 1954.[11] The first few lectures were given by Dr. K. N. (Ken) Dodd, who had used the Manchester University computer. He had written two comprehensive manuals describing both the Mark I and Mark I* computers.[12,13] Produced in the spring of 1953, these manuals would have been very helpful to programmers making the transition from Mark I to Mark I* machines. The November 1954 course lecturers comprised Dodd, Gawlick and G. B. Cook from Fort Halstead's own Senior Superintendent Applied Mathematics (SSAM) branch, Alick Glennie from AWRE, four Ferranti programmers (Conrad Berners-Lee, Mary Berners-Lee, Cyril Gradwell and Erik Robertson) and the senior maintenance engineer from the Manchester University installation (Ted Hodgkinson). Besides basic programming topics, the material included sections on interpretive routines, linear programming, ordinary differential equations and Auto-coding—this last being given by Alick Glennie.

---

[9]Ellson, Allan. 2015. *Ferranti MkI* Commissioning and Installation: Some Recollections*, 14th Oct 2015. Two typed pages. Allan Ellson was a Ferranti engineer who rose to be in charge of commissioning Mark I* computers at Moston and Gem Mill.

[10]Pattison, Gordon., *Playing Chess with Turing: How Computers Came to Fort Halstead.*

[11]Armament Research Establishment Annual Report 1954, p. 18.

[12]Dodd, K.N. 1953a. *The Ferranti Electronic Computer (Parts 1 & 2: the Mark I Model).* Armament Research Establishment, report 10/53. Manual containing 90 pages of text and about the same number of pages of logical diagrams, line diagrams and photos. The machine described became known as the Ferranti Mark I. K. N. Dodd spent time at Manchester, using the University's Ferranti Mark I.

[13]Dodd, K.N. 1953b. *The Ferranti Electronic Computer (Parts 3, 4 7 5: the Mark I* Model).* Armament Research Establishment, report 11/53. Manual containing 101 pages of text and about the same number of pages of circuit diagrams, logical diagrams and line diagrams. The machine described became known as the Ferranti Mark I*.

Fig. 11.1 AMOS, the Ferranti Mark I* at Fort Halstead in 1954. The person at the console is Jim Gawlik, the first ARDE Computing Branch head. Allan Ellson, a former Ferranti engineer, remembers that the computer "was situated in the middle of a wood and was also the first [Mark I*] machine to have the trunking for refrigeration, as shown here" (E-mail dated 20th December 2008 from the former Ferranti maintenance engineer Allan Ellson to the author [SHL]). *Image* © Crown Copyright, MOD

The Ferranti Mark I* installed at Fort Halstead is shown in Fig. 11.1. This image is unusual in showing the only Mark I* installation that had a window looking out onto pleasant wooded parkland—though in the photo it appears more like snowscape than landscape. The Ferranti programmer Joan Travis recalls[14] that: "I visited the Mark I* at Fort Halstead, near Sevenoaks in Kent some spring—but which year I can't remember. I think Mary Lee Woods and Conway Berners Lee [also from Ferranti] were both there at the same time as me. Perhaps that visit was to do with a line printer, but I can't remember. What I do remember is that when it was discovered that we were walking round the grounds at lunch time, admiring the wild

[14]E-mail dated 25th October 2015 from the former Ferranti programmer Joan Travis (neé Kaye) to the author (SHL).

primroses, we were banned from strolling around outside. I also remember that Mark I* was in a very pleasant room".

Gawlik and Berry spent the first two years of AMOS's life writing subroutines, according to Pattison.[15] During its active life, the following applications were typical:

(a) Numerical integration of hyperbolic systems of partial differential equations in three independent variables for fluid flow;
(b) Nonlinear simultaneous parabolic equations for laminar flame propagation;
(c) Thermal transport for high intensity thermal radiation from atomic weapons;
(d) Shock diffraction.

11.4 The Maturing Applications of AMOS

Gordon Pattison has stated that "AMOS was at the heart of just about all of our calculations for numerous weapons R&D projects, design and later assessment. It was absolutely critical to our operations and the origin of our central computing bureau which continued until the late 1970's".[16]

Here are two examples of AMOS projects, taken from Pattison.[17] Both have resonances with current defence interests. In 1957, Gawlik produced the first computerised wargame in the UK, based on a manual game which was already in use. The program, written in AMOS machine code, included classic wargame rules relating to movement, detection, fire and terrain effects. Secondly, the largest single program ever written for AMOS was a battlefield intelligence system, developed for the army between August 1963 and January 1965. It was designed to process and disseminate information, both order of battle and enemy situation.

In the early days there was a modest amount of surplus AMOS time available for outside users. For example under this heading, Oxford University carried out work on the electronic structure of Zirconium.

Of the three Defence-related sites running Ferranti Mark I* installations, Fort Halstead seemed to have been the most outward-looking in terms of sharing software with other Mark I* sites. In 1956 John Berry introduced the Intercode scheme, which was publicised widely.[18] This paper's Introduction described Intercode as "an interpretive simplified coding scheme, designed for 'small' computations only … it also serves as an introduction to proper programming …. A modified version of Intercode has been successfully used on the Mark I* computer at Armstrong

[15]Pattison, Gordon., *Playing Chess with Turing: How Computers Came to Fort Halstead.*

[16]Pattison, Gordon., *Playing Chess with Turing: How Computers Came to Fort Halstead.*

[17]Pattison, Gordon., *Early Computers at the Fort.*

[18]Berry, F.J. 1959. Intercode, A Simplified Coding Scheme for AMOS. *Computer Journal* 2 (2): 55–58.

Fig. 11.2 The PACE 231R analogue computer at Fort Halstead in about 1960. This analogue machine was known locally as BIVAC. *Image* © Crown Copyright, MOD

Siddeley Motors Ltd where, as on AMOS, the computing problems are either very large or very small". More information about Intercode is given in Chap. 15.

There were some problems for which AMOS proved unsuitable. In 1959, Fort Halstead was working on the design of a new mortar for the army and naturally AMOS was used for the ballistic calculations. But AMOS was taking 12 h to compute mortar trajectories and these predicted trajectories did not always agree with actual test firings.[19] It was later realised that the mortar round's fins were introducing wobble. A more rapid means of trajectory-simulation was required—preferably one which would also give a visual indication of flight-path. The PACE (Precision Analogue Computing Equipment), which had been demonstrated at a trade fair in Brussels, allowed such rapid visualisation. Consequently in 1960, Fort Halstead acquired a PACE 231R analogue computer—see Fig. 11.2—which was named BIVAC locally. In no sense did BIVAC replace AMOS: it merely complemented AMOS for certain applications involving simulation and visualisation.

[19]Pattison, Gordon., *Playing Chess with Turing: How Computers Came to Fort Halstead.*

11.5 COSMOS, A Compatible Successor to AMOS

By 1959 AMOS was not providing enough digital computing resources and some of Fort Halstead's compute-intensive jobs were having to be run on Aldermaston's computers. Because of the investment in programs coded for the Mark I*, Fort Halstead decided to develop a new machine that was code-compatible with AMOS but much more powerful. GCHQ had offered Fort Halstead their redundant Mark I* (see also Chap. 10) but Fort Halstead did not have the building space at the time. Ferranti no longer manufactured Mark I* computers and their new machines (Pegasus and Mercury) were software incompatible. Fort Halstead therefore decided to build its own new computer in-house, in their existing electronics section. This electronics section had designed and built the firing circuits for the UK's first atomic bomb, so there was no lack of electronics skill.

The new computer was called COSMOS. It used transistors, not valves, and was parallel (not serial). The COSMOS printed-circuit technology is shown in Fig. 11.3. The machine was projected to be 50 times faster than AMOS, although the performance improvement eventually realised was closer to 60.[20] COSMOS's cycle time was 6.5 μs. Hugh Gawthorp was the designer of COSMOS; his job-title was *Chief of Operational Section, Ministry of Supply*, which conveniently abbreviated to COSMOS. The chief engineers in the COSMOS development were Eric Adshead and Johnny Johnstone and the chief test programmer was Jacqueline Muscott. The COSMOS project was not made public at the time, although it was described in Issue 6 of the in-house journal known as *The Fort*, published in 1961.[21]

One of the design-aims of COSMOS was to provide a programmers' interface that had the look and feel of AMOS. To achieve this convincingly, COSMOS was equipped with an operator's console that mimicked (simulated) the original Ferranti Mark I*'s console. Figure 11.4 shows a prototype display unit mounted behind the AMOS console, during the development phase. The idea of equipping the console of a follow-on computer with the look and feel of an earlier computer was also used by the Instituto Nazionale per le Applicazioni Calcolo (INAC) in Rome, when specifying a 1967 replacement computer for INAC's Ferranti Mark I*—(see Chap. 9).

The stages in the development of the AMOS simulated console for COSMOS are illustrated in Figs. 11.4 and 11.5. Besides being able to run programs written in Mark I* code, COSMOS was also equipped with software that provided a higher-level language interface more in keeping with the evolving expectations of programmers in the 1960s. Fort Halstead's local response to these expectations is described below.

---

[20]Pattison, Gordon., *Playing Chess with Turing: How Computers Came to Fort Halstead.*

[21]Issue 6 of the in-house journal known as *The Fort*, published in 1961. TNAK DEFE 15/943.

Fig. 11.3 Joe Kenny, a PTO (Professional Technical Officer), with printed-circuit boards from Fort Halstead's in-house designed COSMOS computer in about 1961. In the background is a COSMOS cabinet with doors open. *Image* © Crown Copyright, MOD

During approximately the same period that COSMOS was being developed, Fortran was being adopted by many in the UK's physics and engineering community. Unlike Aldermaston, Fort Halstead decided not to adopt Fortran because,

Fig. 11.4 The Fort Halstead Mark I* in about 1961, with an additional prototype display unit mounted behind the normal console. The additional display is a part of the COSMOS project. The square screens on the COSMOS unit are intended to mimic and replicate the two pairs of round screens on the original Mark I*'s console. *Image* © Crown Copyright, MOD

says Gordon Pattison, "the symbols and syntax were not particularly compatible with the written style of mathematical formulae favoured by the Fort Halstead programmers. Therefore, James Gawlik created a new language called MIRFAC (Mathematics in Readable Form Automatically Compiled)[22] "which became the main programming tool at Fort Halstead for 10 years".[23] The final COSMOS installation that replaced AMOS is shown in Fig. 11.6, by which time MIRFAC was in regular use.

"To quote[24]: "MIRFAC's central aim is to offer to the general computer public a language which makes it as easy as possible for the non-specialist to present problems to a computer. In support of this aim a special purpose typing machine–costing very little more than the standard [Friden Flexowriter] product–has been

[22]Gawlik, H.J. 1963. *MIRFAC: A Compiler Based on Standard Mathematical Notation and Plain English. Communication of ACM* 6 (9): 545–547. See also: Gawlik, H.J., and F.J. Berry. 1967. *Programming in MIRFAC*, 2nd ed.; The quotation "MIRFAC's central aim … " comes from the second source.

[23]Pattison, Gordon., *Playing Chess with Turing: How Computers Came to Fort Halstead.*

[24]Gawlik, H.J., *MIRFAC: A Compiler Based on Standard Mathematical Notation and Plain English.* See also: Gawlik, H.J., and F.J. Berry., *Programming in MIRFAC.*

Fig. 11.5 John Berry using a version of the Mark I* console simulator, as attached to an early version of COSMOS. *Image* © Crown Copyright, MOD

developed. The language is thus able to accept for compilation any mathematical equation written in standard text-book notation. Those statements in a problem which do not fall into this category–input, output, jumps and the like–are written in 'plain English' forms designed so as to avoid special conventions".

There was a COSMOS development period at Fort Halstead when AMOS was running in parallel with COSMOS. During January 1967 "all work was transferred to COSMOS and AMOS was switched off at the end of the month. It is a tribute to the realisation of the original aim to make COSMOS compatible with AMOS that it was possible to transfer the workload on to COSMOS .... In the succeeding months, AMOS was moved into storage and has been put up for disposal".[25] As far as is known, the AMOS hardware units were eventually sold for scrap.

[25]Quotation from Fort Halstead's D1 Branch Annual Report for 1967.

Fig. 11.6 The completed COSMOS installation in the summer of 1967. The user in the foreground (Berry) is seated at a Friden flexowriter and is programming in the MIRFAC language. The user in the background (Gawlik) is seated at a special console that simulated (mimicked) the original console of the Ferranti Mark I*. *Image* © Crown Copyright, MOD

11.6 Into the 1980s

COSMOS was switched off in the early 1970s, by which time an ICL 1906A computer had been installed at Fort Halstead in 1974. Gordon Pattison recalls that "By the late 1970's, the Mathematics and Assessment Branch were using computers to play wargames in real time which required the vast majority of the resources of the 1906A, and as a result other branches had started to acquire small digital or hybrid computers for their own purposes. The combination of these factors put an end to the central computer bureau service and the idiosyncratic but highly effective combination of AMOS, COSMOS, Intercode and MIRFAC that had been built up by Gawlik and Berry. The ICL 1906A was followed in due course by other commercially-available computers, for example a Cray 1S in 1983 which was used for warhead design computations using hydrocodes".[26]

[26]Pattison, Gordon., *Playing Chess with Turing: How Computers Came to Fort Halstead.*

References

Berry, F.J. 1959. Intercode, A Simplified Coding Scheme for AMOS. *Computer Journal* 2 (2): 55–58.

Dodd, K.N. 1953a. *The Ferranti Electronic Computer (Parts 1 & 2: the Mark I Model).* Armament Research Establishment, report 10/53.

Dodd, K.N. 1953b. *The Ferranti Electronic Computer (Parts 3, 4 7 5: the Mark I* Model).* Armament Research Establishment, report 11/53.

Ellson, Allan. 2015. *Ferranti MkI* Commissioning and Installation: Some Recollections*, 14th Oct 2015.

Gawlik, H.J. 1963. *MIRFAC: A Compiler Based on Standard Mathematical Notation and Plain English. Communication of ACM* 6 (9): 545–547. See also: Gawlik, H.J., and F.J. Berry. 1967. *Programming in MIRFAC*, 2nd ed.

Pattison, Gordon. 2000. *Early Computers at the Fort.* DERA News, Oct 2000.

Pattison, Gordon. 2015. *Playing Chess with Turing: How Computers Came to Fort Halstead.* DSTL Fellow. Lecture given at Fort Halstead on 11th Dec 2015.

Reiners, C.A. 1953. *Survey of Computing Facilities in the UK.* Published by the Ministry of Supply.

Swann, B.B. 1975. *The Ferranti Computer Department: A History.* This 1975 document was initially circulated privately and marked 'confidential'. A copy is held at the NAHC, catalogue number FER/C30.

Chapter 12
The Mark I* at the Atomic Weapons Research Establishment, Aldermaston

12.1 Preparing for the UK's Nuclear Deterrent

The Atomic Weapons Research Establishment (AWRE) was set up by the Ministry of Supply (MOS) on 1st April 1950, at a former RAF aerodrome at Aldermaston in Berkshire. Much building work was necessary before any significant facilities could be accommodated. Previous to this, plans for the UK's independent atomic weapons programme had been initiated in 1946 when the distinguished mathematician William Penney, later Lord Penney, became Chief Superintendent Armament Research at Fort Halstead—(see also Chap. 11). He set up an Atomic Weapons Section and research was carried on at Fort Halstead and at the Royal Arsenal, Woolwich. In 1951 the first scientific staff arrived at Aldermaston and, in due course, all atomic weapons work was transferred here.

For some years the atomic weapons programme lacked its own computer. To quote an AWRE retrospective review[1]: "In the period 1949–1955, AWRE did not have a digital computer. It was my view in 1949, however, that computers would become important in trying to understand the mathematical physics of atomic weapons, and, for this purpose, I invested scarce man-years in this subject long before there could be any direct return. In the autumn of 1949, I assigned two new S.O's [Scientific Officers] to work with Dr. Wilkes at the University Mathematical Laboratory, Cambridge, to help program for EDSAC I. By 1951, these people were trying out their ideas for doing compressible hydrodynamics on the Manchester computer, the Ferranti Mk. I. During the succeeding time, down to the arrival of our own Ferranti Mk. I* in 1955, we always had one, or sometimes two, S.O.'s working at Manchester. These people did a number of fairly short but useful problems and, in addition, cleared the ground so that we could, when our machine arrived, make the earliest possible use of it on hydrodynamic calculations. During this period also,

[1] *Computer Software at AWRE.* Internal AWRE document, four typed pages, dated 22nd March 1965 by John Corner.

S. Lavington, *Early Computing in Britain*, History of Computing,
https://doi.org/10.1007/978-3-030-15103-4_12

one of them spent a good deal of time with the Fort Halstead example of the Mk. I*, which came into use several months before our own".

The leading Scientific Officer referred to above is A. E. (Alick) Glenny; we shall meet him again later. The second SO is believed to have been K. N. (Ken) Dodd—see also Chap. 11.

On 3 October 1952, the first British nuclear device was successfully detonated off the west coast of Australia in the Monte Bello Islands. The activity at AWRE Aldermaston continued and expanded. The first British hydrogen bomb was tested in May 1957.

12.2 Aldermaston Acquires a Mark I*

According to,[2,3] an order for a Ferranti Mark I* computer for Aldermaston was placed in September 1953 and the machine had been installed and passed its Acceptance Tests by April 1955. Presumably it first arrived on site at the very end of 1954. The Ferranti Mark 1* was installed at the south end of a new office building, in what later became the Model Room. The initial configuration had 384 words of 40 bits of primary storage (on 12 Williams/Kilburn tubes) and 16K words of magnetic drum storage. Besides maintenance engineers, one graduate-level person and one data-preparation assistant were put in charge of administering the AWRE computing service for the programmers.[4]

According to,[5] "the Mark 1* was used mainly for 1D hydrodynamic flow calculations, using the method of characteristics. Programmers had no alternative but to write in machine code". Initial programmer reactions were generally favourable, though a cautionary note was raised about the reliability of the local Aldermaston electricity supply. Here is an extract from an AWRE Progress Report for the period January to June 1955[6]:

"Our Ferranti Mark I* electronic computer (DC 7) was installed during the first months of this year and passed its acceptance tests on March 31st. It was handed over to the mathematicians on April 4th and the scheduled maintenance period by Ferranti, Ltd. (one year) began on that date. Since then the machine has been operated during normal working hours giving a total operating time of 385 h up to

---

[2]*Digital computing facilities at AWRE.* 13-page illustrated booklet published by AWRE Aldermaston in December 1966. Archived document AW2105/43-024192.

[3]Taylor, Jim. 2001. History of Scientific Computing at AWE—Part 1. *Discovery, the Science & Technology Journal of AWE* (2): 52–55.

[4]Graphs on the usage of AWRE computers. Undated but deduce 1964. This is a two-page hand-drawn document.

[5]Taylor, Jim., History of Scientific Computing at AWE—Part 1.

[6]Progress Report of the Senior Superintendent Theoretical Physics Division [John Corner at that time] for January to June 1955. Selected information from this formerly-classified document was extracted for the author in March 2016 by G. P. Libberton, AWE Corporate Archives.

the week ending June 10th; of this 247 h were useful operating time. Most of the remaining time was taken up by checking and preventive maintenance by the engineers and three days (from May 31st to June 2nd) were taken to install a modification, whereby the machine will stop at any predetermined point. Of the 272 h operating time by the mathematicians only 25 h were wasted due to machine faults, giving an operating efficiency of over 90%. The general reliability of the machine has been very good, although for some time after April 4th time was lost due to intermittent "clodding" [see Chap. 16]. However, since then clodding has been very infrequent, and occurs much less often than on the Fort Halstead Computer, presumably due to the fact that our machine room is screened whereas the A.R.D.E. building is not. Trouble has arisen at times due to failure in the main electricity supply to the machine, and following a site power failure on June 9th the machine has been very unreliable during the ten working days to the date of writing. These power failures have a very bad effect on the machine since the prescribed running-down time of one minute no longer takes place. If site power failures occur at all frequently (say, one per fortnight) they may have an extremely adverse effect on the reliability of the machine".

1955 clearly marked the end of AWRE's reliance on the use of computing facilities outside of Aldermaston, particularly at Manchester University and at Fort Halstead. Sir William Penney signalled his thanks to Manchester in a letter (presumably to Professor F C Williams) dated 14th September 1955[7] in which he said:

"I thought I ought to write to you, now that we have finally closed down our contract for the use of the Manchester University electronic computer. You will remember that we first used this in what were the early days of electronic computing and the early days of the Atomic Energy Programme. I feel that we could not come to the end of our use of this machine without my saying how very much we are indebted to you for your help in this matter, and how very useful it has been. Actually we are still indebted to you in another sense because we are now using our own Ferranti Mark I* which in effect is your work. I expect XXX [name withheld] has told you about the remarkably reliable performance this machine has been setting up".

For obvious reasons, no details of the operational life of the Ferranti Mark I* at Aldermaston have been released into the public domain. As an antidote to the grim reality of AWRE's Cold War mission, here's a thought from Joan Travis (neé Kaye), a Ferranti programmer who visited Aldermaston in about 1956: "To get to the Mark 1* at Aldermaston we were driven through heathland with bracken. I think in the same building as the Mark 1* was a computer with mercury delay lines [DEUCE, see below]. The only bit I can actually remember about that was the 'mercury' bit. One weekend when we were working at AWRE Aldermaston the resident Ferranti maintenance engineer, who enjoyed archery, brought in bows and arrows and during some spare time we were practising shooting arrows in a long

[7]Extract from *Penneys Pad*, 14th September, 1955. Selected information from this formerly-classified document was extracted for the author in March 2016 by G. P. Libberton, AWE Corporate Archives.

Fig. 12.1 The Ferranti Mark I* at AWRE Aldermaston. *Image* © Crown Copyright, MOD

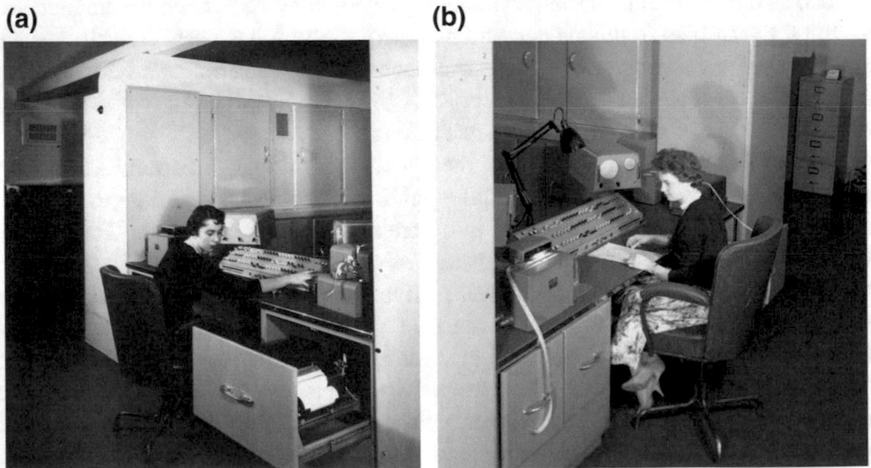

Fig. 12.2 **a** and **b** Two mathematicians at work on the Ferranti Mark I* at Aldermaston. The identity of the persons in the photographs has not been revealed, though it has been suggested that they are Mary Thomas and Florence Rigg. Both women developed reputations as skilled programmers at Aldermaston. *Image* © Crown Copyright, MOD

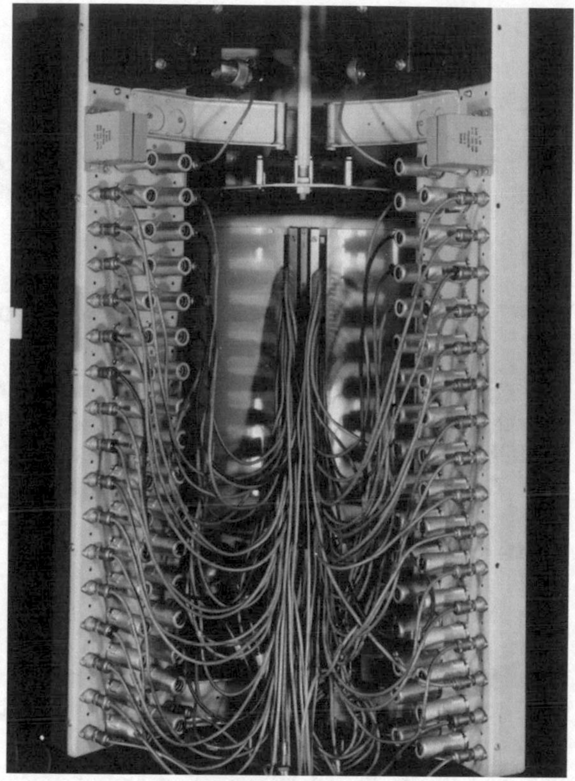

Fig. 12.3 The magnetic drum store on the Ferranti mark I* at Aldermaston *Image* © Crown Copyright, MOD

corridor next to the computer room. The juxtaposition of ancient and modern weapons has stuck in my mind".[8]

An English Electric DEUCE computer was installed in 1956, in a separate room adjacent to the Ferranti Mark 1*. Both DEUCE and the Mark 1* were run for 24 h a day, 7 days a week by scientific staff: there were no operators! Programming was still carried out in machine code.[9] "The computer building at Aldermaston continued to expand and with the completion in 1957 of the East Computer Room an IBM 704 was installed …. Programming was carried out in assembly language and hardware floating point arithmetic was available. The large quantities of data created by these programs were written to one of 10 magnetic tape units. All the 1D codes were rewritten and a start was made to 2D. Mesh methods were now used for hydrodynamics and the coupling of neutronics and hydrodynamics was achieved".[10]

[8]Joan Travis, e-mail to SHL dated 25th October 2015.

[9]Taylor, Jim., History of Scientific Computing at AWE—Part 1.

[10]Taylor, Jim., History of Scientific Computing at AWE—Part 1.

12.3 The End of the Mark I* at Aldermaston

Judging from[11] the Mark I* was being phased out of operation during 1959. By that time Aldermaston had started to acquire some seriously powerful IBM computing facilities, as described below. It is not known what became of Aldermaston's Mark I* after it ceased operation. It was probably sold for scrap.

12.4 What Came Next? the Supercomputer Race

From 1956 onwards AWRE had a policy of continuous renewal of computing resources with, it is tempting to say, no expense spared. Clearly, this government establishment was of major strategic importance in the UK's response to the Cold War. The high-performance computers acquired by Aldermaston over the next ten years are summarised in Tables 12.1 and 12.2, which is based on information.[12,13,14]

The 1951 Ferranti Mark I has been described by retrospective commentators, with some justification, as a *Supercomputer.* In terms of power it was soon overtaken, as illustrated in the Tables below. To a large extent, IBM became the world-wide manufacturer of Supercomputers from the mid-1950s to the mid-1960s until overtaken by the American company CDC. With the exception of the Ferranti Atlas 1, which was inaugurated at the end of 1962, the UK played second fiddle to the USA in the period covered by Tables 12.1 and 12.2. Aldermaston always required the fastest computer available and so, notwithstanding the dollar premium, IBM became AWRE's preferred supplier.

It is worth closing this section by taking an overview of Aldermaston's computing evolution in the period 1955 to 1965. Aldermaston serves as a datum by which to compare the computing resources accessible by all other UK scientific and engineering organisations during that decade. To put it crudely, whilst the likes of Avro and Armstrong Siddeley were trying to urge the last stride out of their Ferranti Mark I* installations as the decade drew to a close, players with more lavish funding were running miles ahead.

---

[11]Graphs on the usage of AWRE computers. Undated but deduce 1964. This is a two-page hand-drawn document.

[12]*Computer Software at AWRE.* Internal AWRE document, four typed pages, dated 22nd March 1965 by John Corner.

[13]*Digital computing facilities at AWRE.* 13-page illustrated booklet published by AWRE Aldermaston in December 1966. Archived document AW2105/43-024192.

[14]*Digital computing facilities at AWRE Aldermaston.* 12-page illustrated booklet produced by AWRE, December 1966.

Table 12.1 The main digital computing facilities at AWRE between 1955 and 1965, together with their storage capacities

Computer	Year installed	Primary storage	Secondary storage	Exchangeable storage
Ferranti Mark I*	Early 1955	383 words, Williams/ Kilburn tubes	16K words drum	None
English Electric DEUCE	Sept. 1956	402 words, mercury delay lines	8K words drum	none
IBM 704	Feb. 1957	Initially 8K words, ferrite cores; upgraded to 32K words	16K words drum, later dropped	10 mag tape decks
IBM 709	July 1959	32K, ferrite cores	?	12 mag tape decks
IBM 7090	Oct. 1960	32K, ferrite cores	?	12 mag tape decks
IBM 7030 (Stretch)	May 1962	96K words, ferrite cores	2M words disk	16 mag tape decks
Ferranti Atlas 2	End 1964	128K words, ferrite cores	4M words disk	14 mag tape decks

By the time the IBM 709 came into full use, the Mark I* and DEUCE were no longer active. The IBM 709 replaced the 704 and, in turn, the 709 was replaced by the 7090. The IBM Stretch and the Ferranti Atlas 2 operated concurrently

Table 12.2 The fixed-point addition times, in microseconds, of the computers shown in Table 10.1, together with a Ferranti Mercury for comparison

	Ferranti Mark I*	English Elec. DEUCE	IBM 704	Ferranti Mercury	IBM 7090	IBM Stretch	Ferranti Atlas 2
Fxpt ADD	1200	Min = 64 Max = 1064	24	60	4.8	1.5	1.6

This is only a crude indication of useful speed, or of useful work-rate. Neither the Ferranti Mark I* nor the English Electric DEUCE had any floating-point hardware. Only the Atlas 2 had what we would now recognise as a multi-tasking Operation System

To quote from the retrospective 1966 AWRE booklet[15]: "It has always been the case that the bulk of the work carried out on the digital computers at AWRE has originated within the establishment itself, and the majority portion of this work has been concerned with the design and improvement of nuclear weapons. Nevertheless, so vast has been the total volume of work through the AWRE computers in the last ten years, that even the minority devoted to non-weapons work has covered a large number of different fields of application".

The 1966 booklet states that tasks in the civil nuclear field were carried out from time to time on behalf of government atomic energy establishments at Harwell, Culham, Winfrith and Risley. More generally, work was undertaken in the areas of

[15] *Digital computing facilities at AWRE.* 13-page illustrated booklet published by AWRE Aldermaston in December 1966. Archived document AW2105/43 -024192.

aerodynamics, astrophysics, plastic flow, optical lens design and geophysical data analysis. Organisations for which computing has been carried out included the Medical Research Council, the Royal Aircraft Establishment Farnborough, the Space Research Management Unit and the Home Office Forensic Unit.

12.5 Software Activity at AWRE

On the software side, AWRE found itself obliged to improve or replace some of the software supposedly provided by the computer manufacturers. Such activity drew criticism from those who thought that AWRE staff had become too involved in writing, or modifying, systems software such as compilers and operating systems—to the detriment of time spent writing programs specific to AWRE's nuclear weapons remit. To head off this criticism, AWRE produced a document[16] that is historically very interesting since it describes trends that were, sooner or later, to hit all scientific computer users by the mid-1960s. It is worth quoting large sections of this document, which now follow:

"1955 – 1961 was a period in which we used computers a great deal and, as time went on, we began to use more sophisticated software, but this software was always supplied by the makers of the machine. In 1955–58 we were using our Ferranti Mk I*, programming this in machine language. The same was true of our use of DEUCE from 1956 to about 1958, but during our later years with DEUCE (1958 – 61) we used it entirely on short jobs for which easier programming became desirable. For this reason, we used an artificial language, together with an interpreter program which the machine used to decode the written instructions into machine language which it could understand. These artificial languages we used were *General Interpretive Program* written by NPL, *Tabular Interpretive Program* written by Bristol-Siddeley Engines [see[17] and Chap. 15], and *DEUCE Alphacode* written by English Electric.

"In 1957 we received an IBM 704. This was a very much faster machine and again it was programmed in what was essentially machine language. As this differed from the language of the Mk I* and DEUCE, we had to re-write most of our programs. This was a considerable labour. It is not precisely true to say that we programmed the 704 in machine language. The difference from truly machine language is that we were allowed to assign storage locations in a symbolic manner which the machine itself later converted to absolute locations by means of a *Symbolic Assembly Program*. Such a program, known as SAP, had been written for the 704 by United Aircraft Corporation, and was in full working order by the time we received our machine, which was the 32nd 704.

"We had been promised that there would be available with the 704 a programming language which would make it very much easier to deal with mathematical

---

[16] *Computer Software at AWRE.* Internal AWRE document, four typed pages, dated 22nd March 1965 by John Corner.

[17] Intercode handbook: Ferranti Mark I* digital computer. Computing Department, Bristol Siddeley Engines Ltd., Ansty. January 1960. 16 typed foolscap pages.

formulae. This *Fortran* did not arrive from IBM (USA) until several months after our 704 had started, and the program was so full of bugs that it could not be put into real use until the end of 1957. By that time, of course, we had had to teach all our programmers SAP language. It is interesting to note the late delivery of the Fortran Compiler Program…. It may be well to notice at this point that this language was later re-designated Fortran I and that the Complier could be used only with the 704.

"In 1959 we replaced our 704 by an IBM 709, which differed in its instruction code, in the manner of handling input and output. Thus all our programs had to be re-written, at least in part, diverting some of our mathematical effort from weapons physics. It was at this point that we resolved never again to endure this rewriting if it could possibly be avoided.

"In 1960 we exchanged our 709 for a 7090 which had an identical instruction code to the 709; therefore no reprogramming was required. The 7090 was so much faster than the previous computer that it forced upon its users the need for 'automatic operation'. A software system written by IBM was available to do this. It included a new version of the simplificd programming language, Fortran II, and the machine language symbolic assembler FAP (the successor of SAP). This automatic operator was widely but not exclusively used: it was suitable mainly for jobs not requiring use of private magnetic tape reels, and was not much used for our big programs. The Fortran II language was used for small programs, but many of the large programs still persisted in the machine language in which they had been written.

"In 1960 we ordered Stretch [aka the IBM 7030], for delivery in 1962. Because of the labour of rewriting programs as we changed from one machine to another, and our increasing stock of exceptionally big programs, we decided that we would have to change our programs to some standard language which would stay current during the life of at least several machines. It was decided near the end of 1960 that it would be most sensible to use Fortran. Unfortunately IBM could not undertake to deliver a Compiler for handling Fortran on Stretch until at least one quarter after the delivery of the machine. Since AWRE intended to put all its programs into Fortran and therefore could not use the machine until its Fortran compiler were available, AWRE decided that it had to write a Compiler of its own. Because Fortran II was an existing language, whereas Fortran IV was, at that stage, only a hope, it was decided that our Stretch Compiler must be for Fortran II.

"Fortunately, AWRE had a man who had been interested in software for many years. This was Mr. Glennie, whom we had seconded to Cambridge as an SO in 1949. He had also worked on our behalf at Manchester during the period 1951–54. During this time he had become interested privately in the writing of languages which would make mathematical programming more easy. He did, in fact, write one such language in 1952 which he christened Autocode. This was the first such language to be written and the name which he gave it became a type name for many other easy-programming languages, especially those written by the Manchester school. Foreseeing that we might have to take some interest in computer software, and knowing that Mr. Glennie had many ideas on this subject which he had never had time to develop (he was our expert on implosion theory), AWRE gave Mr. Glennie leave of absence during the period January to July 1960 to work at Pittsburg with Professor A. J. Perlis.

"It therefore seemed that Mr. Glennie could be a key figure in the writing of a Stretch Compiler for Fortran II. The difficulty still remained of providing him with

Fig. 12.4 Alick Glennie pictured at a NATO Software Engineering Conference in 1968. *Photo credit* Prof. Brian Randell

enough high grade programmers to carry out this assignment in the little more than a year which remained before the machine was scheduled to come into use at AWRE. We were, however, fortunate in having an extremely bright SO [Scientific Officer] programmer called Vaughan-Williams, and we also received important help from IBM who lent us two very good programmers. This team, with some E class help, built up the Compiler for a dialect, called S1, of Fortran II, and did this before Stretch arrived.

"Harwell [the UKAEA's establishment] had decided not to use the S1 version of Fortran II; they preferred to wait for the IBM Compiler which would handle Fortran IV on Stretch. This compiler was several months late, even relative to the IBM schedule, and after Harwell had used it for a short while they decided that it was completely useless for their purpose. In this difficult position for Harwell, AWRE undertook that Mr. Glennie would supervise the writing of a new compiler for Stretch [the S2 compiler] …. Whereas the construction of the S1 compiler had emphasised its delivery date, the design of the new compiler emphasised quick compilation ….."

"It was decided that the Authority should buy an Atlas 2 computer which would be delivered at the beginning of 1965". Atlas 2 was a slimmed-down version of Atlas 1[18] which had first been demonstrated in December 1962.

[18]Lavington, Simon. 2014. *Ferranti Atlas 2 Computers at AWRE Aldermaston and at the CAD Centre*. 22nd Aug 2014. 28-page illustrated report, prepared for the website *Memories of the Ferranti Atlas computer.* See: http://elearn.cs.man.ac.uk/~atlas/docs/Atlas%202%20at.%20AWRE%20and%20CAD%20Centre%20Final.pdf.

The AWRE document[19] ends with an account of how it was necessary to equip Atlas 2 with a Fortran II compiler and a tailored version of the Atlas 1 Operating System (called the Supervisor), how ICT (the successor to Ferranti's mainframe computing activities) provided a Supervisor but that Alick Glennie led a team of AWRE and ICT people who provided the Fortran II compiler for Atlas 2. An overview of all the computing facilities at Aldermaston in 1966 is provided.[20]

It seems incredible to realise that, whilst all this high-performance computing activity was taking place in Berkshire using machines based on transistors and ferrite core RAM, a Ferranti Mark I* computer called FINAC was still performing useful work in Rome until 1966 using thermionic valves and CRT RAM dating from the 1940s. Another much-loved Mark I* called AMOS was active until 1967 at Fort Halstead in Kent.

References

Lavington, Simon. 2014. *Ferranti Atlas 2 Computers at AWRE Aldermaston and at the CAD Centre*. 22nd Aug 2014.

Taylor, Jim. 2001. History of Scientific Computing at AWE—Part 1. *Discovery, the Science & Technology Journal of AWE* (2): 52–55.

[19] *Computer Software at AWRE.* Internal AWRE document, four typed pages, dated 22nd March 1965 by John Corner.

[20] *Digital computing facilities at AWRE Aldermaston.* 12-page illustrated booklet produced by AWRE, December 1966.

Chapter 13
What Came Next?

Whilst the seven Mark I* machines were being manufactured and installed, the original Manchester Mark I had been moved to a spacious new home in 1954, where it thrived. From this base, the University and Ferranti continued to cooperate on designs for new computers. Not surprisingly, the course of these projects did not always run smoothly. By the end of this chapter we will have passed through the highs and the lows, looking ahead to the 1990s and the effective demise of the Ferranti company itself. On the way, there were to be some notable achievements. But in the end, the Mark I and Mark I* machines were all but forgotten in comparison with bigger, better, computers. Thankfully a few historical artefacts do survive for public inspection. At the end of the chapter the locations of these remaining fragments of *Mark I stardust* are listed.

13.1 The New Life at Dover Street

Manchester University's 1951 Computing Machine Laboratory (CML) in Coupland Street had always been intended to be temporary accommodation for the first Ferranti Mark I. It was not spacious enough for the totality of planned computing activities. In any case, the CML was intimately connected with the Department of Electrical Engineering which, having had to share space with the Physics Department for many years, urgently needed its own premises. The new Electrical Engineering building in Dover Street had a basement, three full floors with wings and a top floor—see Fig. 13.1. It was very grand, with a marble-panelled central stairway. The final cost was just over £600,000, nearly twice the original estimate.

Down in the basement were workshops, a large lecture theatre and the AC and DC Electrical Machines laboratories. The machines area was soon to acquire a 10-m length of monorail on which a wonderful chariot powered by a novel linear motor could be demonstrated. The middle three floors contained an assortment of lecture theatres, staff offices, laboratories for digital electronics and control

S. Lavington, *Early Computing in Britain*, History of Computing,
https://doi.org/10.1007/978-3-030-15103-4_13

Fig. 13.1 The Electrical Engineering Department's new building in Dover Street, Manchester. Opened in 1954, the top floor contained two computer suites and offices for the staff of the Computing Machine Laboratory. The University's Ferranti Mark I computer was moved here from its original 1951 premises in Coupland Street, about 350 yards away

engineering, a library and, wonder of all, a tall between-floors void containing high voltage gear on which impressive lightning strikes could be generated.

The top floor at Dover Street was devoted to the Computing Machine Laboratory. On this floor there were two computer suites and several offices for programming staff. The Ferranti Mark I was moved into one of the computer suites: the exact re-installation date in 1954 is uncertain but it is believed to have been in the spring or summer. There is anecdotal evidence that the move had taken place by the time Alan Turing tragically died at his home in Wilmslow in June 1954. The other suite was for hardware research and it was to here that the Meg computer was moved. Designed by Tom Kilburn's group, Meg was ten times faster than the Mark I, consumed less power and had floating-point hardware.[1] It first ran a program in May 1954.

[1](a) Kilburn, T., D.B.G. Edwards, and G.E. Thomas. 1956. The Manchester University Mark II Digital Computing Machine. *Proceedings of IEE* 103 (Part B, Supp. 1–3): 247–268.; (b) Lonsdale, K., and E.T. Warberton. 1956. Mercury: A High-Speed Digital Computer. *Proceedings of IEE* 103 (Part B, Supp. 1–3): 483–490.

Fig. 13.2 The *MEG* research machine, shown in Computer Room 2 in Dover Street in 1954. Meg was faster than the Mark I, consumed less power and had floating-point hardware. Meg was the prototype for the Ferranti Mercury computer

Whilst the hardware group had been busy building Meg, the Computing Machine Laboratory staff had been busy running a computing service and developing innovative software—of which more below. Life was getting busier. The computer's reliability had improved dramatically. Here are the computing service statistics for a 12-month period ending in 1955.[2]

The Mark I's user community in 1955 consisted of programmers from:

9 Departments from Manchester University;
6 Departments from other universities;
3 research associations (such as the British Iron and Steel RA);
9 government establishments;
7 industrial companies engaged in engineering work.

During the year 104 people were trained to use the machine and 66 scientific papers were published based on results obtained with the computer. Of the 100 h of average useful computing time per week in 1955, approximately 12 h were

[2](a) Williams, F.C. 1955. *Internal Report on the Computing Machine Running Costs.* Manchester: University of Manchester.; (b) Williams, F.C. 1956. Introductory lecture: IEE Convention on digital computers. *Proceedings of IEE* 103 (Part B, Supp. 1–3): 3–9.

allocated to CML staff, 30 h to other Manchester University Departments, 13 h to other academic institutions and 45 h to government and industry.

Amongst the industrial users of the Mark I were programmers from Ferranti Moston and Ferranti's London Computer Centre. Mary Berners-Lee recalled in an interview[3] that "as the computer at the university got used more and more, Ferranti got restricted in the hours in which we were allowed to use it. It became evening hours, and then later, the University would only let Ferranti have it between midnight and eight in the morning. And it was decreed by [Ferranti's] Personnel Department that women should not be there at night—to the glee, I remember, of one particular member of the male programming staff, because it meant there was more time on the computer for the men! But of course, for the women, it was a disastrous idea. Unfortunately, the Personnel Manager knew that we were there at night, because I shared the flat with the Assistant Personnel Manager [Joy Badham] … The Personnel Department then suggested that we needed a chaperone, I think—because the maintenance engineers, of course, were always there, and they were men—and suggested we had a tea lady. We managed to override that one as well so we used to get through the night on black coffee! And we did get camp beds, so that you could flake out. And it was awful to be woken up. You'd be in a very deep sleep, and to be woken up—although you were dying to get on to the computer—to be woken up in the middle of the night was not very good, when your turn came! However, it was worth it".

Another outside user was the Meteorological Office. F. H. (Fred) Bushby had attended an EDSAC programming course at Cambridge in the autumn of 1951. By the end of that year Fred Bushby and Mavis Hinds had started to use LEO 1 for weather forecasting. After some while a decision was made to change over to using the Ferranti Mark I at Manchester, which was faster than LEO 1. Here is Mavis Hinds' account of what life was like in Dover Street.[4]

"Since we needed the computer for several hours at a stretch, most of our usage was at night and for some years we used the machine for two nights each alternate week. We stayed at a nearby commercial hotel, made up of several elderly terraced houses, now happily demolished. Sleeping during the day was made difficult by the shouting of the cleaners and the insistence of the electricity-meter emptier, and if we returned during the night the chorus of snores through the thin walls was unbelievable. Occasionally our time off enabled us to sample the delights of Edale or the Peak, or watch a second-grade film at the local cinema. More readily available treats were the sight of sunrise over Manchester from the roof near the computer room or the exhilaration of coping with an old-fashioned Manchester smog in which the

[3]Mary Lee Berners-Lee: An Interview Conducted by Janet Abbate for the IEEE History Center, 12th September 2001. Interview #578 for the IEEE History Center, The Institute of Electrical and Electronic Engineers, Inc. See: https://ethw.org/Oral-History:Mary_Lee_Berners-Lee#About_Mary_Lee_Berners-Lee. The IEEE History Center has a collection of more than 800 oral histories in electrical and computer technology which can be accessed via http://ethw.org/Oral-History:List_of_all_Oral_Histories.

[4]Hinds, Mavis K. 1981. Computer Story. *The Meteorological Office Magazine* 110 (1304): 69–81.

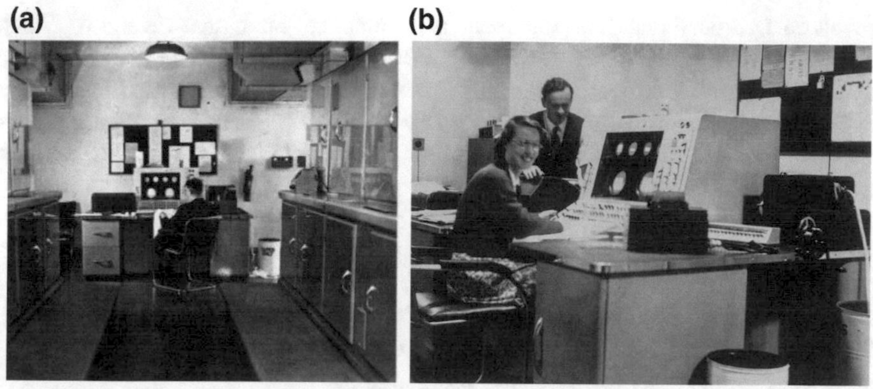

Fig. 13.3 **a** and **b** The Ferranti Mark I, which occupied Computer Room 1 on the top floor of the Dover Street building. The two people in photo **b** are believed to be Mavis Hinds and Fred Bushby of the Meteorological Office

buses were led by a man on foot holding a flare. It was also sometimes necessary to have one member of the party with sufficient athletic prowess to scale the wrought-iron University gate (whilst the others 'kept cave') in order to gain access to the computer building, and several of those who performed this feat have since reached higher directorate level."

Let's now turn to the behind-the-scenes technical work of the Computing Machine Laboratory, in their efforts to make the Ferranti Mark I easier to use.

13.2 The Computer Becomes User-Friendly

The Mark I's original programming system, though compact and efficient, was difficult to learn. Sheila Cooper (neé Gray), who joined Avro in 1954, reckoned that it took a mathematics graduate six months before he or she became a really competent programmer.[5]

Frank Sumner was more forthright about the programming difficulties.[6] Frank had started a Ph.D. in Computational Chemistry in the autumn of 1951, using a Marchant electro-mechanical desk-top calculator to work on molecular energy

---

[5]Sheila Cooper's memories of working with a Ferranti Mark I* computer at Avro Chadderton. Transcript produced by Simon Lavington in April 2017, based on an audio interview of Sheila by Buxton's local historian Vivienne Doyle recorded on 14th April 2016 and supplemented by telephone conversations and e-mails with Sheila in the spring of 2017 and with her daughter Judie Adnett.

[6]Sumner, F.H.(Frank). 1994. Memories of the Manchester Mark I. *Resurrection* 10: 9–13. This article is an edited version of the talk given by Professor Sumner to the Computer Conservation

levels in hydrocarbons. The chemical and numerical analytical details of Frank Sumner's work are vividly described by a fellow user of the Ferranti Mark I, Huw Pritchard.[7] Frank's project was to solve the Hückel molecular orbital eigenvalue problem for some aromatic hydrocarbons by using the contour integration methods though, with the help of Tony Brooker, other numerical methods were later adopted.

Back in 1951, Frank remembers how he got started.[8] One day his Supervisor at Manchester University said "Why don't you go and see Alan Turing? I think he has got some sort of computer over on the other side of campus" Turing's response to Frank was brief. Handing him a copy of the 110-page Programming Manual, Turing said: "Read that. Then go and see Cicely Poppelwell who will give you some time on the machine, and start programming."

"I disappeared for a couple of months, and discovered that the perfect way to write a programming manual is to fill it full of examples, none of which work. That might be an exaggeration - let's say most didn't work. Turing was like that: if he wrote 'k' instead of 't', he knew it was 'k', so why bother to do all that proof reading? So all the programs written in Mark I code [in the manual] had slight errors in them, and by the time you had worked out what the code should have been you had become quite a competent programmer. Anyway, I wrote a program to print out the answers to my energy level problem and took it to the machine. A few months later, it worked!"

The full complexity of writing Mark I programs is explained in Chap. 15. An impression of the difficulties is illustrated in Fig. 13.4. This shows the main section of Christopher Strachey's *Love Letter* code, which was introduced in Chap. 2. As indicated in Fig. 13.4, programmers had to code up each separate instruction as a sequence of the 32 individual 5-track Teleprinter symbols

/	E	@	A	:	S, ...	etc. corresponding to 5−bit patterns:
00000,	10000,	01000,	11000,	00100,	10100, ...	etc. in 'backwards binary'.

Phew!

Memorising the teleprinter notation was not the only obstacle. The Ferranti Mark I had no automatic detection of arithmetic overflow, or indeed no hardware support for floating-point arithmetic. Users had to be aware of the range/scale of their numerical calculations. Finally, users had to shuffle code and data between the fast (but small) primary store and the slow (but large) drum secondary store and be aware of the physical memory addressing rules.

Society's North West Group at The Museum of Science and Industry in Manchester on 21st October 1993.

[7]Pritchard, H.O., and J. Mol. 2001. Computational Chemistry in the 1950s. *Graphics Modelings* 19 (6): 623–627. Huw was appointed Assistant Lecturer in the Chemistry Department at Manchester University in the autumn of 1951.

[8]Sumner, F.H.(Frank)., Memories of the Manchester Mark I. *Resurrection* 10: 9–13.

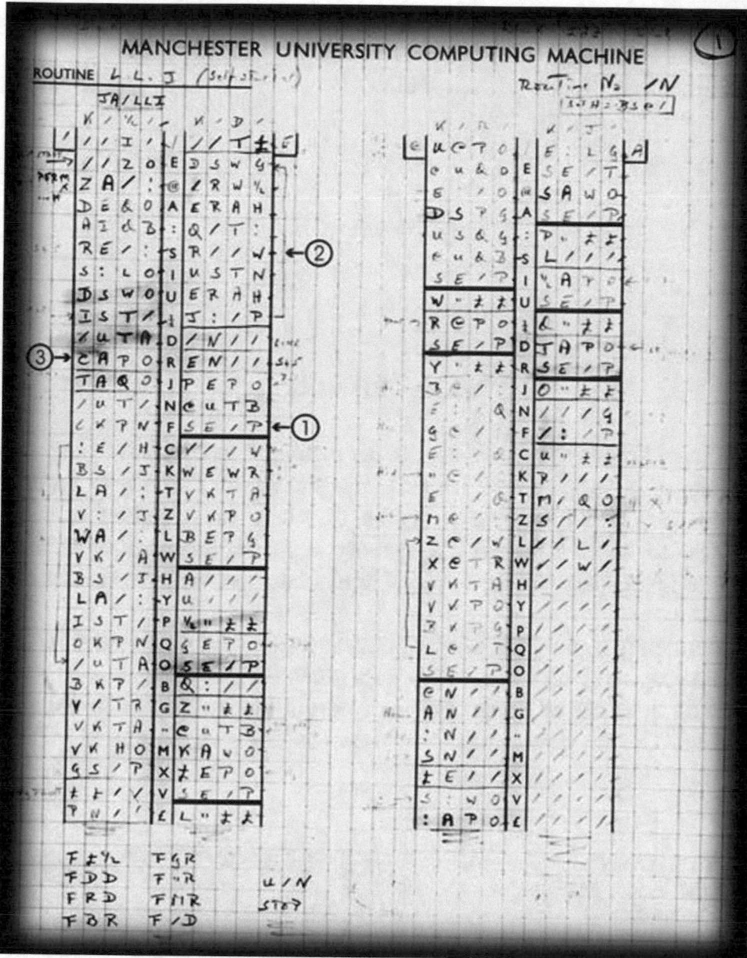

Fig. 13.4 The first page of Mark I code for Christopher Strachey's *Love Letters* program. Some letters generated by this program are quoted in Chap. 2. The code is explained in detail (Zielinsky, Siegfried, and David Link., There Must Be an Angel: On the Beginnings of the Arithmetics of Rays)

In 1954 Tony Brooker, inspired by some earlier research carried out at Manchester by Alick Glennie, introduced his *Mark I Autocode* programming language.[9] This greatly simplified the written form of a program and solved the two

[9]Brooker, R.A. 1955. *An Attempt to Simplify Coding for the Manchester Electronic Computer. British Journal of Applied Physics* 6: 307–311. This paper describes the Mark I Autocode. Previous to Brooker's work, Alick Glennie who was an MOS external user of the machine had developed his personal automatic coding scheme in the summer of 1952. Glennie's system, which he called *autocode*, was very machine-dependent and was not released to other users.

problems of scaling and memory management. The Autocode is best illustrated by example. This short Autocode sequence prints the root mean square (RMS) of the floating-point variables v1, v2, … v100.

$$\begin{aligned}
&\mathrm{n1} = 1\\
&\mathrm{v101} = 0\\
&\mathrm{2v102} = \mathrm{vn1}\ \times\ \mathrm{vn1}\\
&\mathrm{v101} = \mathrm{v101} + \mathrm{v102}\\
&\mathrm{n1} = \mathrm{n1} + 1\\
&\mathrm{j2}, 100 \geq \mathrm{n1}\\
&{}^{*}\mathrm{v101} = \mathrm{F1}(\mathrm{v101})
\end{aligned}$$

As explained more fully in Chap. 15, the symbol * in the above program causes the printing of a variable and F1 calls the square root library subroutine. The Autocode user did not have to worry about where the variables or subroutines existed because the entire Mark I's memory could be imagined as what Brooker called a one-level store and the addressing mechanisms were taken care of behind the scenes.

The complete programmer's description of Mark I Autocode occupied just two sides of foolscap, with an example program given on a third page. New users could be trained in a matter of hours. Program design and testing were comparatively simple tasks. On the down side, Autocode programs ran between five and fifty

Fig. 13.5 R. A. (Tony) Brooker in the 1960s. Tony joined the Computing Machine Laboratory staff in 1951. In 1954 he developed *Mark I Autocode*, thereby making the Ferranti Mark I "possibly the easiest machine to program in Britain"

times slower than equivalent machine code programs and the Autocode was unsuitable for very large problems.

Nevertheless, the popularity of the Mark I Autocode grew rapidly from 1955 onwards. The Computing Machine Laboratory set up a postal service, through which remote users mailed their program manuscripts. The Autocode script was then punched on paper tape, the program run and the printed output returned to the user—usually with a turn-round time of less than a week. Early users of the postal service included (from 1956) the Central Instruments Laboratory of Imperial Chemical Industries (ICI) and the Theoretical Physics Branch of the Atomic Energy Research Establishment at Harwell.

At ICI, the postal service was soon replaced by an early attempt at a digital link. Autocode programs were punched on paper tape, transmitted via the ICI sales Telex network to a convenient terminal in Piccadilly, central Manchester. These tapes were physically transported the mile or so to Dover Street. "Later in the day output tapes with error diagnostics and occasionally results would be returned via the Telex network to the originating engineer anywhere in the ICI world-wide network of factories and design offices".[10] "A key factor in this intriguing early example of remote access computing was the recruitment of Bernard Richards from Tony Brooker's team at the University to staff the Manchester relay station. Bernard subsequently joined the Central Computing Facility of I.C.I. when we installed a Mercury Computer". Bernard Richards had been Alan Turing's second research student, working on morphogenesis.

It was said that, with Tony Brooker's Autocode, the Ferranti Mark I "became possibly the easiest machine to program in Britain".[11] It was certainly influential: similar styles of autocode were developed for the Ferranti Pegasus (in 1957), the Ferranti Mercury (in 1958) and for the Elliott 803 (in 1960).

13.3 Goodbye Mark I, Goodbye FC

The end of the Ferranti Mark I.

For the last couple of years of its life, the Ferranti Mark I at Manchester University was performing well. Huw Pritchard[12] recalls that "by 1956 the machine would often run for a full 24 h without a fault". Maintenance on the machine was

---

[10]Thomas, G.E.(Tommy). 1998. *The Significance of Brooker's Autocodes in Taking the Early Manchester Machines into the Market (1954–1960).* Unpublished paper which first appeared on Thomas's personal website in about 1998 (but is no longer there). It is reproduced here: http://curation.cs.manchester.ac.uk/computer50/www.computer50.org/mark1/gethomas/manchester_autocodes.html.

[11]Campbell-Kelly, Martin. 1980. Programming the Mark I: Early Programming Activity at the University of Manchester. *Annals of the History of Computing* 2 (2): 130–168.

[12]Pritchard, H.O., and J. Mol., Computational Chemistry in the 1950s.

discontinued in December 1958. The last entry in the Log Book, dated 24th December, says “The end of Mark I—RIP”. Then below this, another hand has added “Removed in June 1959 for burial”.[13] After lengthy negotiations with the DSIR and with the Science Museum at South Kensington, which sadly came to nought, the Mark I was dismantled in June 1959. The main logic cabinets and the console were given to Ferranti to form the basis of a projected company museum. The rest was sold for scrap. J. L. (Johnny) Mudge remembers the last remnants "being thrown out of the window” at the very end of September.[14] Within a few days the first section of the MUSE project, the prototype Fixed Store, was being moved into the Mark I’s empty room (see below).

When the main part of the Mark I was dismantled, it was placed in the West Side storage section of Ferranti’s Lowerhouse works, Oldham, “and set up just as it was at the University”. The West Side storage was discontinued by Ferranti in 1962, whereupon the computer was disposed of as scrap.[15]

Negotiations with Ferranti for the replacement of the Mark I had begun in November 1955. The replacement was a Ferranti Mercury which was delivered late in 1957 to the room in Dover Street formerly occupied by Meg (Fig. 13.6). Mercury started useful work in February 1958. The cost of this computer (approximately £100,000) was met partly from Mark I earnings and from design know-how supplied by the University to Ferranti and partly by a grant of £48,000 from the University Grants Committee (UGC, the successor to DSIR). This Mercury was returned to the UGC free of charge in January 1963 upon commencement of the Atlas computing service and it was then re-installed by the UGC at Sheffield University.

A word should be said about Mark I computer earnings, ie the income from hiring out time to outside users at £20 per hour. For the year 1955 the earnings amounted to about £20,000. The computer’s running costs (maintenance, Computing Machine Laboratory salaries, data-preparation equipment, etc.) were met partly out of these earnings, partly from contributions from general University of Manchester funds and partly by the DSIR who had taken over ownership of the Mark I from the MOS in 1952. The remainder of the Mark I earnings were used to support research into computer design and to build up a fund to permit eventual purchase of a better computer.

---

[13]Some surviving log books from the Manchester University’s Ferranti Mark I computer, 1951–58, are preserved in: NAHC/MUC/2/C6.

[14]J. L. Mudge, series of e-mails to Simon Lavington, May/June 2018. Johnny Mudge became a Ph. D. student for R. L. Grimsdale in October 1958, working in Dover Street on the MUSE/Atlas Fixed Store. He later joined the staff and then resigned in 1967 to take up a job in California.

[15]*Manchester University computer Mk I: results of enquiries made regarding the whereabouts of the first Ferranti Mk I computer.* Single-page typed memo, prepared by the Ferranti Archivist in July 1965. To this was later added a two-page typed note headed *Mk I* computer at Leicester Museum.* Undated, but the contents suggest June 1973. The author is not given but it was probably Charles Somers, the Ferranti Archivist at the time.

Fig. 13.6 A Ferranti Mercury computer, installed in 1957 in Computer Room 2 on the top floor of Dover Street. Four magnetic drum cabinets line the right-hand wall. This machine provided Manchester University's computing service until 1963, when the service was transferred to Atlas

The end of F. C. Williams' involvement.

This is the point to say 'goodbye' to Professor F. C. Williams, who had guided the Mark I project from the earliest days. But even in 1946 at the time of his appointment to the University referees had commented that, whilst Williams was an undoubted expert in electronics, the post of Professor of Electrical Engineering could really benefit from a candidate with broader electrical interests.[16] One notable 'broader interest' soon emerged.

By the mid-1950s 'FC', as he was familiarly called by his departmental colleagues, became interested in heavy-current engineering and, in particular, in variable-speed AC motors. It seems likely that the person responsible for Williams' change of direction was E. R. (Eric) Laithwaite. Eric Laithwaite was born in 1921 and served in the RAF from 1941–46 when he rose to the rank of Pilot Officer. He then entered Manchester University, graduated in 1949 and was taken on as a research student working for Williams and Kilburn. At that time the University's prototype Mark I computer was beginning to run a few useful tasks and it appears that Laithwaite became interested in how to arrange access to a library of permanent (read-only) subroutines. His Master's thesis was a proposal for a system based on electro-mechanical selection of a chosen card from a stack of marked (eg punched)

[16]University of Manchester, file VCA/7/46. Vice-Chancellor's correspondence in relation to the appointment of a Professor of Electro-Technics, autumn 1946. Contains letters from referees. F. C. Williams was one of three candidates interviewed; the others were R. W. Sillars and Arnold Tustin.

cards, each one to be read by a flying-spot scanner.[17] Each card would contain code for (part of) a subroutine. Eric, however, had friends in the Lancashire textile industry near his home town, Atherton, and for some time had had the idea of driving shuttles electrically. He therefore changed research from computing to an investigation of linear induction motors, with FC acting as his Supervisor.

FC "saw Laithwaite's experimental motors and wondered what would happen if the rotor was restrained to move at an angle from the direction of field travel. This then started a whole range of really innovative variable-speed induction motor designs (spherical, logarithmic, phase-shift, pole-change), mostly driven and invented by FC and his unlimited imagination".[18] The result was that scientific publications and patents started to appear in 1957.[19] From that point onwards, Williams' name began to be associated with variable-speed AC motors rather than with digital electronics.

FC's public swansong in the computing arena was a live interview on BBC Television for a programme called *The Brain in the Box,* part of the BBC's *Eye on Research* series.[20] Figure 13.7 shows the scene in Computer Room 1 in Dover Street.

Eric Laithwaite left Manchester in 1964 to take up a Chair at Imperial College, where he continued his research into linear motors. Eric died in 1997. F. C. Williams continued to lead the Manchester group, researching variable-speed motors. "This work eventually moved into the field of diesel-electric railway traction and brushless alternators. His invention of the consequential *pole-change* traction drive showed the usual Williams unique ingenuity. A further interesting offshoot was a unique mechanical automatic gearbox relying on an oscillating torque transfer".[21] FC died of cancer at the relatively young age of 66 in 1977.

---

[17]Laithwaite, E.R. 1950. *An Automatically Accessible Three-Dimensional Library for use with Digital Computing Machines*. M.Sc. thesis, University of Manchester, Faculty of Science.

[18]E-mails from Graham (Mac) McLean to Simon Lavington, 10th and 12th July 2018. Mac graduated in Electrical Engineering from Manchester University in 1960 and worked for F. C. Williams, first as a research student and then as a colleague, until Williams' death in 1976. Mac continued to do research into electrical machines at the University until 1990, since when he has continued to act as a consultant to industry.

[19]Williams, F.C., E.R. Laithwaite, and L.S. Piggott. 1957. Brushless Variable-Speed Induction Motors. *Proceedings of IEE* 104 (14): 102–118.

[20]*The Brain in the Box,* part of the *Eye on Research* series of BBC television programmes and first broadcast on the evening of Tuesday 28th January 1958. Robert Reid was the interviewer. This half-hour production used standard OBU equipment based on three cameras, with film inserts done at the base studio. The programme started in the Mark I's computer room in Dover Street and then went to the adjacent Mercury computer room. The occasion is described in: *Chaos in the Computer Laboratories—Mercury goes on TV*, an anonymous illustrated article which appeared in the Manchester University Engineering Society's magazine, Lent Term 1958, pp. 14 and 15.

[21]E-mails from Graham (Mac) McLean to Simon Lavington, 10th and 12th July 2018. Mac graduated in Electrical Engineering from Manchester University in 1960 and worked for F. C. Williams, first as a research student and then as a colleague, until Williams' death in 1976. Mac continued to do research into electrical machines at the University until 1990, since when he has continued to act as a consultant to industry.

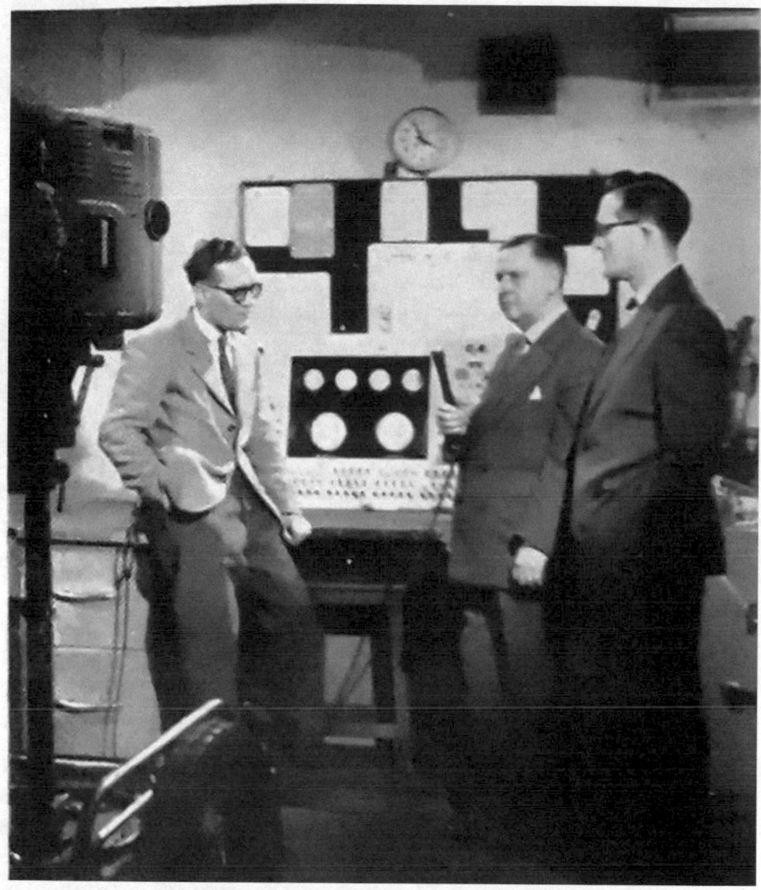

Fig. 13.7 F. C. Williams and Tom Kilburn rehearsing for a BBC half-hour live broadcast on computers in January 1958. The broadcast effectively marked the end of FC's 12-year public association with computer design at Manchester University

13.4 Ferranti and Supercomputers

Back in 1958, a Ferranti Mercury computer had taken over as the Computing Machine Laboratory's main resource. The original Ferranti Mark I was removed from Dover Street in 1959 and, in its space, a new research computer was taking shape. The new project, called MUSE, was Tom Kilburn's answer to an implied challenge from America. In December 1956 the design objectives for two powerful computers, the Univac LARC and the IBM STRETCH, had been presented at the Eastern Joint Computer Conference in the USA. In January 1957 the UK's Atomic Energy Research Establishment and the Atomic Weapons Research Establishment

Fig. 13.8 The prototype Atlas machine in Computer Room 1 at Dover Street in 1960. A room for magnetic tape decks leads off in the right background

issued a joint report stating that "by 1960 the AEA would need a computer comparable to LARC and STRETCH".[22]

The word *Supercomputer* began to be used to describe machines like LARC and STRETCH. By 1957 both NRDC and DSIR's Brunt Committee believed that the UK should be designing its own Supercomputer. This triggered two years of technical and political debate, during which representatives from industry, academia and the government argued about who should be responsible for building such a Supercomputer and who should finance the project. The full, messy, story is recounted in a collection of papers.[23]

From 1957 onwards, Tom Kilburn's MUSE research group went ahead with hardware development, backed by £50K from an internal research fund. In 1958 Ferranti decided to join in. Specifically, the company planned a two-pronged approach to developing large computers. Ferranti would initiate its own medium-sized business data-processing computer (later to be called Orion) and, at the same time, it would cooperate with Manchester University on a larger scientific supercomputer (later to be known as Atlas).

---

[22]Many retrospective articles on the Ferranti Atlas project are collected together here: http://elearn.cs.man.ac.uk/~atlas/ In particular, reference may be made to: (a) Lavington, Simon. 2012a. *NRDC and The Case for a British Supercomputer, 1956–1960*, 13.; (b) Lavington, Simon. 2012b. *Timeline of the MUSE/Atlas Project at Manchester University, 1955–1971*, 19.; (c) Lavington, Simon., et al. 2016. *Tony Brooker and the Atlas Compiler*, 29.

[23](a) Lavington, Simon., *NRDC and The Case for a British Supercomputer, 1956–1960*.; (b) Lavington, Simon., *Timeline of the MUSE/Atlas Project at Manchester University, 1955–1971*. (c) Lavington, Simon., et al., *Tony Brooker and the Atlas Compiler*.

By this time Peter Hall had replaced Brian Pollard as the Manager of Ferranti's Computer Department. From 1959 onwards, Tom Kilburn and Peter Hall exercised joint control over the MUSE/Atlas project—both in respect of development carried out at the University and at Ferranti's West Gorton facility. The terms of the agreement, which is set out in full in,[24] were as follows: (i) the first machine (the production Atlas) was to be installed at Manchester University, who would have half the operating time; (ii) Ferranti would run this machine at their own cost and would sell time to outside users in industry, commerce and academia; (iii) Ferranti would pay Manchester University £10K per annum plus a percentage of the income gained by selling time on the computer; (iv) the main MUSE/Atlas team of between 12 and 20 academics was increased in due course by the addition of between 30 and 40 Ferranti employees.

By the end of the Atlas project, approximately 50 people had been involved in innovative hardware or software design. Of these, just under half were Ferranti employees. Special mention should be made of David Howarth, who joined Ferranti in 1960. Together with a small team that at no time numbered more than six people, Howarth became the principle designer and implementer of the Atlas Supervisor (Operating System).

Prior to this, from September 1959 Tony Brooker and colleagues were planning the writing of high-level language compilers for Atlas. For this work, Brooker conceived the Compiler Compiler.[25] On the hardware side, by September 1960 the outline design of all nine Atlas sub-units had been completed and the detailed logical design of 80% of these units had been finished. By the end of January 1961 the Pilot Model of Atlas was dismantled and moved to West Gorton and the empty room at Dover Street was re-furbished ready to receive the first Atlas production sub-unit in June. On 7th December 1962 Sir John Cockcroft officially inaugurated the completed first production Atlas in Dover Street, accompanied by press coverage. It was said to be "the most powerful computer in the world". The justification for this claim, together with performance figures for other supercomputers of the early 1960s, is given in Chap. 18.

Peter Hall estimated that the total cost of the Atlas project had been £2 m, of which Manchester University had originally put in £50K (plus academic staff wages and overheads). In May 1959 NRDC had loaned Ferranti £300K towards the cost of Atlas development, to be recovered from sales. Ferranti had paid off their loan by March 1963.

At the end of the Atlas agreement with Ferranti in 1972, the University's Computer earnings fund stood at approximately £300K. The Manchester Atlas was switched off on 30th September 1971.

---

[24](a) Lavington, Simon., *NRDC and The Case for a British Supercomputer, 1956–1960.*; (b) Lavington, Simon., *Timeline of the MUSE/Atlas Project at Manchester University, 1955–1971.* (c) Lavington, Simon., et al., *Tony Brooker and the Atlas Compiler.*

[25](a) Lavington, Simon., *NRDC and The Case for a British Supercomputer, 1956–1960.*; (b) Lavington, Simon., *Timeline of the MUSE/Atlas Project at Manchester University, 1955–1971.* (c) Lavington, Simon., et al., *Tony Brooker and the Atlas Compiler.*

Fig. 13.9 The first production Ferranti Atlas installation in Computer Room 1 in Dover Street in 1962. When it was inaugurated in December 1962, it was said to be "the most powerful computer in the world"—a position it did not hold for many months because American machines such as the CDC 6600 were on the horizon

The Atlas project was a success for the academics. Many scientific papers were published and 15 Atlas patents were filed between October 1957 and February 1962. The picture was not so rosy for Ferranti. Marketing Supercomputers proved difficult and, in particular, no progress could be made in America, Europe or Australia. Only three full Atlas systems were delivered, together with three smaller Atlas 2 variants.

Meanwhile, Ferranti's other large computer project, Orion, was having some problems.

The Orion design was started at West Gorton in 1958, based on novel ballot-box logic circuits called *neurons*. Orion's instruction set was derived from an enhanced version of the instruction set used for the Pegasus/Perseus machines. The *neuron* development at West Gorton got behind schedule, risking penalties in unfulfilled customer orders. So in 1961 Peter Hall authorised Ferranti's Bracknell Laboratory to implement the same Orion instruction set but using different, well-proven, circuitry. In the event, West Gorton's Orion 1 version was first delivered to a customer in March 1963. The Bracknell version, known as Orion 2, was first delivered to a customer in September 1964. The two Orion versions were software-compatible but Orion 2 was on average about five times faster. It is believed that a total of about 19 Orion computers (both variants) were sold.

Throughout the 1950s, the Ferranti company had been active in many areas other than computers. It was still basically a huge family-run organisation with world-wide representation. By 1963 Ferranti's West Gorton and Bracknell sites were both involved in the final stages of two significant computer projects, Atlas and Orion, both consuming significant resources. The company management must have begun to wonder whether its deep involvement with large computers was altogether wise.

Table 13.1 Share of installed base of British computers, as in January 1963 and before the take-over by ICT of Ferranti's mainframe computer interests

Company	Value, £m	Percentage share
Ferranti	£6.620	24.5
ICT (including EMI)	£6.614	24.5
Elliott (including NCR)	£6.238	23.1
English Electric	£3.565	13.2
Leo	£1.740	6.4
AEI	£1.140	4.2
STC	£0.540	2.0
Others	£0.570	2.1
	£27.027	100.0

The data is based upon figures given in a contemporary document issued by the organisation *Computer Consultants*. The figures are quoted on page 77 of: *A History of the Ferranti Computer Department*. Bernard Swann. 98-page typed manuscript, produced in 1975 and circulated privately. Catalogue item NAHC/FER/C30 in the National Archive for the History of Computing

13.5 From Ferranti via ICT and ICL to Oblivion

Table 13.1 shows that, at the start of 1963, Ferranti's share of the UK's computer market seemed healthy. But behind the scenes three factors were becoming clear. Firstly, the Atlas experience was proof that Ferranti was struggling to compete in the area of large scientific computers. Secondly, Ferranti was still failing to take full advantage of the rapid growth of the business data-processing market. Thirdly opportunities were opening up in industrial automation and process control, for which small and robust computers were needed.

In 1963 the biggest changes were occurring in the speed at which computers were being applied to business and commerce. The electro-mechanical punched-card machinery market had collapsed between the autumn of 1961 and spring 1962 "with devastating suddenness".[26] One reason for the sudden collapse was the popularity of a new IBM product, the 1401, that had first appeared in Britain in 1960. During 1961–2, IBM installed nearly one hundred of its 1401 computers, which accounted for about a third of the computer sales in Britain during that period. Anticipating this American competition, International Computers and Tabulators (ICT) had been formed in 1959 from the amalgamation of the two UK punched-card equipment manufacturers, namely the British Tabulating Machine Co. (BTM) and Powers-Samas. ICT continued to sell BTM 1200 series computers up to 1963. In 1962 the company started to deliver the first of well over a hundred ICT 1301 computers, thereby out-selling both Ferranti and Elliott in this area.

[26]Campbell-Kelly, M. 1989. *ICL: A Business and Technical History*. Oxford: Oxford University Press. ISBN 0-19-853918-5.

ICT, through its two ancestor companies, had many years of experience in the provision of office machinery based on electro-mechanical punched-card equipment. Critically, ICT understood the day-to-day data-processing needs of the business community. By 1963, ICT needed to expand its computer manufacturing base whereas Ferranti was minded to pull out of the large computer area. Discussions took place, as a result of which ICT took over the West Gorton site and Ferranti's mainframe computer interests in 1963. The Atlas and Orion machines were still promoted, but under the ICT banner.

This still left Ferranti with the military and control computers such as Argus and some substantial computer-related R&D sites at Bracknell, Wythenshawe, Edinburgh, etc. Thus, Ferranti continued as a competitor to Elliotts in the military sector and in the process control market where the Ferranti Argus series was applied with increasing success.

In 1968, with government encouragement, ICT was merged with all the UK's remaining mainframe computer manufacturing companies to form ICL. The timeline of these historical amalgamations is shown in Fig. 13.10. ICL had some success with its 1900 series and 2900 series of computers. But by the 1980s ICL was suffering from a lack of in-house semiconductor technology that would have given it access to low-price chips. In October 1981 ICL reached an agreement with Fujitsu, whereby the Japanese company provided ICL with semiconductor chips and design tools. Eventually Fujitsu acquired ICL and in June 2001 Fujitsu decided to dispense with the name ICL.

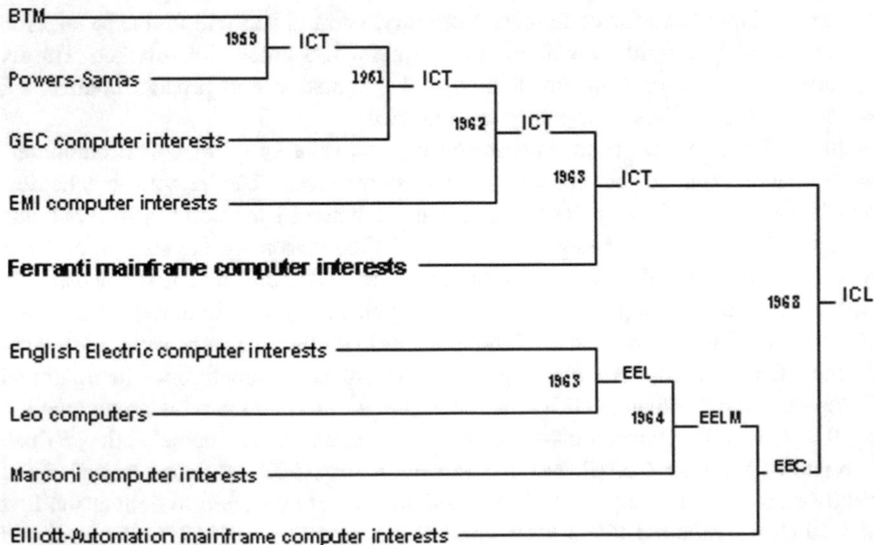

Fig. 13.10 The take-overs and mergers in the British computer industry between 1959 and 1968, showing how Ferranti's mainframe interests became amalgamated with ICT, which itself became ICL. This still left Ferranti with an independent interest in smaller computers for military and industrial control applications. Diagram adapted from (Campbell-Kelly, M., ICL: A Business and Technical History)

Meanwhile, Ferranti was successful in supplying small and medium-sized computers for defence projects and for industrial process control and automation. Then in 1987 Ferranti purchased International Signal and Control (ISC), a US defence contractor based in Pennsylvania. The company subsequently changed its name to Ferranti International plc. Unknown to Ferranti, it eventually transpired that ISC's business primarily consisted of illegal arms sales. The resulting financial and legal difficulties forced Ferranti into bankruptcy in December 1993. Thus ended the name of a pioneering supplier of British computers. It is small comfort to note that some former Ferranti divisions live on today under the umbrella of BAE Systems and Integrated System Technologies plc.

13.6 Any Star Dust Left Today?

For many people, computer history begins in the 1980s with the first Personal Computers. Is it therefore more that can be reasonably expected that any artefacts from 1950s computers have survived? The thread of preservation is tenuous. We should be surprised and grateful if anything but dust has survived from the days of the first Ferranti machines.

Summary data about the two Mark I and seven Mark I* computers is presented in Table 13.2, which gives the dates on which each of them was finally switched off. Some contemporary effort was made to preserve significant parts of three of the

Table 13.2 A reminder of the nine Ferranti Mark I and Mark I* computers whose life and times are covered in previous chapters

Ferranti's DC number and model	Customer, site	Also known as	Approx. delivery date	Approx. final switch-off
DC1 Mk I	Manchester University	–	Feb. 1951	Dec. 1958
DC2 Mk I	Toronto University	FERUT	May 1952	End of 1966
DC3 Mk I*	GCHQ, Cheltenham	CLEOPATRA	Sept. 1953	1959/1964
DC6 Mk I*	Shell Labs., Amsterdam	MIRACLE	June 1954	Dec. 1961
DC5 Mk I*	ARDE, Fort Halstead	AMOS	July 1954	Jan. 1967
DC4 Mk I*	INAC, Rome	FINAC	Jan. 1955	1966
DC7 Mk I*	AWRE, Aldermaston.	–	Early 1955	1959
DC8 Mk I*	A. V. Roe Ltd., Chadderton	–	Early 1956	1965
DC9 Mk I*	Armstrong Siddeley, Ansty	–	Oct. 1957	1964

machines in Table 13.2: the Manchester Mark I, FINAC and the Avro Mark I*. As far as can be determined, in only the Avro case has anything of significance survived to this day.

As described in Chap. 6, Avro donated the hardware of their Mark I* computer (but minus the power supply equipment) in 1966 to the new Museum of Technology at Leicester. In the event, the Museum felt unable to display the artefacts but put them in storage in a disused Victorian sewage pumping station. There the pieces languished in conditions that were far from ideal.

The Avro Mark I* artefacts were 're-discovered' by Jane Pugh of the London Science Museum in 1973, at a time when she was looking for exhibits for a planned Computing Gallery in London. In the summer of 1973 Jane Pugh and representatives from Leicester Museum, Ferranti Ltd., the Royal Museum of Scotland and Manchester University went to Leicester to examine the computer's remains. They had deteriorated to the point at which a satisfactory restoration of the unified assembly would have been very difficult. It was decided to divide up the artefacts and distribute them to various interested museums and collections. As the years passed, some of the Avro artefacts became further distributed and a few, it is feared, have vanished.

In principle it should today be possible to see units from the Avro computer at any of the following locations:

The Computer History Museum, Mountain View, California;
The Science Museum, South Kensington, London;
The National Museums Scotland, Edinburgh;

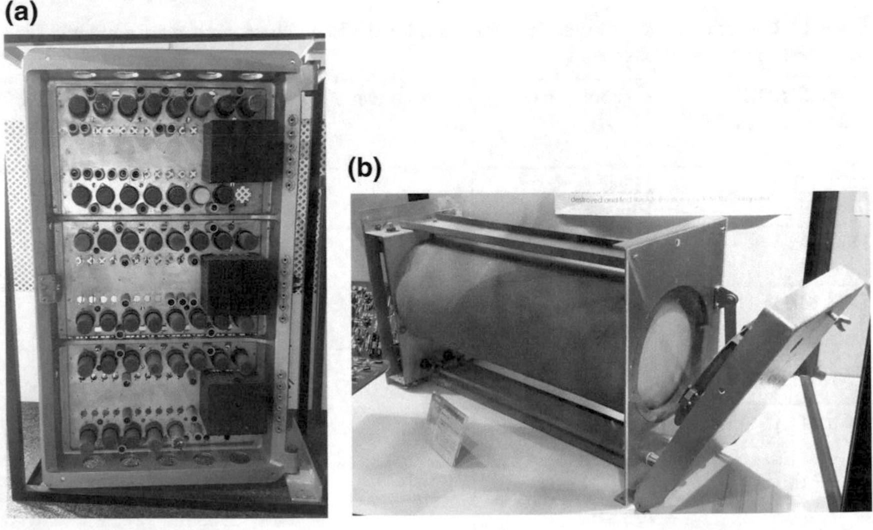

Fig. 13.11 **a** and **b** Two artefacts from the Avro Mark I* computer, as displayed at The National Museum of Computing, Bletchley Park. Photo **a** shows a logic door from the central processing unit. Photo **b** shows half of a twin Williams/Kilburn CRT storage unit. In both cases, the paintwork is not the original

Fig. 13.12 A working replica of the 1948 Small-Scale Experimental machine *(the Baby)* at the Science and Industry Museum in Manchester. This exhibit was created in 1998 and has been giving live demonstrations ever since. This photo, taken in 2018, shows some project volunteers and (centre) Chris Burton of the Computer Conservation Society, the project leader

The Museum of Science and Industry, Manchester;
The National Museum of Computing, Bletchley Park;
The School of Computer Science, University of Manchester;
The Leicester Museum of Science and Technology, Abbey Pumping Station.

In most cases, however, the artefacts are held in storage so permission to view has to be obtained in advance. At the time of writing, the easiest way to see two artefacts (a logic door and a Williams/Kilburn storage unit) on public display is to visit The National Museum of Computing, Bletchley Park. Fig. 13.11 shows photographs of what you might see.

A more vivid impression of the basic technology underlying the Ferranti Mark I and Mark I* machines is available at the Science and Industry Museum in Manchester. Here, a full-scale working replica of the 1948 Small-Scale Experimental Machine (the *Baby*) has been demonstrated to the public by an enthusiastic team of volunteers since 1998. A photograph of the exhibit is given above (Fig. 13.12). This image may be compared with the original Baby's control panel, shown in Chap. 1, Fig. 1.10.

At the time of writing (2018) there is no publicly-available code simulator for the Ferranti Mark I or Mark I* computers. However, a simulator for the Small-Scale Experimental Machine (Baby) is available here: https://www.davidsharp.com/baby/

This provides a flavour of the coding conventions for early Manchester and Ferranti computers.

David Link[27] has written a Ferranti Mark I emulator in connection with his installation based on Christopher Strachey's Love Letters program (see Chap. 2). He is currently using the emulator to produce a demonstration of Strachey's Draughts program and then, when this is complete, to release the emulator for public access.

The source code for several original Ferranti Mark I programs still exists. It is to be hoped that enthusiasts might get one or more of these to run on a modern simulator. A particularly interesting challenge would be to breathe life into the original source code for Tony Brooker's Mark I Autocode.

References

Brooker, R.A. 1955. An Attempt to Simplify Coding for the Manchester Electronic Computer. *British Journal of Applied Physics* 6: 307–311.

Campbell-Kelly, Martin. 1980. Programming the Mark I: Early Programming Activity at the University of Manchester. *Annals of the History of Computing* 2 (2): 130–168.

Campbell-Kelly, M. 1989. *ICL: A Business and Technical History*. Oxford: Oxford University Press. ISBN 0-19-853918-5.

Hinds, Mavis K. 1981. Computer Story. *The Meteorological Office Magazine* 110 (1304): 69–81.

Kilburn, T., D.B.G. Edwards, and G.E. Thomas. 1956. The Manchester University Mark II Digital Computing Machine. *Proceedings of IEE* 103 (Part B, Supp. 1–3): 247–268.

Laithwaite, E.R. 1950. *An Automatically Accessible Three-Dimensional Library for use with Digital Computing Machines*. M.Sc. thesis, University of Manchester, Faculty of Science.

Lavington, Simon. 2012a. *NRDC and The Case for a British Supercomputer, 1956–1960*, 13.

Lavington, Simon. 2012b. *Timeline of the MUSE/Atlas Project at Manchester University, 1955–1971*, 19.

Lavington, Simon., et al. 2016. *Tony Brooker and the Atlas Compiler*, 29.

Lonsdale, K., and E.T. Warberton. 1956. Mercury: A High-Speed Digital Computer. *Proceedings of IEE* 103 (Part B, Supp. 1–3): 483–490.

Pritchard, H.O., and J. Mol. 2001. Computational Chemistry in the 1950s. *Graphics Modelings* 19 (6): 623–627.

Sumner, F.H.(Frank). 1994. Memories of the Manchester Mark I. *Resurrection* 10: 9–13.

Thomas, G.E.(Tommy). 1998. *The Significance of Brooker's Autocodes in Taking the Early Manchester Machines into the Market (1954–1960).*

Williams, F.C. 1955. *Internal Report on the Computing Machine Running Costs*. Manchester: University of Manchester.

Williams, F.C. 1956. Introductory lecture: IEE Convention on digital computers. *Proceedings of IEE* 103 (Part B, Supp. 1–3): 3–9.

Williams, F.C., E.R. Laithwaite, and L.S. Piggott. 1957. Brushless Variable-Speed Induction Motors. *Proceedings of IEE* 104 (14): 102–118.

Zielinsky, Siegfried, and David Link. 2006. There Must Be an Angel: On the Beginnings of the Arithmetics of Rays. In *Variantology 2, On deep time relations of Arts, Sciences and Technologies*, 15–42. Published Cologne: Konig.

[27]Zielinsky, Siegfried, and David Link. 2006. There Must Be an Angel: On the Beginnings of the Arithmetics of Rays. In *Variantology 2, On deep time relations of Arts, Sciences and Technologies*, 15–42. Published Cologne: Konig.

Chapter 14
Appendix A. Baby's Conception: The Back Story

Since the whole of the Ferranti Mark I developments sprang from activity at Manchester University, it is interesting to examine more closely the factors that might have influenced the design of the 1948 Small-Scale Experimental Machine (the SSEM, or Baby computer). We start with some pre-history.

14.1 Computer History: The Wider Context

An underlying property of a stored-program computer is that it is *universal* rather than *problem-specific*. The general principles of universal computers may have been evident to the Cambridge mathematician Charles Babbage in the 1850s, as he wrestled with the design of his *Analytical Machine*. The concepts then lay dormant for many years until the 1930s when they were given a formal and novel mathematical description by Alan Turing, at that time a 23-year old Fellow of Kings College Cambridge. Turing's formalisation of a universal machine[1] was worked out in the context of mathematical logic, to help answer Hilbert's so-called *Entscheidungsproblem* which can be expressed as: *Is mathematics decidable?*

Turing's research supervisor at Cambridge, Max Newman, was impressed by the originality of his approach and took steps to promote Turing's academic career. Turing spent two years at the Institute of Advanced Study at Princeton where he obtained a Ph.D. on some implications of Gödel's Theorem. His Ph.D. supervisor at Princeton was John von Neumann. The relevance of all this to the non-mathematical world lay in the links formed between Max Newman, John von Neumann and Alan Turing—links that were to have some indirect bearing on the development of practical computers at Manchester in the late 1940s.

---

[1]Turing, A.M. 1937. On Computable Numbers, with an Application to the Entscheidungsproblem. *Proceedings of the London Mathematical Society,* Series 2 42 (1): 230–265.

S. Lavington, *Early Computing in Britain*, History of Computing,
https://doi.org/10.1007/978-3-030-15103-4_14

The notion of a universal machine only started to have tangible relevance in 1945, when the immediate post-war strategic and technological conditions in the USA and in the UK were ripe for it. Particularly, there was a need for fast, general-purpose computers. The challenge was taken up by engineers skilled in high-frequency electronics who had been released from war-time projects such as radar.

In America, stored-program computer activity had begun at the Moore School of Electrical Engineering, University of Pennsylvania, in 1945. There a huge special-purpose electronic calculator called ENIAC (Electronic Numerical Integrator and Computer) had recently been built. Even before ENIAC came into operation, the designers had been thinking about a more flexible successor, a true stored-program machine to be called EDVAC (Electronic Discrete Variable Automatic Computer). The ideas were written up by John von Neumann in June 1945 in a 101-page document entitled *First draft of a report on the EDVAC*. By 1946 copies of this report had been distributed widely on both sides of the Atlantic. It was to become the inspiration for several British computer projects.

In July/August 1946 Professor Max Newman sent David Rees, a 28-year-old Lecturer in the Mathematics Department at Manchester University, to America to attend an eight-week course of lectures at the Moore School on *The theory and techniques for design of electronic digital computers*. It is believed that Rees, who was perhaps looking for something more mathematical, was not inspired by the lectures. Rees' Manchester colleague Jack Good remembers that "I did not see Rees' informal report on the Moore School seminars. I recall being somewhat put out by this though Rees told me they would not interest me much".[2] In contrast, Maurice Wilkes from Cambridge University also attended the Moore School lectures and derived much useful material from the EDSAC information. It inspired Wilkes to initiate his own EDSAC computer project at Cambridge.

The EDVAC ideas were taken up by American academic institutions (for example at the Institute of Advanced Study at Princeton University and at the Massachusetts Institute of Technology), by government establishments (for example the National Bureau of Standards and the Ballistics Research Laboratory), and in due course by entrepreneurial companies (for example the Eckert-Mauchley Computer Corporation and Engineering Research Associates). In true American style, the projects were much more ambitious than those in the UK and the early commercial successes were financially more significant.

In the UK about eight research groups began designing and building prototype stored-program computer hardware. The locations, the names of the resulting computers and the dates when they first ran a program are listed.[3]

---

[2] *Early Notes on Electronic Computers*. I. J. Good. 78 typed and hand-written pages mostly covering the period 1947–8, with Good's retrospective introduction dated 23rd March 1972, and Good's covering letter to SHL dated 7th April 1976. Catalogue NAHC/MUC/2/A4.

[3] Lavington, Simon. 1980. *Early British Computers*. Manchester University Press. This is out of print but has helpfully been made available at: http://ed-thelen.org/comp-hist/EarlyBritish.html. The locations, the names of the resulting computers and the dates when the first British computers

First in the field was Alan Turing's Automatic Computing Engine (ACE) project at the National Physical Laboratory. The ACE computer was definitely shaped by Turing's own creativity. Furthermore, the ACE project was initially promoted by the Director of NPL as being the main (and only?) British National Computer.[4] For the other UK projects which were slower to get under way but quicker to mature, much inspiration was derived from ideas coming from America via John von Neumann's EDVAC Report and from von Neumann's subsequent project at the Institute of Advanced Study, Princeton.

At this stage, 1945/46, there was no cost-effective computer storage in the world that had been proved to work at electronic speeds. Devising suitable memory technology was the single most important problem faced by all implementers of universal computers.

We will see that computer *construction* activity at Manchester University was rather slow to get off the ground. However, enthusiasm for the possible *mathematical applications* of universal computers at Manchester dated from the autumn of 1945, when Max Newman left his war-time assignment at Bletchley Park and became the Fielden Professor of Mathematics at the University. He arranged that two of his Bletchley Park colleagues, I. J. (Jack) Good and David Rees, joined him. Good and Rees were both Cambridge graduates, Good having obtained his Ph.D. At Manchester they were appointed respectively as Lecturer and Assistant Lecturer in the Mathematics Department.

---

ran a program are: From 1946 onwards: at the National Physical Laboratory. (Pilot ACE, 1950); • From 1947 onwards: (i) at Cambridge University. (EDSAC, 1949). (ii) at Elliott's Borehamwood Research Labs. (Elliott 152, 1950). (iii) at TRE Malvern. (TREAC, 1953). (iv) at the Post Office's Dollis Hill Laboratory. (MOSAIC, 1953). (v) at Birkbeck College, London. (ARC/SEC/APE(X)C, 1948–52). • From 1948 onwards: at Manchester University. (SSEM 1948 & Mark I 1949). • From 1949 onwards, at the caterers J Lyons & Co., London. (LEO, 1951).

[4]*Proposed electronic calculator.* A. M. Turing. 48 typed pages. Undated but assumed written at the end of 1945. This Report was presented as paper E881 at a meeting of the NPL Executive Committee held on 19th March 1946. Turing attended the meeting to explain his proposals, which were accepted 'with enthusiasm'. The initial plans envisaged that ACE should be the Mathematics Division's highest-priority project. Subsequent internal difficulties with the implementation of the computer's storage and electronic circuitry led Sir Charles Darwin, the NPL Director, to seek help from external sources such as TRE, the Post Office's Research Laboratory and the Universities of Cambridge and Manchester, all to no avail even though the ACE was promoted by Darwin as being "the highest priority of any work that was being done for the Department of Scientific and Industrial Research". DSIR funded NPL and similar national establishments. See for example *Notes on a visit to the NPL to discuss the Automatic Computing Engine with the Director.* R. A. Smith (Superintendent, TRE Physics Division), TRE Internal memo D4070, 25th November 1946.

14.2 Max Newman's Plans for a Computing Machine Laboratory

At Manchester in the summer of 1946 the University established a Royal Society Calculating Machine Laboratory (RSCML), with a Senate Sub Committee to help the administration. This was in response to a Royal Society grant obtained by Max Newman. The award of £35,000 (equivalent to about £1 million today) was for salaries (£3000 over five years) and construction (£20,000) of a computing machine. Newman's intention was that his computer would be used for work in algebra and topology, probably as a way of exploring the role of computing on the development of Mathematics itself.

At that point there was no RSCML premises. The word 'calculating' in the title had been changed to 'computing' by 1948. In due course a physical Computing Machine Laboratory was built using the £20,000. It was equipped in 1951 with a Ferranti Mark I machine paid for by a Ministry of Supply grant, as described in Chap. 2.

It is interesting to construct a time-line of Professor Max Newman's 1945 plans for this Laboratory and to relate these to F. C. Williams' 1948 plans to build a practical computer at Manchester. The story is quite revealing, when set in the immediate post-war context. We start with the chronology shown in Table 14.1.

The thread in Table 14.1 starts in the autumn of 1945 when Max Newman set up a group of three mathematicians (himself, Jack Good and David Rees) at Manchester to investigate universal computers with the intention that, in due course, they could equip a Computing Machine Laboratory with a machine to be used for pure mathematical research. As Max Newman explained to John von Neumann in a letter dated February 1946[5]:

"By about eighteen months ago I had decided to try my hand at starting up a machine group when I got out [of war-time work at Bletchley Park]. It was indeed one of my reasons for coming to Manchester that the set-up here is favourable in several ways. This was before I knew anything of the American work, or of the scheme for a unit at the National Physical Laboratory. Later I heard of the various American machines, existing and projected, from Hartree and Flowers.… I am of course in close touch with Turing.… I don't expect to get started, even on paper, in less than a few months and then the first thing will be for the whole unit [Newman, Good and Rees] to try and learn something in detail about what has already been done. I am, of course, a complete beginner … what I should most like is to come out [to Princeton] and talk to you … I also hope to get hold of a good circuit man, though they are both rare and not procurable when found".

Looking at Table 14.1 we may deduce that electronic capability at Manchester had shrunk dramatically in September 1946 with the departure of Professor Willis Jackson and his immediate staff. We return to the story of Willis Jackson below in

[5]Letter dated 8th February 1946 from Max Newman to John von Neumann. This is reproduced here: http://www.alanturing.net/turing_archive/archive/m/m14/M14.php.

Table 14.1 Timeline of Max Newman's computing activity at Manchester, 1945–46

Date	Newman's Computing Machine Lab group, etc., at Manchester	Williams' storage group at TRE	Some events elsewhere
June 1945	Max Newman is released from war work at Bletchley Park		von Neumann writes *First draft of a report on the EDVAC*
29th Sept. 45	Max Newman becomes Prof of Maths at Manchester. I. J. Good and D. Rees appointed to Maths Dept. "to help with the development of an electronic computer"		A. M. Turing joins NPL
7th Mar. 46	Newman applies for Royal Society grant to set up a Computing Machine Laboratory		
19th Mar. 46			Turing's ACE Report goes to NPL Exec Committee
May 46			Turing is joined at NPL by mathematicians Wilkinson and Woodger
16th May 46	Newman awarded RS grant		
25 May–7th July 46		Williams in USA; sees attempts at digital storage	
June 46			First Report on IAS computer at Princeton. See[a]
Summer 46	Newman and Good spend a week at NPL, talking to Turing about ACE		
July–Aug. 46	Rees attends Moore School lectures Tells Good that "they would not interest you much"		Moore School lectures. Wilkes also attends
7th Aug. 46		Williams proposes research on storage	
Sept. 46	Prof Willis Jackson and his research group leave the Department of Electro-Technics, which is now without any electronics activity	Williams and Kilburn involved in storage research at TRE	Wilkes starts the Cambridge EDSAC project, inspired by the EDVAC Report

(continued)

Table 14.1 (continued)

Date	Newman's Computing Machine Lab group, etc., at Manchester	Williams' storage group at TRE	Some events elsewhere
Oct.–Dec. 46	Newman visits Princeton. Learns details of the IAS computer project		
Oct. 46		NPL sends Williams a copy of Turing's ACE Report	
22nd Nov. 46		Williams and Uttley visit NPL; Turing explains his ACE design	
Dec. 46		Kilburn attends ACE lectures at the Adelphi, London	
11th Dec. 46		Williams files first storage patent	
Dec. 46 or Jan. 47	Newman gives "two or three lectures" on the IAS computer to Mathematics staff at Manchester		
25th Dec. 46		Williams becomes Prof of Electro-Technics	

[a]*First IAS Report.* 28th June 1946: "Prepared in accordance with the terms of Contract W-36-034-ORD-7481 between the Research and Development Service, Ordnance Department, US Army and the Institute for Advanced Study". A Second edition is dated 2nd September 1947. See: http://grch.com.ar/docs/p1/Apuntes/eng/Logical%20Design%20of%20an%20Electronic%20Computing%20Instrument.pdf

Sect. 1.3. Manchester had not recruited what Newman would call 'a good circuit man' until the arrival of F. C. Williams at the end of 1946. F. C. Williams was one of three people interviewed for the Chair of Electro-Technics. There is some evidence[6] that the 35-year-old Williams, whose last academic appointment had been from 1936–39 as an Assistant Lecturer, was not the first preference of the Appointing Committee. Williams, who was called *FC* by his colleagues and *Freddie* by close personal friends, turned out to be an excellent choice.

Little hard evidence has survived to show how Max Newman's computing plans may have matured in the year following F. C. Williams' arrival but in Table 14.2 we give some pointers. It is frustrating that, today, no explicit documentary

[6]Conversation with Professor D. B. G. Edwards, 17th April 2012. Professor Edwards explained that, in pursuit of his own research into early computing activity, he had been allowed access to confidential files in the Registrar's Department, University of Manchester.

Fig. 14.1 M. H. A. (Max) Newman, at about the time when he became the Fielden Professor of Mathematics at the University of Manchester in 1945

evidence has been discovered that reveals Newman's own thoughts on computing in the period January 1947–January 1948. This might be because Newman was busy with his teaching and research responsibilities as Head of the Mathematics Department at Manchester. He was, after all, a 50-year old with well-established on-going research into combinatorial topology. The inference is that Newman handed the pursuit of ideas for computer design to his 36-year old colleague Irving John (Jack) Good. It was Jack Good who seems to have first coined the name *Baby Machine* in the summer of 1947, as discussed below in Sect. 14.6. There is, frustratingly, little surviving evidence to show that Jack Good ever had meaningful discussions with either F. C. Williams or Tom Kilburn before December 1947. Perhaps this was because, prior to December 1947, the research preoccupations of both Williams and Kilburn were focussed on oscilloscopes and soldering irons—in other words, the practical realisation of their novel ideas for digital memory technology. More on Jack Good's ideas is given in Sect. 14.6.

Table 14.2 Timeline of computing events at Manchester, January 1947–June 1948

Date	Newman's Computing Machine Lab group at Manchester	Williams' digital storage group at Manchester, etc.	Some events elsewhere
7th–10th Jan. 47			Harvard symposium. Turing attends
Feb. 47	Good issues an internal memo listing symbols and abbreviations, based on the IAS computer		H. D. Huskey (SEAC, USA) starts his year at NPL
3rd Mar. 47	Husky's first Manchester visit: Newman "plans to copy Princeton's CPU and get Princeton to make their [Manchester's] magnetic wire I/O"	Huskey's first visit: Williams has "2 lines of CRT storage at prf of 125Kc/s, using the anticipation-pulse scheme"	
1st Apr. 47		Williams writes to A. M. Uttley (TRE): "our longest memory period so far is 4½ h …"	
Apr. 47		P. W. W. Woodward and A. M. Uttley give lectures at TRE on Computers and Computing	
9th Apr. 47		Two copies of the TRE course notes sent to Williams	
1st May 47		Formal start of Uttley's own computer project at TRE	
4th May 47	Good issues 8-page internal memo entitled *The Baby Machine*. Gives an instruction set based loosely on the Princeton IAS proposal		
10th June 47	Good gives an outline register-level block diagram of a Baby Machine		
30th Sept. 47			Turing leaves NPL. Ted Newman and Clayden start work at NPL on electronics for a Pilot ACE

(continued)

Table 14.2 (continued)

Date	Newman's Computing Machine Lab group at Manchester	Williams' digital storage group at Manchester, etc.	Some events elsewhere
Autumn 47		Demonstration of storing 2048 digits 'for a matter of hours'	
1st Dec. 47		Kilburn's Report: "*A storage system for use with binary digital computing machines*"	
Dec. 47	Good writes notes on programming issues and instruction-set design. Rees, Newman, Wilkinson and Harry Huskey contribute thoughts	10th Dec. Huskey's second visit. Evidence that Williams and Kilburn discuss ideas with Newman, Good, etc.	
4th Mar. 48	Newman gives a Royal Society talk on automatic digital computers	Williams gives a Royal Society talk on *A cathode ray tube digit store*	Wilkes gives a Royal Society talk on EDSAC
Apr. 48	I. J. Good leaves Manchester. D Rees leaves at the end of the summer 1948		NPL makes Electronics Section a separate unit. Pilot ACE moves forward
4th June 48		Tootill draws his block diagram of "first experimental machine"	
21st June 48		The SSEM (Baby) runs its first program	

Much of the evidence for the left-hand column of Table 14.2 comes from Jack Good's surviving papers.[7] These indicate that, although Newman's computing ambitions at Manchester may at first have been inclined towards Alan Turing's ACE design,[8] he switched towards John von Neumann's IAS Princeton design after Newmans' visit to Princeton in the autumn of 1946. The internal (i.e. register-level) architecture of the ACE and IAS computers were radically different. When the American engineer Harry Huskey visited Manchester on 3rd March 1947 he met

[7] *Early Notes on Electronic Computers.* I. J. Good. 78 typed and hand-written pages mostly covering the period 1947–8, with Good's retrospective introduction dated 23rd March 1972, and Good's covering letter to SHL dated 7th April 1976. Catalogue NAHC/MUC/2/A4.

[8] Turing's ACE explanation, assumed to have been given to Newman and Good at their meeting with Turing at NPL in the summer of 1946, is held at the Newman Archive at St. Johns College, Cambridge: http://www.cdpa.co.uk/Newman/MHAN/ box 5 folder 6 item 6, *Outline account of computing machine ACE.* (16 typed pages).

Fig. 14.2 I. J. (Jack) Good, who was a Lecturer in the Mathematics Department at Manchester University from September 1945 until April 1948, after which he returned to GCHQ

both Williams and Newman.[9] It was quite clear to Huskey that Newman proposed to follow the IAS computer, though Huskey hints that any machine built at Manchester might use F. C. Williams' CRT storage system rather than the RCA Selectron storage tubes intended for the IAS machine. (The IAS machine finally used Williams/Kilburn tubes after the Selectron failed to live up to expectations). Significantly, Newman is reported as "intending to have Princeton fabricate their input and output equipment" (consisting of a magnetic wire recording system).

When interviewed in 1976,[10] the 80-year-old Newman was understandably but frustratingly vague about the progress of his computing machine project. After dealing with Newman's wartime work at Bletchley Park where he had led the Colossus project, the interviewer Dr Chris Evans then asks: "And then did you concern yourself with, um, computing machinery after that?"

Newman responds: "I did, yes, I then, it was at that time that I was appointed Professor of Mathematics at Manchester. And they, er, then Williams came along there and, er, started up one of the very first universal, general-purpose computing machine projects well, in the world, really, and it was a very small machine to start with – it had only a thousand digits storage in all in the first machine and I was associated with it at that stage but it soon, it was, they were concerned with actually

[9]Huskey, H.D. 1947. *The State of Computing in Britain and the US*. 14 typed pages plus 3 pages of correspondence. See: http://www.alanturing.net/turing_archive/archive/l/l01/l01.php.

[10]Interview of M. H. A. Newman, conducted by Dr. Chris Evans (NPL) in 1976. This is interview 15 in a series of interviews of computer pioneers sponsored by the Science Museum.

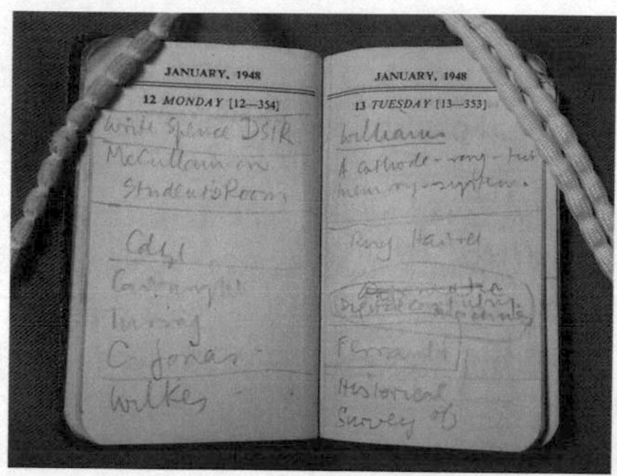

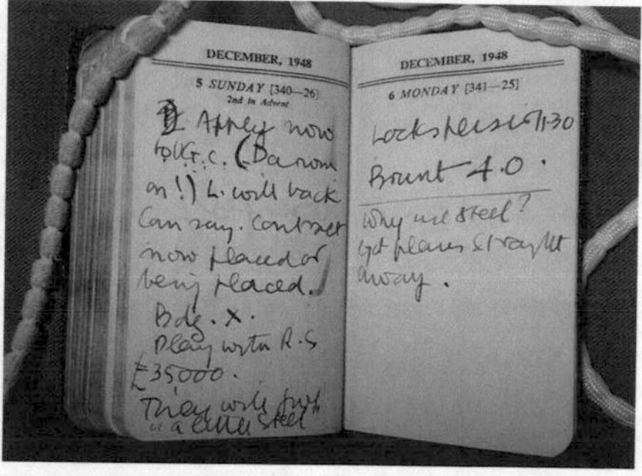

Fig. 14.3 Pages from Max Newman's pocket diary for 1948. *(left)*: January 12th and 13th; *(right)*: December 5th and 6th. Names mentioned on the left include Turing, Wilkes, Williams and Hartree. Names mentioned on the right are Darwin, Lockspeiser and Brunt

designing the machine, not the use of it, and it was quite clear to me that this was a job for an engineer and so I, I went, left it, and went back to mathematics".

Max Newman's small pocket diary for 1948 still survives.[11] Being used by Newman simply for appointments and occasional cryptic notes, the diary offers only brief glimpses of Newman's contacts with people and events connected with computer development. During January and February 1948 there are diary mentions of Williams, Hartree, Turing, Wilkes and Booth, in the period when Newman was planning the line-up of speakers for the March 1948 Royal Society Symposium—see for example Fig. 14.3. Then during mid-March to the end of April in the aftermath of the Royal Society Symposium, Newman appears to be involved in discussions with Blackett and Williams on the future of the Computing Machine Laboratory and the possibility that the CRT store might give Manchester a lead. One may deduce that, at this point, there is the chance that Turing might be enticed to Manchester now that Good had left and David Rees was not showing any interest.

In July 1948 there are cryptic diary entries which appear to indicate liaison with Williams about an announcement of the June 1948 computer's successful operation and a letter to Nature. Then there appears to be a long diary gap (for summer vacation, followed by Newman's ill health?) until relevant entries resume in early December. One can deduce that during the autumn the Williams/Kilburn computer activities had proceeded much faster than Newman had thought possible. There were hasty decisions to be made about financing a new building to prepare for the arrival at the University of a Ferranti production machine. The entries for December 5th and 6th 1948 reveal evidence of the plans being made for a new (temporary) building, now that Ferranti had just been given the go-ahead—see Fig. 14.3.

[11]Appointments Diary for 1948, item 4-13-1 in the Newman Archive at St. Johns College Library, Cambridge. See the catalogue here: http://www.cdpa.co.uk/Newman/MHAN/view-box.php?Box=4. The diary is a Letts Pocket Diary, measuring about 4" × 3". In the front and in the back are lists of the staff of the Mathematics Department at Manchester, for academic years 1947/48 and 1948/49.

The building, which necessitated special permits for the steel used in its construction, was to be financed from Newman's Royal Society grant. The support of the University Grants Committee (UGC) was to be sought for on-going Computing Machine Laboratory funding.

After the resolution of these December 1948 issues, which are more fully discussed in Chap. 2, Newman seems to have become detached from meaningful involvement with computing at Manchester University.

Returning to 1947, notice in Table 14.2 that in May of that year Jack Good produced an 8-page internal memo entitled *The Baby Machine,* which contained an instruction set based loosely on the Princeton IAS proposal. This is discussed in Sect. 14.6. A technical description of instruction sets and their idiosyncratic evolution at Manchester after June 1948 is presented in Chap. 15. It is not clear whether Kilburn ever saw Jack Good's May 1947 document. Questioned about this in 1975, Kilburn could not recall ever having seen it.[12] Certainly, the instruction set finally chosen by Kilburn for the June 1948 computer was different from the one proposed by Good and at Manchester the now-familiar name *Baby* was not associated with the June 1948 computer until many years later. When the June 1948 design was formally described in print[13] it was called the *Small-scale Experimental Machine*, abbreviated as SSEM.

We now give the necessary technical background to Table 14.2. The intention is to explore the influences upon computing activity at Manchester from four sources during the period 1946–1948:

Willis Jackson and redundant Bletchley Park equipment;
NPL and Alan Turing's ACE computer;
Albert Uttley and computer design at TRE;
Jack Good's interpretation of John von Neumann's Princeton computer.

The order of presentation reflects the time-line of Table 14.2. Readers with little appetite for technical details can safely skip to the conclusions in Sect. 14.7.

14.3 Willis Jackson and Max Newman's BP Hardware

In his February 1946 letter to John von Neumann, Max Newman had said: "It was indeed one of my reasons for coming to Manchester that the set-up here is favourable in several ways". Why was Manchester University's set-up 'favourable'

---

[12]SHL had many conversations with Tom Kilburn in the period leading up to the publication in 1975 of the booklet *A History of Manchester Computers,* published by the National Computing Centre. (Second edition published in 1998 by the BCS, ISBN: 0-902505-01-8). F. C. Williams and M. H. A. Newman were also interviewed by SHL during this period. All three pioneers checked and approved the text of the booklet.

[13]*Universal high-speed digital computers: a small-scale experimental machine.* Proc. IEE, Vol. 98, Part 2, No. 61 Feb. 1951, pp. 13–28.

to the type of computing machine that interested Newman? Possibly because Manchester was already home to three distinguished people who were well-disposed towards the construction of such a machine: Douglas Hartree FRS (Professor of Applied Mathematics since 1929), Patrick Blackett FRS (Professor of Experimental Physics since 1937) and Willis Jackson (Professor of Electro-technics since 1938).

Hartree and Blackett had national influence: they were certainly well-regarded in Ministry of Supply circles. Jackson's more modest influence lay within the Department of Scientific and Industrial Research but he also had good links with the Manchester engineering company Metropolitan-Vickers, who had built Hartree's Differential Analyser (an analogue computer used primarily for solving differential equations) in 1935. It is probable that both Hartree and Blackett were aware of the importance of Newman's war-time work as leader of the Colossus group at Bletchley Park. There was a close personal relationship between the three: Blackett was Best Man at Douglas Hartree's wedding in 1923 and at Max Newman's wedding in 1934.

Back in August 1945 Prof Willis Jackson was asked to advise Newman about what redundant Bletchley Park electrical equipment would be useful for computing machinery, if the equipment were to be transferred to Manchester. In a memo to the relevant authorities dated 8th August 1945, Newman wrote[14]:

> After going round the equipment with me, Professor Jackson thinks the proper request for us to make is for the material of two complete Colossi; and in addition a few thousand resistances and condensers off other machines (these I understand from Maile are useless to the PO and to commercial firms).
>
> We should like the counter racks and the 'bedsteads' (tape racks) to be in working order but the rest could be dismantled so far as is necessary to make the circuits unrecognisable.
>
> If a punch and reader (creed or Teletype) were available it would of course be most valuable.
>
> I shall be able to let you know before long when we shall be able to take the stuff away.

It is not quite clear how much equipment from the above ambitious shopping list finally arrived in Manchester, and when. From a November 1945[15] letter it seems that the Ministry of War Transport was being asked to take "three large items and some small stuff" to Professor Newman at Manchester later that year. The dimensions of the large items, of total weight 7 ton, were given as:

4 ft × 2 ft 6 in. × 7ft.
2 ft × 2 ft 6 in. × 7 ft.
4 ft × 2 ft 6 in. × 7 ft.

[14]Internal memo dated 8th August 1945 from M. H. A. Newman to DD(A) [probably an administrative section within Bletchley Park].

[15]Internal memo dated 8th November 1945 from M. H. A. Newman to Mr. Reiss [probably an administrator within Bletchley Park].

What was the exact nature of the components that reached Manchester? Is the '7 ton' an accurate estimate of the weight? It is worth pointing out that much of the equipment in use at Bletchley Park during the war was electro-mechanical (relays, rotary contactors, etc.). Colossus, however, was electronic. The Mark II versions contained up to 2400 thermionic valves (tubes) per machine as well as relays, plug-boards (jack-fields), switches, etc. Westinghouse *Westat* 50 V power supplies and passive components (resistors, capacitors, etc.) completed the component list. The whole Colossus assembly was mounted on 7 ft high steel racks, some of which were the standard Post Office variety (19 in. wide) and some of which were wider. The valve-types used in Colossus were:

EF36: audio-frequency pentode; octal base with top-cap grid connection;
GT1C: Thyratron (gas-filled triode); 5-pin base;
807: medium-power beam tetrode; 5-pin base with top-cap anode;
L63: AF triode amplifier; octal base.

It is worth noting that the Manchester University computers (Baby, Mark I) constructed in the period 1948–1949 by Williams and Kilburn made no use of any of the above valve types. In addition to the Cathode Ray Tubes which formed the main memory and central registers, Williams and Kilburn employed the well-proven valves in common use at TRE, particularly:

EF50 (VR91): RF pentode; 9-pin base connection
EA50 (VR92): miniature signal diode.

Noting the difference between AF (audio frequency) and RF (radio frequency), it is unlikely that any of the Colossus valves would have been of any interest to Williams and Kilburn, even if TRE had not been supplying components to Manchester from 1947 onwards free of charge. There remains the possibility that re-cycled Post Office 19" racking (i.e. bare metal) and the commodity Westinghouse *Westat* 50 V power supplies could have had a use in the early Manchester computers —but such hardware could also have been supplied by TRE. Indeed, in a letter from Newman to the Royal Society dated 22nd December 1948[16] Newman states: "All materials needed for construction work on the prototype [the SSEM or Baby computer] have been given without cost by the Ministry of Supply (TRE), and it is reasonable to suppose they will continue to give all necessary electrical material".

The fate of the (seven ton of?) Bletchley Park equipment that arrived in Manchester at the end of 1945 has not yet been discovered. In the autumn of 1946 Willis Jackson and his team quit Manchester, leaving the Electro-Technics Department severely depleted and especially so in the area of electronics. Albert Cooper, the Department's Chief Technician, was reported to be "disgusted" with Jackson's move because it seemed to him that "the whole of the Department was

[16]Letter dated 22nd December 1948 from M. H. A. Newman to Professor D. Brunt at the Royal Society. See NAHC/MUC/2/C/2.

disappearing into Jackson's van".[17] It is not obvious that Jackson's departure was in fact a loss to Newman because Jackson, then or subsequently, seems to have had no particular interest in constructing computers. Jackson's personal research interest was in the dielectric properties of materials. Nevertheless, Jackson may have taken some of the ex-Bletchley Park electronic equipment with him to Imperial College.

14.4 Alan Turing and the ACE Connection

In 1972 Jack Good recalled that: "In the summer of 1946 Newman and I visited NPL, where Turing was designing the logical control of the ACE, and Turing explained it to us in detail. We stayed for a week".[18] Looking at Jack Good's surviving notes, which mostly date from 1947 and 1948, there is no mention of ACE. As shown in Sect. 14.6, from 1947 Newman's group was following the Princeton star and they had abandoned any thought of following the ACE project.

From the centre column of Table 14.1 it is clear that both Williams and Kilburn were aware of Alan Turing's ACE design. Indeed, the hypothetical computer included as an illustrative system in Tom Kilburn's December 1947 storage report is based on the ACE's structure—that is to say, it has a number of distributed *sources* and *destinations* but no central accumulator.[19]

The inclusion of the ACE architecture in Kilburn's December 1947 Report is curious. If anything, it hints that Kilburn was unfamiliar with the details of the Princeton hardware or with Jack Good's interpretations of the Princeton register-level structure. Instead, perhaps Kilburn had been relying on notes he had taken at the December 1946 ACE lectures at the Adelphi in London.

We know from contemporary correspondence[20] that Williams and Kilburn were not temperamentally inclined to follow Alan Turing's ideas. Indeed, when Williams

---

[17]Broadbent, T.E. 1998. *Electrical Engineering at Manchester University; The Story of 125 years of Achievement*. Published by The Manchester School of Engineering, University of Manchester. ISBN 0–9531203-0-9.

[18]*Early Notes on Electronic Computers*. I. J. Good. 78 typed and hand-written pages mostly covering the period 1947–8, with Good's retrospective introduction dated 23rd March 1972, and Good's covering letter to SHL dated 7th April 1976. Catalogue NAHC/MUC/2/A4.

[19]A good analysis of the unique nature of the ACE design is given in: Carpenter, B.E., and R.W. Doran. 1977. The Other Turing Machine. *The Computer Journal* 20 (3): 269–279. A diagram of Kilburn's hypothetical computer, based on the general ACE structure, is in: Kilburn, T. 1947. *A Storage System for use with Binary Digital Computing Machines*. Typed foolscap document, consisting of 52 pages of text, 32 pages of diagrams and one page with three photos. A paper copy is held in Manchester—see NAHC/MUC/2/A1. For an electronic copy plus comments, see: http://curation.cs.manchester.ac.uk/computer50/www.computer50.org/kgill/mark1/report1947.html.

[20]In 1947 NPL sent a draft contract between NPL and the University of Manchester to F. C. Williams. The purpose was to get Williams' group to help build NPL's ACE computer. Williams declined the contract because: (a) he did not need additional funding for his own research; (b) the wording of the NPL contract appeared restrictive compared with normal academic practice; (c) Turing's ACE design was incompatible with Williams' own storage project. See:

was contemplating submitting the first Manchester storage paper to the Proceedings of the Institution of Electrical Engineers, he had doubts about including anything on Kilburn's hypothetical computer. TRE's official clearance was required for journal publication and, in a letter to the TRE Director[21] written in February 1948, Williams says: "It had occurred to me that the sections on the hypothetical machine tended rather to cloud the issue, but I was in some doubt whether our objective in making a digit store, and the requirements for such a store, would be clear without some reference to a hypothetical machine …. The problem of splitting the report into two papers is rather a serious one because we just do not know enough about the hypothetical machine to be able to write a paper about it alone".

From February 1948 onwards, there is no detectable influence of Turing's ACE architecture in any of the Manchester computer designs. The spotlight had turned to architectures based on a central accumulator register, as exemplified by the IAS computer project at Princeton.[22] This is confirmed by a retrospective historical summary given in October 1948,[23] where Max Newman says that "the most suitable type [of computer] seemed to be that which was being made at Princeton under John von Neumann's direction".

NAHC/MUC/1/B1a. Furthermore, Turing and Williams had clashed during the meeting at NPL with TRE staff in November 1946. A. M. Uttley, who was at this meeting, recalled in conversation with SHL that there was no love lost and that "sparks flew between Turing and FC".

[21]Letter dated 18th February 1948 from F. C. Williams to the TRE Director R. A. Smith. See: NAHC/MUC/1/B1a.

[22]*First IAS Report.* 28th June 1946: "Prepared in accordance with the terms of Contract W-36-034-ORD-7481 between the Research and Development Service, Ordnance Department, US Army and the Institute for Advanced Study". A Second edition is dated 2nd September 1947. See: http://grch.com.ar/docs/p1/Apuntes/eng/Logical%20Design%20of%20an%20Electronic%20Computing%20Instrument.pdf.

[23]*A status report on the Royal Society Computing Machine Laboratory.* M. H. A. Newman. A typed report prepared for the University of Manchester Senate Sub-Committee on the Royal Society Computing Machine Laboratory, 15th October 1948. See NAHC/MUC/2/C/2. Those present at this meeting included Newman, Blackett, Williams and "Mr. Turing by invitation". This report includes some interesting points: (a) Confirms that Newman's original plan was to build a machine based on the Princeton model, buying in a number of RCA Selectron tubes for the memory; (b) Confirms Kilburn as the designer of 'the first automatic general-purpose computing machine to have actually worked'; (c) States that Kilburn is to be appointed as a staff member of the Computing Machine Laboratory from 30th November 1948 to 29th September 1951, with the status of University Lecturer. (d) Confirms that Princeton intends to switch to Williams Tubes; (e) States that a 3000 ft^2 temporary building should be erected near the existing Physics Building [in Coupland Street]. This would contain the machine (1200 ft^2), workshop, staff rooms and would operate for three years until a much larger building could be constructed on the other side of Oxford Road [actually in Dover Street, which went eastwards from Oxford Road whereas Coupland Street went westwards], hopefully to be ready by 1953. More on the new Dover Street building is given in Chap. 13.

14.5 Albert Uttley and the TRE Connection

Events at the Telecommunications Research Establishment (TRE) had no influence on Max Newman, Jack Good or David Rees at Manchester. In contrast, F. C. Williams had naturally remained in close contact with TRE and, in particular, with A. M. (Albert) Uttley who was to lead TRE's own computer design project. In the spring of 1947 Uttley organised a TRE course of lectures on digital computers and sent Williams two copies of the hand-outs.[24] It is reasonable to believe that Uttley's hand-outs represent, for Williams and Kilburn, the first down-to-earth written description of a stored-program computer that they had encountered. Here are some extracts from Albert Uttley's lecture hand-outs:

"To make a start, we will sketch some of the probable requirements of an electronic digital computer, drawing largely on the ideas of Professor Von Neumann who is engaged on the design of such machines in America".

"*1. Memory*. It will be necessary to store thousands of numbers and orders; the store may be likened to a town containing numbered streets, each street containing numbered houses, each house containing either a number or an order".

Uttley goes on to define by means of a simple diagram an address x as "a number containing complete information of street number and house number". He then introduces a list of orders, or instructions, thus:

"*2. The code for orders*. Von Neumann considers that any order can be expressed by means of 20 binary digits. The first 12 of which state the address of the number that has to be operated on, and the last 8 of which specify the type of operation.... Remembering from lecture 1 the need of a Register R and an Accumulator A, together with the idea of a house, x, then Von Neumann believes that the minimum requirements are as follows".

Uttley goes on to list a repertoire of instructions for transfer of data, arithmetic operations and control transfers. Here are examples of Uttley's notation and his explanations:

x → A load an operand from memory into the accumulator.
A → x store the contents of the accumulator to memory.
A + x → A add an operand into the quantity in the accumulator.

More examples of candidate instruction sets are given in Sect. 14.6.

To put Uttley's notes in context, here is an extract from the covering letter he wrote to Williams on 9th April 1947[25]:

---

[24]*Lectures I and II in Electronic Digital Computer*, TRE Lecture Series, typescripts, pp. 6 (1947). See NAHC/MSC/D3. See also NAHC/MUC/1/B1a, for correspondence between Williams and Uttley and others at TRE and between TRE and NPL during 1947–48.

[25]From the correspondence in NAHC/MUC/1/B1a it is clear that F. C. Williams and A. M. Uttley were in regular communication in 1947. For example there is a letter from FC to Uttley dated 1st April 1947 when FC tells Uttley of "our longest memory period so far is 4½ h during which 10^9 regenerations were formed without error … this is quite the most promising result to date". Uttley

"Macfarlane and Woodward [Uttley's colleagues at TRE] have started work on mathematical requirements, and I now have a rough mental picture of the whole device. I have been giving a series of talks to familiarise people with digital computers; it has been well attended, I enclose a couple of copies. Of course it is just a 'Tiny Tots' affair, but I thought you and Kilburn might like a copy.

"I intend that we build a 16 CRT Computer with our own fair hands, and let the memory grow as rapidly as Industry can implement a contract. I presume that both you and I will want at least 256 storage units".

The formal project start-date for what became the TREAC computer was 1st May 1947.[26] TREAC was the UK's first *parallel* (as opposed to *serial*) computer.[27] Its gestation period was rather long—TREAC did not run its first program until 1953—and its 'zero/one/error' style of circuitry made it quite different from the Manchester computer designs even though TREAC employed Williams/Kilburn electrostatic CRT storage for its primary memory. In short, by the start of 1948 Williams and Kilburn were ploughing a different furrow from TRE. Nevertheless, it may have been Uttley at TRE who first presented them with an introduction to basic computing terminology in the Princeton style.

14.6 Jack Good and the Princeton Connection

Following Max Newman's visit to John von Neuman's group at the Institute for Advanced Study at Princeton, Jack Good seems to have been a possible channel through which the Princeton's computer ideas may have reached Williams and Kilburn during the period from May to December 1947. The evidence, which is sparse, is contained in the collection of typed and hand-written notes made by Jack Good,[28] which are accompanied by his retrospective comments of 1972.

Good's evolving thoughts on computer design began in 1946, though the surviving written evidence does not start until February 1947. Max Newman visited Princeton in the autumn of 1946, during a three-month sabbatical. The state of

---

replied to this letter on 9th April 1947, saying that "Electronic Computing is now on the DCD programme as an MOS project … We start Research on Digital Computing on May 1st ".

[26]From the correspondence in NAHC/MUC/1/B1a it is clear that F. C. Williams and A. M. Uttley were in regular communication in 1947. For example there is a letter from FC to Uttley dated 1st April 1947 when FC tells Uttley of "our longest memory period so far is 4½ h during which 10^9 regenerations were formed without error … this is quite the most promising result to date". Uttley replied to this letter on 9th April 1947, saying that "Electronic Computing is now on the DCD programme as an MOS project … We start Research on Digital Computing on May 1st".

[27]Carter, R.H.A., and A.M. Uttley. 1953. The Telecommunications Research Establishment Parallel Electronic Digital Computer, Chap. 10. In *Faster than Thought,* ed. B.V. Bowden. Published by Pitman.

[28]*Early Notes on Electronic Computers.* I. J. Good. 78 typed and hand-written pages mostly covering the period 1947–8, with Good's retrospective introduction dated 23rd March 1972, and Good's covering letter to SHL dated 7th April 1976. Catalogue NAHC/MUC/2/A4.

Princeton's Institute for Advanced study (IAS) computer at this time may be deduced.[29] Good remembers that "on Newman's return, he gave two or three lectures on the Princeton plans, together with some ideas of his own concerning notations for programming". Good's contemporary notes and his retrospective comments show that Good spent the period February to May 1947 "proposing some notional conventions for programming" in the Princeton style.

Of interest is Good's evolving ideas on the number of, and purpose for, central registers. This includes a brief mention of D, a Dubbing Table. In 1972 Good explained that "The term 'dubbing' refers to the changing of address by means of an additive (when using relative addresses). The term was probably invented by Newman or von Neumann". No reference to a dubbing register has so far come to light in any other contemporary documents. The de-dubbing process is briefly mentioned by Good on page 5 of his 1947 notes, but is then dropped. Following terminology introduced in the late 1960s,[30] Jack Good's Dubbing register might possibly now be called a *Name Base* register.

If anything, the brief appearance of a D register in Good's notes simply indicates that Good was juggling with several notions at the conceptual level.

In a retrospective comment, Good explains that by May 1947 "Kilburn was still working on the main memory unit, the Williams Tube, and he had asked for suggestions for a small number of basic instructions (or operations)". Good responded with an eight-page typed report dated 4th May 1947 and headed *The Baby Machine*. The register-level architecture envisaged by Good at this stage had the following features:

(a) A 64-word memory, 32 words being reserved for instructions and 32 for numbers;
(b) A proposed instruction set of 12 orders.

No explanation is given of the symbols used by Good and there are some inconsistencies with his earlier notations. However, once one realises that the computer described by Good in May 1947 has a one-address instruction format and 2's complement number representation (as, for example, had von Neumann's IAS computer at Princeton), the meaning is clear.

Seven of the pages in Jack Good's 8-page *Baby* Report describe the solution of five specimen problems. His follow-up notes of 10th June 1947 seem to represent his last formal ideas on computer architecture before he left Manchester University in April 1948 to re-join GCHQ. Good's contribution to GCHQ's involvement with stored-program computers in the early 1950s is described in Chap. 10.

---

[29]*First IAS Report.* 28th June 1946: "Prepared in accordance with the terms of Contract W-36-034-ORD-7481 between the Research and Development Service, Ordnance Department, US Army and the Institute for Advanced Study". A Second edition is dated 2nd September 1947. See: http://grch.com.ar/docs/p1/Apuntes/eng/Logical%20Design%20of%20an%20Electronic%20Computing%20Instrument.pdf.

[30]Kilburn, T., D. Morris, J.S. Rohl, and F.H. Sumner. 1968. A System Design Proposal. In *Proceedings of IFIP Congress*, 76–80, Edinburgh, Section D.

Table 14.3 Register-level symbols suggested by Jack Good in February 1947

Good's 1947 notation	Good's 1947 explanation	Comment, based on other contemporary evidence
A	Accumulator	Notation used by Princeton in June 1946, but not by Turing
B, B1, B2, etc.	Temporary storage	Not a notation used at Princeton or by Turing
C	Control	Used by Princeton and others
CC	Conditional control	
D	Dubbing table	In 1972 Good explains this retrospectively as follows. "The term 'dubbing' refers to the changing of address by means of an additive (when using relative addresses). The term was probably invented by Newman or von Neumann". See also below
R	Register (arithmetic)	A working register, also specified by Princeton in June 1946
S	Selectron	Assumed (by Princeton and by Good at this time) to be the computer's main memory technology (RAM)
T	Tree	As in *address decoder, instruction decoder,* etc.

Good's notations were evidently evolving during the spring of 1947. For example, registers B and D were soon abandoned as Good developed his ideas for a simpler computer

Good's notes start again in December 1947, coincident with Tom Kilburn's Report[31] giving evidence that the CRT store was proving itself worthy of being taken seriously. Evidence for interaction between Good and Williams and/or Kilburn is sparse, comprising three instances:

(i) From p. 46 of Good's original notes it appears that Newman had been talking to Williams in December 1947 and was able, in conversation with Good, to inform Good that at the moment Williams was committed to "32 bits/word but may go to 40 bits …".

(ii) On 10th December 1947 Good writes: "saw Kilburn & Tootill and it happened that Huskey (from Moore School [actually, from SEAC by then], returns 1/1/48), Wilkinson & Williams turned up.

(iii) On 27th February 1948 Good writes: "Williams & Kilburn asked me whether one was likely to return often in a short time to:

  (a) a single word,
  (b) a single stretch of words (of length 32 words say),
  (c) adjacent words.

[31]Kilburn, T., *A Storage System for use with Binary Digital Computing Machines.*

Table 14.4 A *Baby Machine's* instruction set, as proposed by Jack Good in May 1947

Good's original notation	Typical modern assembler mnemonic	Explanation and comment
x → A	LDA x	Load the contents of memory location x into the accumulator
A + x → A	ADD x	Add the contents of memory location x into the accumulator
− x → A	LDN x	Load the negative of memory location x into the accumulator
A − x → A	SUB x	Subtract the contents of memory location x from the accumulator
A → *R*	TAR	Transfer contents of accumulator to register R
R → A	TRA	Transfer contents of register R to the accumulator
A → x	STO x	Store the contents of the accumulator at memory location x
x → R	LDR x	Load the contents of memory location x into register R
l	SHL	Shift the contents of the accumulator arithmetically one place left
r	SHR	Shift the contents of the accumulator arithmetically one place right
C → x	JMP	Jump unconditionally to memory location x
CC → x	JPOS	If A $\geq$ 0, transfer control to memory location x

The 'arrow' notation was also used by John von Neumann's computer design group at Princeton and by Albert Uttley's computer design group at the UK's Telecommunications Research Establishment (TRE)

This conversation clearly relates to the strategy for refreshing lines in an electrostatic memory, something that was of crucial concern to Williams and Kilburn at that time.

By way of summary, Tables 14.3, 14.4 and 14.5 chart Jack Good's evolving notations, in comparison with others in use at the time. Figure 14.4 shows Jack Good's only-known register-level diagram of a computer, dated 10th June 1947. For comparison, Fig. 14.5 shows the diagram used by Williams and Kilburn to explain their June 1948 Small-scale Experimental Machine and its enhancement 10 months later.

In conclusion, Jack Good's 1947 notes on computer design were potentially available to Williams and Kilburn in much the same way as Alan Turing's ACE design notes and Uttley's TREAC design notes were also potentially available in 1947. The modest Williams/Kilburn computer that emerged in the spring of 1948 was closer to Good's (and hence Princeton's) ideas than to ACE or TREAC but the Williams/Kilburn design also had an original flavour of its own.

Table 14.5 A comparison of register-level notations used by Jack Good in June 1947 and Tom Kilburn in June 1948

Good's notation	Good's explanation	Comment, using modern terminology	Notation used by Kilburn in 1948
Tree	(for the memory). Has 12 inputs and 4096 outputs	Memory address decoder	*(See comment below)*
IT	Instruction tree. Has 6 inputs and 64 outputs	Function-bits decoder	*(See comment below)*
CR	Control register. Contains one-word instruction	Instruction register	PI, present instruction
AC	Act Counter. Contains the house number of the current or next instruction	Program counter	CI, control instruction
M	Memory. "A set of a few hundred cathode ray tubes, Selectrons or mercury tanks. The words in M are regarded as belonging to a number of 'houses' numbered consecutively say from 1 to 4096"	Main memory.	S, 32 lines of Williams Tube storage each line containing 32 digits
W	Works. The circuits which perform the possible types of instruction	ALU	*(By implication, the subtractor circuit plus the A storage tube)*
A	Accumulator	Accumulator	A
R	Arithmetic register	Working register	–

Kilburn's 1948 computer did not use a specific term for a decoding network
The address-bits, s, and function-bits, f, of the current instruction were held in a flip-flop register called the *Staticisor* whilst being decoded

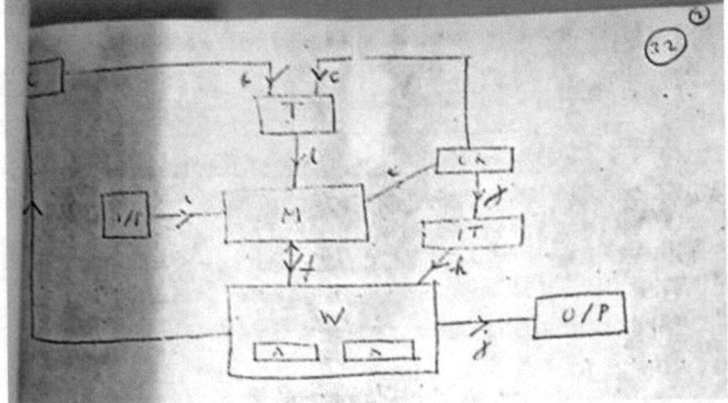

Fig. 14.4 Jack Good's only-known register-level diagram of a computer (10th June 1947). The box in the top left says AC (Act Counter). T = Tree; CR = Control Register; IT = Instruction Tree; W = Works; A = Accumulator; R = Arithmetic Register

The way the Small-scale Experimental Machine then evolved during the next 15 months, leading to the Ferranti Mark I, was unique—especially in the hardware support provided for addressing structured data. Of course Kilburn, who gradually

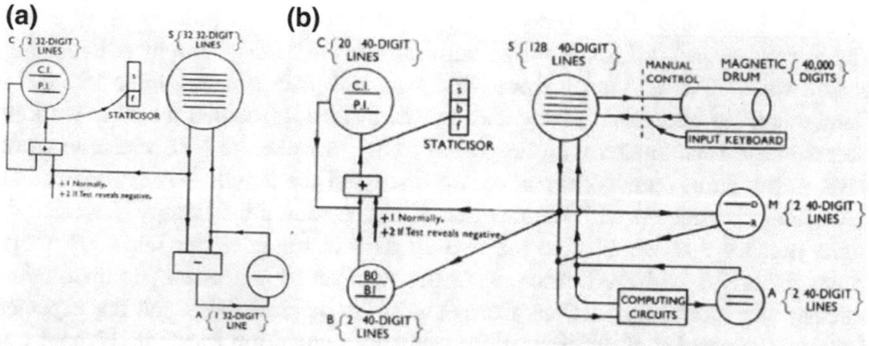

Fig. 14.5 **a** and **b** The diagrams used by Williams and Kilburn to explain the June 1948 Small-scale Experimental Machine *(left)* and the April 1949 Manchester Mark I computer *(right)*. The register notation is as shown in Table 14.5. Note that all central registers as well as the main RAM memory were implemented using Williams-Kilburn electrostatic storage—shown as circles in the diagram. The right-hand diagram represents the enhancements that had taken place to the SSEM over 10 months, including longer word length, expanded instruction set with multiplication, index registers (denoted as B0, B1 above) and a magnetic drum. Drum transfers and paper tape input/output were manually controlled in April 1949 but were under program control by October 1949

Table 14.6 The instruction set for the Small-scale Experimental Machine (SSEM) which first ran a program on 21st June 1948

Function bits (least-sig. digit on left)	Kilburn's original notation	Modern equivalent	Comment
000	s, C	JMP S	Set the Control Register to the contents of address S (absolute indirect unconditional jump)
100	c + s, C	JRP S	Add the contents of address S to the Control Register (relative indirect unconditional jump)
010	−s, A	LDN S	Set the Accumulator equal to the negated contents of address S
011	a, S		Copy the contents of the Accumulator to address S
001 or 101	a − s, A	SUB S	Set the new value of the Accumulator equal to the former contents minus the contents of address S
011	Test	CMP	If a < 0, add 2 to Control Register; else, add 1 as normal
111	Stop	STP	Halt and sound a steady hooting (i.e. buzzing)

Note that the control register was incremented (by 1 or 2) before the start of every order, so the programmer used JMP and JRP to point to an instruction one before the intended destination line

took over leadership of computer design at Manchester, was influenced by others. There were several visitors. Tom Kilburn remembers[32] that: "The most interesting visitor was Wilkie Wilkinson from NPL, who had in fact been a tutor of mine at Cambridge. He came up one day and he was thrilled to see this machine working because he'd been landed with the job of trying to make the NPL machine work. I remember during our conversation we discussed the length that a word should have, and we came up with the fact that 32 digits were a bit skimpy. I wanted 40 digits because that would give me two 20 digit instructions per line ....". Other topics discussed with Wilkinson were the number of addresses per instruction, whether to partition instructions from data in the main memory and the expected run-time frequencies of multiplication instructions and drum transfers. If there was such a thing as a distinct Manchester brand of computer architecture, it evolved from the autumn of 1948 onwards.

14.7 Conclusions on Conceptions

The early Manchester computer story was re-told by F. C. Williams in 1974.[33] He talked of the "uncertainty and excitement" of a time when "there was no storage system in existence that was known to work with the reliability and freedom from errors that would be required" for a stored-program computer. He recalled that in March 1948 Professor Max Newman of Manchester University had said that "the machines now being made in America and in this country will be universal—*if they work at all*". When Williams gave two lectures to Canada's Atomic Energy research team at Chalk River[34] in August 1949 he said: "When I first entered the field of computers, fresh from the field of radar after the war, I was prepared to believe you could do a lot with electronics because I had had some experience of it, but the stumbling block, the thing I didn't see, was how the programme was organised to be put onto these [computing] machines with a finite amount of effort and of space in the machine".

In his 1974 retrospective account, Williams adds that "we had much assistance from other bodies …With [the 1948 CRT] store available, the next step was to build a computer around it. Tom Kilburn and I knew nothing about computers, but a lot about circuits. Professor Newman and Mr. A. M. Turing in the Mathematics

[32]Kilburn, Tom. 1990. From Cathode Ray Tube to the Ferranti Mark I. *Resurrection the Journal of the Computer Conservation Society* (2).

[33]Williams, F.C. 1975. Early Computers at Manchester University. *The Radio & Electronic Engineer* 45 (7): 327–331. Presented at a Royal Society Colloquium on 12th November 1974.

[34]On 19th and 22nd August 1949 F. C. Williams gave two lectures in Canada at the National Research Council's Atomic Energy project, Research division, Chalk River, Ontario. These lectures were typed up from a wire recording and bound as Report LT-24, 14th Sept. 1949, *High speed universal digital computers.* (28 typed pages and 16 figures). This Report essentially describes the CRT storage system, the SSEM and the factoring program. See NAHC/MUC/1/D5.

Department knew a lot about computers but substantially nothing about electronics. They took us by the hand and explained how numbers could live in houses with addresses and how if they did they could be kept track of during a calculation. In addition, Professor Newman had a grant from the Royal Society. The collaboration was fruitful …".

In the light of Sects. 14.3, 14.4, 14.5 and 14.6 above, Williams' 1974 retrospective story of Baby's conception is a little misleading. In 1948 any expert in electronics would have needed rather more than the simple analogy of numbers living in houses before being able to produce a working computer. It is more reasonable to deduce that, although Williams and Kilburn were late-comers on the computing stage, they were aware of Turing's design activity going on at NPL, Uttleys' activity at TRE and, through Newman and Good, von Neuman's activity at Princeton. The architecture of the Baby computer must have been influenced to some extent, in positive or negative ways, by all of these prior projects. In hindsight, perhaps the most obvious positive indirect influence was that of Princeton via Jack Good (see Sect. 14.6)—though it is doubtful that this was as clear in 1948. In the event, Princeton's Institute of Advanced Study (IAS) computer was not operational until early 1951,[35] using Williams/Kilburn tubes because the RCA Selectrons proved unsuitable and unavailable.

As an aside, it is worth quoting a story from Julian Bigelow's IAS paper,[36] since it conveys a contemporary American view of the impression that the Baby machine, and in particular its novel storage system, made on the Princeton engineers:

"In June [1948] we received a preprint copy of a report, the IEE paper,[37] by Williams and Kilburn from Manchester, and it was read quickly by everybody in our group. The consensus was that this looked like the most promising scheme [for primary storage] yet, and arrangements were made for me to go to Manchester for a few weeks to visit Williams, while at IAS Jim Pomerene would start an effort to see whether the Williams scheme could be made to work in an experimental rig in our laboratory. My visit to Manchester (also to Cambridge and London) was a delightful experience. F C Williams was a true example of the British 'string and sealing wax' inventive genius, who had built a primitive electronic computer from surplus World War II radar parts strictly on his own inspiration – in the middle of which were two cathode-ray tubes storing digits in serial mode – the 'Williams memory'. I can remember him explaining it to me, when there was a flash and a puff of smoke and everything went dead, but Williams was unperturbed, turned off

[35]Bigelow, J. 1976. Computer Development at the Institute for Advanced Study. In *International Research Conference on the History of Computing*, Los Alamos, 10th–15th June 1976. The papers were later published in: Metropolis, N., J. Howlett, and G.-C. Rota (eds.). 1980. *A History of Computing in the Twentieth Century*. Academic Press. The IAS paper is pp. 291–310.

[36]Bigelow, J., Computer Development at the Institute for Advanced Study. The papers were later published in: Metropolis, N., J. Howlett, and G.-C. Rota (eds.)., *A History of Computing in the Twentieth Century*. The IAS paper is pp. 291–310.

[37]Williams, F.C., and T. Kilburn. 1949. A Storage System for use with Binary Digital Computing Machines. *Proceedings of IEE* 98 (Part 2, No. 30): 183 ff.

the power, and with a handy soldering iron, replaced a few dangling wires and resistors so that everything was working again in a few minutes. When I 'phoned Princeton from Manchester after having been away about three weeks, Pomerene reported to me that an experimental Williams memory tube was already working there and storing some 16 digits. So I came home".

Apart from its novel storage system, what was unique about the 1948 Baby? Simply its appearance as the first working example of a universal computer, running the first program. From this small beginning, computer science innovations followed at Manchester over the years that were indeed novel, including Index Registers and Virtual Memory. For better or worse, a subtle *not invented here* fog (of tobacco smoke?) arose between Tom Kilburn's collaborative computer projects with Ferranti Ltd. and the work of other companies and other research groups. It is a personal opinion that this self-imposed Manchester fog lasted at least until the 1970s. Perhaps this fog has unwittingly coloured previous analyses of the early history of computing?

References

Bigelow, J. 1976. Computer Development at the Institute for Advanced Study. In *International Research Conference on the History of Computing*, Los Alamos, 10th–15th June 1976.

Broadbent, T.E. 1998. *Electrical Engineering at Manchester University; The Story of 125 years of Achievement*. Published by The Manchester School of Engineering, University of Manchester. ISBN 0–9531203-0-9.

Carpenter, B.E., and R.W. Doran. 1977. The Other Turing Machine. *The Computer Journal* 20 (3): 269–279.

Carter, R.H.A., and A.M. Uttley. 1953. The Telecommunications Research Establishment Parallel Electronic Digital Computer, Chap. 10. In *Faster than Thought,* ed. B.V. Bowden. Published by Pitman.

Huskey, H.D. 1947. *The State of Computing in Britain and the US*.

Kilburn, T. 1947. *A Storage System for use with Binary Digital Computing Machines*.

Kilburn, T., D. Morris, J.S. Rohl, and F.H. Sumner. 1968. A System Design Proposal. In *Proceedings of IFIP Congress*, 76–80, Edinburgh, Section D.

Kilburn, Tom. 1990. From Cathode Ray Tube to the Ferranti Mark I. *Resurrection the Journal of the Computer Conservation Society* (2).

Lavington, Simon. 1980. *Early British Computers*. Manchester University Press.

Metropolis, N., J. Howlett, and G.-C. Rota (eds.). 1980. *A History of Computing in the Twentieth Century*. Academic Press.

Turing, A.M. 1937. On Computable Numbers, with an Application to the Entscheidungsproblem. *Proceedings of the London Mathematical Society,* Series 2 42 (1): 230–265.

Williams, F.C. 1975. Early Computers at Manchester University. *The Radio & Electronic Engineer* 45 (7): 327–331.

Williams, F.C., and T. Kilburn. 1949. A Storage System for use with Binary Digital Computing Machines. *Proceedings of IEE* 98 (Part 2, No. 30): 183 ff.

Chapter 15
Appendix B. Mark I and Mark I* Software Details

We introduce the notations and instruction sets for the Ferranti Mark I and Mark I* computers and survey the more important systems software associated with these two types of machine. It will be shown that there was a big culture gap between Mark I and Mark I* installations such that, though the underlying hardware was very similar, the surrounding software was in practice not directly compatible.

15.1 Notations and the 5-Bit Teleprinter Code

For the Small-Scale Experimental Machine (*Baby*) in June 1948, the Manchester University engineers keyed in instructions and data to the computer manually and read results by visual inspection of patterns of dots on a cathode ray tube (CRT) screen. (The electronics behind the dots is explained in Chap. 16). This primitive method of input/output was made easier on or about 24th October 1948[1] when Alan Turing introduced the engineers to the 5-bit teleprinter code—of which he'd had plenty of experience during the war at Bletchley Park. Turing taught the engineers to divide strings of digits into 5-bit groups, each of which could be made to print a visible character on a teleprinter—see Table 15.1. Once the character code had been learned by heart, reading and writing programs became somewhat simpler. There was one twist to the story: the computer's monitoring tubes presented data serially, with time flowing from left to right and hence with the least-significant digit on the left of a string of digits. This 'backwards binary' view was preserved by programming systems in the Ferranti Mark I, but not in the Mark I* computer.

[1]*Digital Computer—notes on design and operation.* This is a quarto-sized laboratory notebook kept by G. C. Tootill whilst working with F. C. Williams and Tom Kilburn at Manchester University. The first entry is dated 4th June 1948 and the last is 28th November 1948. See National Archive for the History of Computing, document NAHC/MUC/2/C3.

S. Lavington, *Early Computing in Britain*, History of Computing,
https://doi.org/10.1007/978-3-030-15103-4_15

Table 15.1 The Teletype character code used for the Ferranti Mark I

Decimal equivalent of normal binary value	'Backwards binary' bit-pattern	Printed symbol, when teleprinter is in Letter shift	Printed symbol, when teleprinter is in Figure shift
0	00000	/	0
1	10000	E	1
2	01000	@	2
3	11000	A	3
4	00100	:	4
5	10100	S	5
6	01100	I	6
7	11100	U	7
8	00010	½	8
9	10010	D	9
10	01010	R	+
11	11010	J	–
12	00110	N	.
13	10110	F	
14	01110	C	
15	11110	K	
16	00001	T	
17	10001	Z	
18	01001	L	
19	11001	W	
20	00101	H	
21	10101	Y	
22	01101	P	
23	11101	Q	
24	00011	O	
25	10011	B	
26	01011	G	
27	11011	"	
28	00111	M	
29	10111	X	
30	01111	V	
31	11111	£	

This is an adaptation of the standard Murray code, arranged so that, in letter-shift, a printable character results from every code combination

Table 15.1 shows the particular character codes used for the Ferranti Mark I. These symbols are essentially the standard Murray teleprinter code, except that the combinations 00000, 01000, 00100, 00010, 11011, 11111 which in true Murray code are represented by *no effect, line feed, space, carriage return, figure shift, letter shift* are now re-assigned the symbols /, @, :, ½, ", … £ as follows:

no effect	line feed	space	carriage return	figure shift	letter shift
/	@	:	½	"	£

These symbols, sometimes known as stunt characters, were chosen (by Alan Turing?) so as to enable the upper case of a typewriter to be used throughout. Whilst awkward to modern eyes, the presentation of Table 15.1 did mean that Mark I programmers and engineers had a direct and mutually-understandable representation of information inside the computer.

15.2 Instruction Format and Programmer-Accessible Registers

Normal Ferranti Mark I instructions were 20 bits long:

10 bits		3 bits		1	6 bits	
Address, n		B		Spare	Function (Op code)	
0	9	10	12	13	14	19

The B digits denoted eight index registers, called B lines. By convention, the contents of B0 was always zero. With the above division, addresses in the primary store could be represented by two teletype characters.

Programmers always wrote the functions (op codes) as two characters, of which the left-most character was either / or T (giving a least-significant 0 or 1 in 'backwards binary') assuming index register B0 was implied in an instruction. A list of functions and their corresponding character representation is shown later in Table 15.2.

The Ferranti Mark I's normal word length was 40 bits, with 20-bit instructions. The original Manchester terminology referred to *lines* rather than words. Hence a long line was 40 bits, a short line 20 bits, and in the contemporary literature a line usually meant a short line. Control-transfer addressing was to short lines (ie half-words).Operand addressing was to a 40-bit line (word) whose digits were made from a pair of 20-bit lines (half-words).

Table 15.2 The instruction set for the Ferranti Mark I computer

Programmers' code	Backwards binary	Function	Comment
//	000000	h is obeyed as a magnetic instruction	H is the pattern set up on the handswitches
/E	010000	s′ = am	Store M
/@	001000	am′ = am + i, where i = 63 if s = 0; else, $2^i \leq s < 2^{i+1}$	Indicates the position of the ms '1' in S. See Note (a)
/A	011000	s′ = am; AM cleared	Store and clear
/:	000100	s is obeyed as a magnetic instruction	S contains a drum or I/O transfer Control Word—(see below)
/S	010100	s′ = al	Store (ls part of) Acc
/I	001100	am′ = al; al′ = am	Swap halves of Acc
/U	011100	s′ = al; al′ = am; AM cleared	40-bit move and clear
/½	000010	a′ = a − d × s, (unsigned)	Multiply & subtract (unsigned)
/D	010010	a′ = a − d × s, (signed)	Multiply & subtract (signed)
/R	001010	am′ = am + (number of 1s in s)	Sideways add into m ('population count')
/J	011010	am′ = am + s	Add (upper)
/N	000110	a′ = a + d x s, (unsigned)	Multiply and add (unsigned)
/F	010110	a′ = a + d x s, (signed)	Multiply and add (signed)
/C	001110	d′ = s; d treated as an unsigned no.	Load multiplicand (unsigned)
/K	011110	d′ = s; d treated as a signed number	Load multiplicand (signed)

(continued)

Table 15.2 (continued)

Programmers' code	Backwards binary	Function	Comment
/T	000001	c′ = s if b ≥ 0, else c′ = c + 1	Conditional indirect absolute jump on the last-referenced B line
/Z	010001	s′ = h	Copy the digits set up on the handswitches into s
/L	001001	Stop if switch /L is set on the console	Known as a '*dummy Stop*'
/W	011001	al′ = 20 random digits in the ls end	Hardware random number generator
/H	000101	c′ = s if a ≥ 0, else c′ = c + 1	Conditional indirect absolute jump
/Y	010101	*unassigned*	
/P	001101	c′ = s + 1	Unconditional indirect absolute jump. See Note (b)
/Q	011101	c′ = c + s	Unconditional indirect relative jump
/O	000011	c′ = c + s + 1 if b ≥ 0, else c = c + 1	Conditional indirect relative jump on the last-referenced B line
/B	010011	*unassigned*	
/G	001011	Stop if switch /G is set on the console	Known as a '*dummy Stop*'
/"	011011	*unassigned*	
/M	000111	c′ = c + s + 1 if a ≥ 0, else c = c + 1	Conditional indirect relative jump
/X	010111	*unassigned*	
/V	001111	Hoot	Pulse the console's Audio amplifier
/£	011111	*unassigned*	
T/	100000	al′ = s; am cleared	(unsigned)

(continued)

Table 15.2 (continued)

Programmers' code	Backwards binary	Function	Comment
TE	110000	*unassigned*	
T@	101000	all' = 65th line of tube addressed by S; alm & am are cleared	Load drum address of page
TA	111000	s' = al; al is cleared	Store (ls end of) Acc & clear
T:	100100	a' = 0	Clear acc
TS	110100	*unassigned*	
TI	101100	al' = s + al, any carry being added into am	Add (unsigned)
TU	111100	*unassigned*	
T½	100010	a' = s with sign-extn into am	Load acc (signed)
TD	110010	a' = extended s OR a	Logical OR (sign-extended into ms end)
TR	101010	a' = extended s AND a	Logical AND (sign-extended into ms end)
TJ	111010	a' = extended s NEQ a	Logical NEQ (sign-extended into ms end)
TN	100110	a' = a − (sign-extended) s	Subtract (signed)
TF	110110	a' = − (sign-extended) s	Load acc negatively (signed)
TC	101110	a' = a + (sign-extended) s	Signed addition
TK	111110	a' = 2 x (sign-extended) s	Arithmetic shift up by one place
TT	100001	b'= s	Load B (*This instruction may itself be modified*)
TZ	110001	s' = b	Store B (*This instruction may itself be modified*)
TL	101001	b' = b − s	B subtract (*This instruction may itself be modified*)

(continued)

Table 15.2 (continued)

Programmers' code	Backwards binary	Function	Comment
TW	111001	Same action as TL	
TH	100101	unassigned	
TY	110101	unassigned	
TP	101101	unassigned	
TQ	111101	unassigned	
TO	100011	b′ = s	Load B (*This instruction is **not** itself modified*)
TB	110011	s′ = b	Store B (*This instruction is **not** itself modified*)
TG	101011	b′ = b − s	B subtract (*This instruction is **not** itself modified*)
T″	111011	*Same action as TG*	B subtract (*This instruction is **not** itself modified*)
TM	100111	*unassigned*	(*This instruction, if used, is **not** itself modified*)
TX	110111	*unassigned*	(*This instruction, if used, is **not** itself modified*)
TV	101111	*unassigned*	(*This instruction, if used, is not itself modified*)
T£	111111	*Dummy order*	(*This instruction is not itself modified*)

Notes

(a) (for the /@ instruction): for non-zero numbers, the position of the most-significant logical 1 will be an integer in the range 1–39. For the special case of s = 0, the arbitrary number 63 is chosen as the indicator

(b) (for the /P instruction): For example, if the decimal value of the ten least-significant digits of the number in the operand-address is 54 then the instruction in line 56 will be the next to be obeyed

The computer had the following programmer-accessible registers:

A 80-bit double-length accumulator. The sections of this are accessible as:
AL the less-significant 40 bits of A
ALL the least-significant 20 bits of L
ALM the most-significant 20 bits of L
AM the more-significant 40 bits of A

D 40-bit multiplicand register

B one of 8 modifier (index) registers. B0, by convention, contains zero.

QA the most-significant bit of A

QB the Q digit. This is the most-significant digit of the B line which was last operated upon by one of the B group of instructions.

C 10-bit Control register (Program Counter). C is normally incremented at the conclusion of the present instruction, unless the instruction is a control-transfer order for which the jump-to condition has been satisfied.

H the hand-switches on the operator's console

P the 65th line of the CRT specified by the current address. The contents of this line may be taken as the address of this page when the page is held on the magnetic drum backing store.

The instruction set allowed for 64 orders, of which only 50 were defined. Table 15.2, which is spread over four pages, gives the complete picture. Further explanations follow in Sect. 15.3.

Below is an 8-line code sequence that illustrates just how difficult it was to 'read' Mark I programs, let alone write them. The sequence places in address `/C` the scalar product of two 18-element vectors whose fixed-point values are stored in the following addresses (inclusive):

Vector (i) in lines `/N, @N, ..., LN`
Vector (ii) in lines `/F, @F, ..., LF`

A diagram of how the primary store was addressed is given later.

Table 15.3 A code sequence to form the scalar product of two 18-element fixed-point vectors

Line address in primary store	Machine code	Explanation
`//`	`L///`	Number of elements in the vectors
`/E`	`IST/`	Entry point; set round-off
`/@`	`//QO`	Set B7
`/A`	`/NUK`	Add product
`/:`	`/FUF`	... to partial sum
`/S`	`A:QG`	B7 := B7 − 1 (ie adjust counter)
`/I`	`A:/T`	Test for last cycle, jumping back to line `A` if more to do
`/U`	`/C/A`	Transfer result to address `/C`

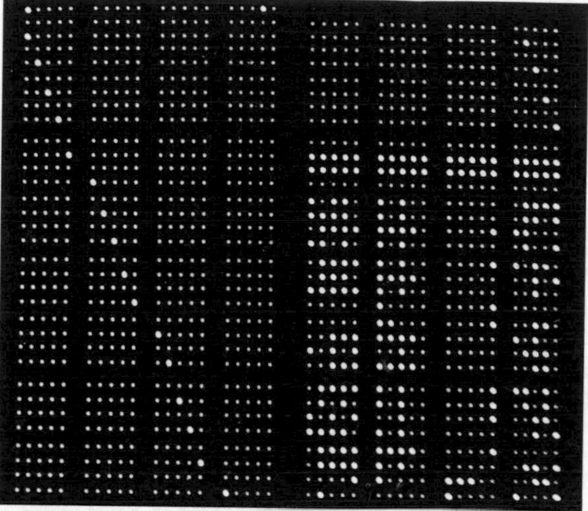

Fig. 15.1 The contents of a Williams/Kilburn storage tube, as displayed on the console monitor. The display shows two columns, each with 32 20-bit lines, with an extra line at the top of the first column. The left-hand column and part of the right-hand column in this example are displaying powers of two (in 'backwards binary') in successive 40-bit words. The code displayed on the right is Scheme A PERM

15.3 Further Explanation of the Mark I Instruction Set

Alan Turing issued the first Programmers' Handbook in March 1951.[2] This magnificent 110-page foolscap manual contains some errors and is not easy to understand. A much better starting-point for the modern reader is,[3] which also gives an account of the Mark I's early programming systems, several example programs and some interesting background on the programming philosophy at Manchester University.

The contents of a Ferranti Mark I's 40-bit word may be regarded as a 2's complement signed number or as an unsigned number. It will be seen in Table 15.2 that two versions of some of the arithmetic instructions existed: signed and unsigned. Signed numbers may be regarded as either integers or fractions. Most applications programmers worked with fixed-point signed fractions in the range $-½ \leq x < ½$. Floating-point arithmetic was performed by software.

---

[2]*Programmers' handbook for Manchester Electronic Computer Mark II*. Anon and undated, but known to have been written by Alan Turing and issued in March 1951. 110-page typed foolscap manual. (The qualification 'Mark II' was used by Turing to distinguish this computer from its research predecessor at the University of Manchester. The nomenclature 'Mark II' refers to the production version which was later called the Ferranti Mark I computer). Two sets of Errata sheets were issued, probably by Cicely Popplewell according to Dai Edwards, respectively on 13th and 28th March 1951. A second edition of the manual was issued by Tony Brooker in October 1952. Subsequent editions followed: a third was issued in September 1953 and a supplement to this in January 1956. A transcription of Turing's first edition, with helpful comments, is available at: http://curation.cs.manchester.ac.uk/computer50/www.computer50.org/kgill/mark1/progman.html

[3]Campbell-Kelly, Martin. 1980. Programming the Mark I: Early Programming Activity at the University of Manchester. *Annals of the History of Computing* 2 (2): 130–168.

Multiplication was normally done in two steps: (i) the setting of D; (ii) the multiplication proper following afterwards. There was no hardware divide. There was no general shift instruction (apart from `TK` in Table 15.2) so multiplication was used instead. In the monitor display photograph of Fig. 15.1 one can see a handy list of the powers of two.

Control transfers were all indirect. Absolute transfers go to line n + 1 where n is the address contained in the line addressed by the instruction. Relative transfers skip n instructions.

The normal instructions for manipulating the B-lines (ie the index registers) were `TO`, `TB`, and `TG`; these instructions were not themselves modified. For the occasional programming trick, the alternative, modified, versions `TT`, `TZ`, and `TL` could be used.

There are two unusual instructions in Table 15.2. These are `/R` (the sideways add that gave the number of binary 1s in a word) and `/W` (the hardware random number generator). In the words of Geoff Tootill, these orders came under the category of "the requirements placed upon us by Professor Newman and Mr. Turing".[4] One possible application of the *sideways add* is in pattern-matching. If individual bits in a word are used as indicators of the presence or absence of certain properties in a (large) set of objects so as to produce a descriptor of the set, then the general characteristics of sets can be compared by comparing the sum of bits in their descriptors.

The *random number generator* was expressly included by Alan Turing, who provided Tootill with a possible circuit diagram as part.[5] Turing's circuit was similar to an arrangement already developed at GCHQ Eastcote.[6] It is very unlikely that Turing intended the Ferranti Mark I's random number generator to be used for serious cryptographic purposes, if only because of the difficulty of ensuring strict physical security of the installation. Rather, it probably chimed with his own thoughts about writing programs that simulated the apparently random elements in nature and in human reasoning. In any case, the particular Mark I hardware was, it seems, not always correctly adjusted.[7] There were at least three known, non-serious,

---

[4]Tootill, G.C. 1949. *Informal Report on the Design of the Ferranti Mark I Computing Machine*. Approximately 30 typed pages plus diagrams. See NAHC/MUC/C4. The report includes a two-page Appendix on a possible circuit for generating random numbers, written by Alan Turing.

[5]Tootill, G.C., *Informal Report on the Design of the Ferranti Mark I Computing Machine*.

[6]Donald Horwood worked at GCHQ Eastcote after the end of the war on the development of a random number generator and high-speed checker. The relevant correspondence with SHL is in the Bodleian Library and is described at K4 in this catalogue: http://www.ourcomputerheritage.org/CatK.pdf. It is believed that many Donald Duck units were produced, for use with one-time pads. Interestingly, GCHQ advised the Department of National Savings on ERNIE and the original ERNIE random-number hardware was based on Donald Duck.

[7]In Campbell-Kelly (1980) is an anecdote about deliberate mis-adjustment for demonstration purposes. Also Anthony Ralston, later to become Director of the Computing Centre and Professor of Mathematics the State University of New York at Buffalo, spent seven weeks during the summer of 1953 as an MIT vacation student at Ferranti Moston. His task was to test the random number generator. In an e-mail to Simon Lavington dated 27th October 2015, Olaf Chedzoy

uses of the Random Number Generator instruction at Manchester: (i) Christopher Strachey's *Love Letters* program; (ii) Martin Wingstedt's *Greyhound* dog-racing program; (iii) FINAC playing pretend Mozart music starting with random 20 bit numbers, based on a program written by D. G. Prinz.

Input and output from/to paper tape equipment was organised by a 20-bit control word via what was called a *magnetic instruction*—see `/:` in Table 15.2. A full description of the control word is given below, when we describe the drum store. For paper tape input/output, the control word has bit 14 set to 1. The interpretation of bits 10–13 were then as follows:

Bits 10–13	Symbol	Meaning
00011	O	Input a character from the paper tape reader—(see note (a))
00001	T	Output a character (see note (b))
10011	B	Test output (see note (c))
10001	Z	Teleprinter space
01001	L	Teleprinter carriage return
11001	W	Teleprinter line feed
00101	H	Switch teleprinter to Figure Shift (see note (d)
10101	Y	Switch teleprinter to Letter Shift (see note (d))
11111	£	No action

Notes

(a) During input, a 5-bit character is transferred from the paper tape reader and the tape is moved on to the next character-position. The logical OR is performed between this 5-bit character and the five ms digits of the accumulator, the result being placed in the five ms digits of the accumulator.

(b) When a character is output, the five ms digits of the accumulator may be routed according to the setting of a three-position switch on the operator's console. The three possibilities are:

(i) send the character to the paper tape punch;
(ii) send the character to the teleprinter;
(iii) send the character to both devices.

(c) for option (b) above, the last 5-bit character that was sent to the output equipment is ORed with the ms five digits of L and the result is placed in the five ms digits of L. This enables a programmer to check that the correct five digits were last sent to the output devices. However, this instruction does not check that the output equipment itself was working correctly.

---

remembers that Ralston "discovered that there was a small but noticeable bias towards the production of '1's. However, he also discovered that this could be overcome by using a 'not equivalent' function on two random numbers, and the result was then unbiased". Tony Ralston's paper giving the mathematical background is: Ralston, Anthony. 1980. Random Number Generation on the Ferranti Mark I. *Annals of the History of Computing* 2: 270–271.

(d) The teleprinter remains in either the letter-shift or the figure-shift state until a counter-command is received. In letter-shift, the teleprinter prints the symbols of the Ferranti Mark I's code as given in Table 15.1. When in figure-shift, the decimal digits 0–9 can be printed, along with: +, −.

Once one of the input/output instructions had been initiated, the computer continued to obey other instructions until a further input or output instruction was encountered. At this point the machine paused until the first was complete.

A Bull lineprinter was later added to the Ferranti Mark I at Manchester—and indeed to several of the Mark I* installations—as an upgrade. This upgrade necessitated the inclusion of an extra Williams/Kilburn tube as a buffer store. As a humorous aside, Frank Sumner remembered that: "If things went wrong it could be physically dangerous. A piece of metal could fly off the fast printer, so an old army tin helmet was available to put on before going into the room containing the printer".[8]

15.4 Storage Organisation: Primary and Secondary Memory

The primary memory initially consisted of 8 random-access Williams Tubes, extensible to 16 tubes. Each tube held 1280 bits, arranged as 64 lines of 20 digits each. The 64 lines were referred to as a *page*. A photograph of a Ferranti Mark I's tube as displayed on the console monitoring screen is shown in Fig. 15.1.

When loading a 40-bit quantity into AL from memory, the contents of S went into ALL and the contents of (S + 1) went into ALM.

Using the symbols of Table 15.1, programmers addressed the eight Williams/Kilburn tubes thus, where each tube holds 64 20-bit lines or a *page*:

Tube 0	Tube 1		Tube 7
// /E	/@ /A		/C /K
E/ EE	E@ EA		EC EK
@/ @E	@@ @A	...	@C @K
..	..		..
..	..		..
..	..		..
£/ £E	£@ £A		£C £K

[8]Sumner, F.H.(Frank). 1994. Memories of the Manchester Mark I. *Resurrection* 10: 9–13. This anecdote probably refers to the period from 1954 onwards, when the Ferranti Mark I had been moved to the new Electrical Engineering Department in Dover Street.

The Ferranti Mark I's magnetic drum had a maximum of 256 tracks of 2560 digits each. In practice, the number of tracks available to programmers was usually less than this because the drum was, at least to begin with, one of the most unreliable parts of the whole Mark I computer. The situation improved dramatically for the Mark I* version. Unusually, the drum was synchronised to the CPU's clock so that, in principle, several drums could be added. No Ferranti Mark I or Mark I* system ever had more than one drum attached. Some Mark I*s had a magnetic tape deck attached from time to time.

Each drum track contained two pages—ie $2 \times 64 \times 20 = 2560$ bits. The 65th line in each half-track contained fixed information specific to that track. The contents of a 65th line was in fact the control-word (see below) required to read from the half-track into the first tube of primary store. It was arranged as follows:

digits $0 \rightarrow 7$ contain the number of that track;
digits $10 \rightarrow 14$ contain 00000, which is the bit-pattern for read from this track;
digits $15 \rightarrow 19$ contain 00000, thus identifying the first tube.

Digits $0 \rightarrow 7$ therefore give the address of a page of primary storage when that page is stored on the drum. At Manchester University this concept was later to lead to the idea of a page's Virtual Address and the hardware-assisted memory-management of virtual memory on the Manchester/Ferranti Atlas computer.

Transfers to and from the magnetic drum used the `//` or `/:` instruction with an appropriate Control Word. The general format of the 20-bit Control Word was:

8 bits		2		5		5	
Track number		Spare		Function		Tube number	
0	7	8	9	10	14	15	19

The function-bits were designated as follows:

Bit 10: if = 0, then the left-half drum track in a two-tube drum transfer goes to the first tube.
If = 1, then the right-half track in a two-tube drum transfer goes to the first tube.
Bit 11: one-tube transfer (=0) or two-tube transfer (=1). For one-tube transfers, bits 15–19 give either of the column-numbers of the two columns in the desired tube. For two-tube transfers, the pair of tubes specified must be an 'odd-plus-even' pair.
Bits 12 and 13:

00 = read: transfer data, including the 65th line, from drum to CRT;
01 = write: transfer data, excluding 65th line, from CRT to drum;
10 = compare data on specified CRT, including 65th line, with data on the specified half-track. If data is equivalent, `c' = c + 3;` if there is disagreement, `c' = c + 1` as usual at the end of this instruction.
A programmer typically issues this check after a *read* transfer.
11 = compare data as above, but exclude the 65th line. A programmer typically issues this check after a *write* instruction.

Bit 14: if = 0, the control word specifies a drum transfer, as above.
If = 1, the control word specifies an input/output transfer from/to paper tape equipment, etc. In this case, bits 0 to 9 of the control word are ignored. See later for interpretation of the five function-bits for paper tape transfers.

It took about 36 ms for reading and checking transfers and about 63 ms for write transfers. Track-selection was electronic for reading (the more frequent operation) and by electro-mechanical relay for writing. Drum transfers caused a 'hesitation' in the regular rhythm of the Ferranti Mark I's fetch-execute cycle. In other words, computation was suspended whilst the drum transfer took place.

15.5 Basic System Software

A modest amount (well, tiny by modern standards!) of system software was provided with each Ferranti Mark I computer. The essential part of this, what we might today call a *loader* or an *assembler*, was a set of input routines stored permanently on drum tracks that had been write-protected by a manual switch. The basic section was known as PERM, which during normal operation was permanently resident in the primary store. PERM contained useful constants and a sequence of instructions for bringing down a routine from drum.

The bootstrap procedure at the start of a day was firstly to clear all central registers and the primary store by means of switches on the console (see Sect. 15.7). The computer was then set to 'auto', at which point the instruction at store line zero was obeyed. This was all zeros, which caused the handswitches on the console to be taken as a magnetic instruction (see Table 15.2). The track specified on the handswitches was brought into primary store and control now took the next instruction from line 1, which was the start of the Input routine. As explained,[9] the Input routine read a paper tape on which was punched pairs of instructions, each pair preceded by a warning character and the address of the long line into which the instruction-pair was to be placed. A final sequence caused the program (usually just one of many routines) to be written to the drum. Thus, each subroutine in a larger program could be read in and placed on the drum. Finally a master routine was brought from the drum and entered.

[9]*Programmers' handbook for Manchester Electronic Computer Mark II*. Anon and undated, but known to have been written by Alan Turing and issued in March 1951. 110-page typed foolscap manual. (The qualification 'Mark II' was used by Turing to distinguish this computer from its research predecessor at the University of Manchester. The nomenclature 'Mark II' refers to the production version which was later called the Ferranti Mark I computer). Two sets of Errata sheets were issued, probably by Cicely Popplewell according to Dai Edwards, respectively on 13th and 28th March 1951. A second edition of the manual was issued by Tony Brooker in October 1952. Subsequent editions followed: a third was issued in September 1953 and a supplement to this in January 1956. A transcription of Turing's first edition, with helpful comments, is available at: http://curation.cs.manchester.ac.uk/computer50/www.computer50.org/kgill/mark1/progman.html.

As time went by, several versions of the Input routine existed at Manchester—see Table 15.4. Further details are given in Campbell-Kelly[10] and (for Gradwell).[11] Toronto University devised their own Input and Directory scheme for FERUT in November 1952—see Chap. 3. A related, but much improved, Input routine was developed for the Ferranti Mark I* computer as is explained later.

At the suggestion of John Bennett, Mary-Lee Woods, one of the Ferranti programmers based in the Tin Hut at Moston, adapted Scheme A to give a primitive debugging facility called *Stop and Print*. Mary-Lee recalls: "What I did was to have a different routine-changing sequence. I still had to leave the powers of two there, because everybody used them, but instead of having Turing's routine-changing sequence, I had my little program. It took in data from a punch tape, little by little, as the [main] program went, and the punch tape would say, "Substitute for a certain instruction in the main program a jump into my program." Then I would be able to use the whole of the Perm to bring down another routine, which took in from tape what was needed to be printed out, and printed them out, and then returned to the master program".[12] Thus, the state of the accumulator, etc., could be printed at critical points in a program's execution.

More details of the primary store layout at run time is given,[13] where it is explained that a Mark I program consisted of a master routine and a group of subroutines. When a subroutine was called, it was normally brought down from the drum, over-writing the master routine. When a subroutine ended it would bring down a fresh copy of the master routine and re-enter this at the instruction following the subroutine call. Scheme A had a five-instruction routine-calling sequence in PERM.

Over the years a reasonable library of subroutines was collected by the Computing Machine Laboratory at Manchester University. For Scheme A, the library contained about 50 sub-routines, divided into mathematical (14), input/output (13), and miscellaneous (23).[14] The *miscellaneous* category included certain sub-routines used in connection with Turing's Formal Mode programming environment (see below). The Scheme A sub-routines were dated between July 1951 and September 1952 and were variously written by R. A. Brooker, A. E. Glennie, N. E. Hoskin, R. K. Livesley, C. M .Popplewell and A. M. Turing. In Chap. 3 we

---

[10]Campbell-Kelly, Martin., Programming the Mark I: Early Programming Activity at the University of Manchester.

[11]E-mail dated 11th November 2015 to Simon Lavington from Joan Travis (neé Kaye), who worked for Ferranti Moston.

[12]Mary Lee Berners-Lee: An Interview Conducted by Janet Abbate for the IEEE History Center, 12 September 2001. Interview #578 for the IEEE History Center, The Institute of Electrical and Electronic Engineers, Inc. See: https://ethw.org/Oral-History:Mary_Lee_Berners-Lee#About_Mary_Lee_Berners-Lee.

[13]Campbell-Kelly, Martin., Programming the Mark I: Early Programming Activity at the University of Manchester.

[14]Campbell-Kelly, Martin., Programming the Mark I: Early Programming Activity at the University of Manchester.

Table 15.4 The various Input schemes used at Manchester University

Name	Date first introduced	Author(s)	Comments
Scheme A	Spring 1951	Alan Turing	4 pages of code
Scheme B	Spring 1952	Tony Brooker and Alick Glennie	2 pages, including a Directory
Input G	1952	Cyril Gradwell	Avoided use of checksum in Scheme A. Gradwell worked for Ferranti at Moston.
Scheme C (TELEINPUT)	Early 1953	Tony Brooker	Optional Directory
TELEINPUT F	1954/5	Donald Gillies	Allowed flexibility of routine's page-allocation. Gillies was working for NRDC on the aircraft Flutter problem

describe how a paper tape copy of the Scheme A library, apparently totalling about 9000 lines of code[15] was taken to Canada in the summer of 1952, in preparation for the installation of the second Ferranti Mark I computer (FERUT) at Toronto University.

Somewhat smaller subroutine libraries were later created for Scheme B and for Scheme C. In due course the library at Manchester included a small number of larger routines or complete programs of general interest, including those for matrix manipulation, Runge-Kutta solutions, Bessel functions and analytical differentiation. Meanwhile, Ferranti programmers at Moston were adding their own library subroutines.

For Scheme B, Tony Brooker developed FLOATCODE, an interpretive system for floating-point arithmetic. This is described with an example.[16] In 1951, previous to FLOATCODE, D. G. Prinz and others had developed an interpretive floating-point system for use by Ferranti programmers at Ferranti Moston.[17] After 1954, floating-point operations at the University were usually carried out in the context of Brooker's Autocode system, described in Sect. 15.9.

---

[15]Gotlieb, Calvin C. 1954. The Cost of Programming and Coding. *Computers and Automation* 25: 14 ff.

[16]Campbell-Kelly, Martin., Programming the Mark I: Early Programming Activity at the University of Manchester.

[17]Bennett, J.M., D.G. Prinz, and M.L. Woods. 1952. *Interpretative Sub-routines.* In *Proceedings of ACM National Conference*, 81–87, Toronto, Sept. 1952.

15.6 Alan Turing's Formal Mode: An Embryonic Operating System?

In the 1950s programmers booked sessions (say of half-an-hour) on the computer, during which debugging took place manually and interactively using the console's displays and handkeys (see Fig. 15.2). In the early days of the Ferranti Mark I such sessions were made more taxing if there was an intermittent hardware error. Even when the computer was working perfectly there was a tendency for programmers to get the most out of a half-hour's session by 'playing the handkeys' and making quick modifications to code in primary store. The trouble with this ad hoc method of debugging was that, after the event, there was usually no written record of changes.

Alan Turing wrote a series of sub-routines under Scheme A known collectively as *Formal Mode* which provided a well-defined interface between programmer (the person at the console) and the machine at run time. At the centre of *Formal Mode* was a sub-routine called ACTION. The program passed messages to the programmer via ACTION and then halted; the programmer responded to ACTION messages (usually only by setting handkeys or loading a paper tape) and then restarted the machine. ACTION printed out the state of the handkeys and continued with the program. After the debugging session the user thus had a printed record of all man-machine interactions during the (half-hour) session.

Formal Mode also had some behind-the-scenes useful routines, such as:

INTERFERE: enabled the programmer to alter single lines of store;
BURSTS B: ran a given sub-routine repeatedly until consistent results were obtained, as verified by printing a checksum;
COPY R OUT: produced a reloadable dump of the relevant part of the magnetic drum at the end of a BURST B.

Formal Mode was frequently used by Alan Turing. Most other Mark I programmers found it excessively fussy.[18] Figure 15.2 shows some computer output fixed to a sheet of Alan Turing's hand-written notes of 24th May [1954], illustrating the efforts he was making to develop his morphogenesis program shortly before his tragic death on 7th June 1954.

[18]Campbell-Kelly, Martin., Programming the Mark I: Early Programming Activity at the University of Manchester.

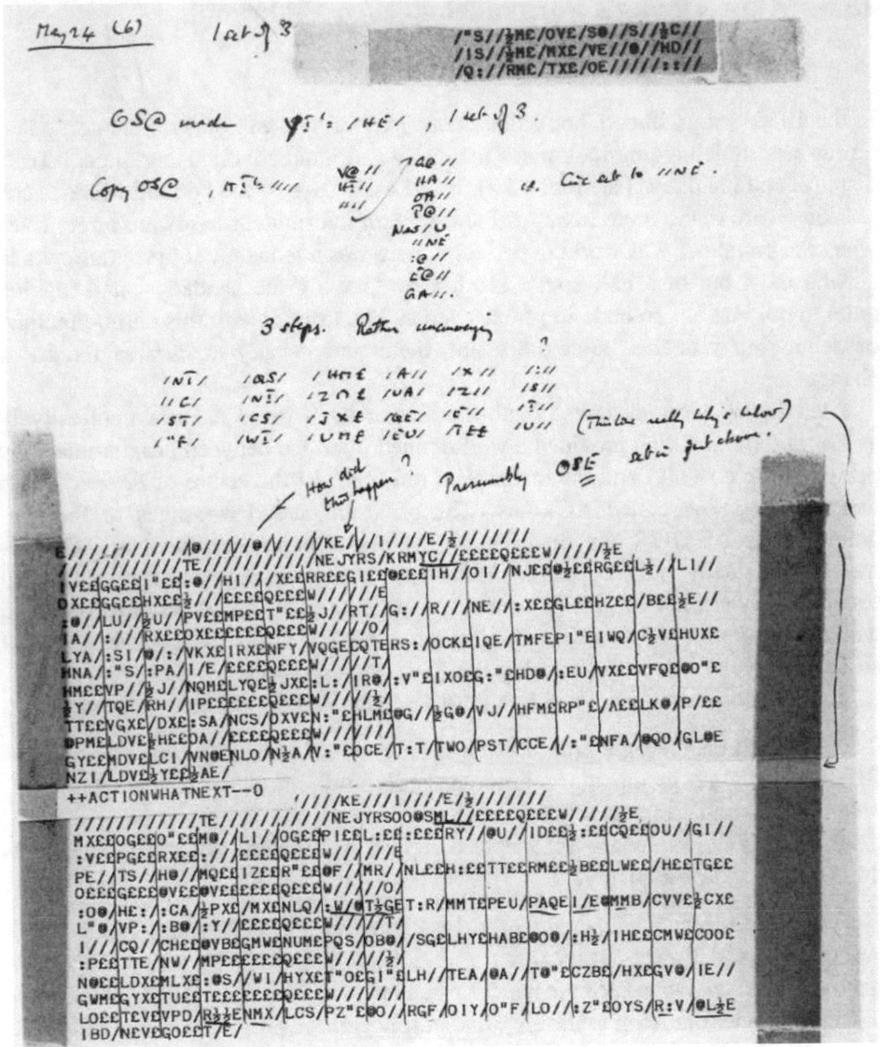

Fig. 15.2 Two fragments of output from one of Alan Turing's morphogenesis programs during development. The ++ACTIONWHATNEXT heading comes from the *Formal Mode* system

15.7 The Mark I Operator's Console

The console contained two larger screens, each of which could be switched to display the contents of a tube of the primary memory (CRT store), together with four smaller screens and several switches—see Fig. 15.3.

The smaller screens showed, respectively, the eight B lines, Control (Program Counter), the 80-bit Accumulator and the 40-bit Multiplicand register.

The main switches on the console included the following:

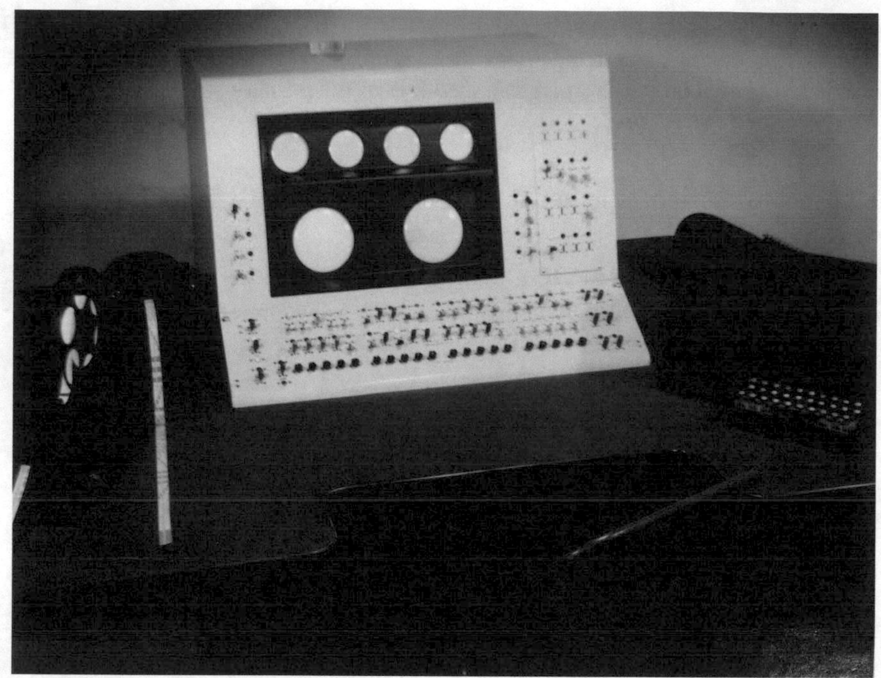

Fig. 15.3 The console of the Ferranti Mark I computer at Manchester University

/G, /L	these determined whether each of the `/G` and `/L` instructions caused the computer to halt.
KAC, KBC, KCC, KMC, KSC	these switches (keys) allowed, respectively, the accumulator, the entire B store, the control register (Program Counter), the D lines and the entire primary memory to be cleared (ie set to zero) manually. A further switch called KEC has the same effect as pressing all five of the previous switches. KEC stood for 'Key Everything Clear'.
KLC	line-clear
Write	this switch enabled an operator to inhibit the writing current to the drum, thus inhibiting all write-transfers

The machine could be switched to operate at *single-shot, semi-continuous* or *continuous* rates. *Semi-continuous* implied a slow clock-rate and *continuous* implied the standard full-speed rate.

A row of 20 handkeys (hand-switches) was provided, allowing for example a 20-bit binary quantity to be set up in preparation for use with either the `/Z` or // instruction. A row of 20 manual instruction switches was also provided. These enable an instruction to be set up and obeyed repeatedly, for diagnostic purposes. Finally, 20 'Typewriter' buttons, which together with an *erase/insert* switch, enabled digits to be inserted into, or removed from, a specified line of the primary store.

15.8 Instruction Speeds for the Ferranti Mark I

The digit frequency was 100 kHz, giving a nominal 10 μs digit-period. 24 digit-periods (240 μs) was known as a *beat.* Computational instructions normally took an integral number of beats, as follows:

Simple arithmetical and logical functions	5 beats (total 1.2 ms)
Multiplication instructions	9 beats (total 2.16 ms)
Most other instructions	4 beats (total 0.96 ms)

The `/W` instruction took 5.8 ms to generate a 20-bit random number. Floating-point addition by software took about 90 ms.

15.9 Automatic Coding and Autocodes

In the summer of 1952 Alick Glennie developed his personal automatic coding scheme which, although the first *autocode* in the field, was very machine-dependent. Ordinary users at Manchester had to wait until the spring of 1954 when Tony Brooker released the Mark I Autocode.[19] This was much more user-friendly and addressed three problems:

simplifying the written form of arithmetic statements;
taking care of the scaling of variables;
taking care of the transfer of information between primary and secondary store.

The Mark I Autocode is best introduced by a short sequence which prints the root mean square (RMS) of the floating-point variables v1, v2, … v100.

```
n1 = 1
v101 = 0
2v102 = vn1 x vn1
v101 = v101 + v102
n1 = n1 + 1
j2, 100 ≥ n1
*v101 = F1(v101)
```

[19]Brooker, R.A. 1955. An Attempt to Simplify Coding for the Manchester Electronic Computer. *British Journal of Applied Physics* 6: 307–311.

The 5-track paper tape for this sequence would have been prepared on a special teleprinter in the Computing Machine Laboratory which had been adapted with appropriate printable symbols (illustrated here in the type font *Times New Roman*). In the Autocode convention, n1 is an integer and v1 etc are floating-point numbers. The symbol * causes the printing of a variable to ten decimal places on a new line and F1 signifies the intrinsic function *square root*. The Autocode user does not have to worry about the distribution of information between primary (CRT) and secondary (drum) storage because the entire memory can be imagined as what Brooker called a *one-level store*. The calling of library sub-routines is also hidden from the user.

The manner in which Brooker's Mark I Autocode was integrated with standard Scheme C sub-routines is covered,[20] where an assessment of the Autocode's place in history is also given. It is concluded that, with the Autocode, the Ferranti Mark I "became possibly the easiest machine to program in Britain". It was certainly influential: similar styles of autocode were developed for the Ferranti Pegasus (in 1957), the Ferranti Mercury (in 1958) and for the Elliott 803 (in 1960).

15.10 Instruction-Set Differences Between the Ferranti Mark I and Mark I*

The Mark I had 26 instructions that operated, in the broadest sense, on the contents of the accumulator—see Table 15.2. The Mark I* had only 13 such accumulator instructions, including the following 'specials': sideways add, standardise. There was no logical OR, and no hardware random number generator instruction. The Mark I had 15 unassigned instructions and 8 B instructions, some of which were duplicates and some of which were alternatives which allowed the instruction itself to be modified. The Mark I* rationalised this situation by having only two unassigned instructions and only four B instructions, none of which may itself be modified.

Numbers in the Mark I* were written down and displayed in conventional binary, rather than the 'backwards binary' of the Mark I. Further points of operational detail between the two types of computer are now listed.

(a) When addressing operands in an ascending sequence, the Mark I* incremented addresses from page to page throughout the memory, whereas the Mark I treated ascending addresses as modulo the page of the first-occurring address in the sequence.

[20]Campbell-Kelly, Martin., Programming the Mark I: Early Programming Activity at the University of Manchester.

(b) The Ferranti Mark I's jumps were all indirect, whereas in the Mark I* the jump-to address was given in the instruction itself. Also, the Mark I added +1 to the jump-to address, whereas the Mark I * did not.
(c) The Mark I distinguished between unsigned and signed arithmetic operations. In the Mark I*, only signed arithmetic is performed. Both machines used the two's complement representation for negative numbers. In the Mark I, the binary point for signed quantities was one place from the end; in the Mark I* it was *two* places from the left-hand (more-significant) end, so that data was held in fractional form—see examples below.
(d) The Mark I* only used the so-called magnetic instruction for drum transfers. Input/output from/to paper tape was performed by the separate instructions `F` and `$` and not (as in the Mark I) by using an appropriately-configured magnetic instruction.
(e) The layout of the Mark I*'s console was different, as can clearly be seen from photographs such as those in Chap. 8 (Fig. 8.11) and Chap. 11 (Fig. 11.1).

15.11 Number Representation and the Instruction Set for the Ferranti Mark I*

The character code for the Mark I* was chosen to be (a) easier to remember and (b) containing an ordered alphabetic collating sequence. Numbers in the Mark I* were written down and displayed in conventional binary, rather than the 'backwards binary' of the Mark I. The teleprinter symbols corresponding to each five-bit pattern were as shown in the Table 15.5.

Table 15.5 The 5-bit teleprinter code for the Ferranti Mark I*

Decimal equivalent	Binary	Teleprinter symbol
0	00000	f
1	00001	£
2	00010	1
3	00011	O
4	00100	@
5	00101	:
6	00110	$
7	00111	A
8	01000	B
9	01001	C
10	01010	D
11	01011	E
12	01100	F
13	01101	G
14	01110	H

(continued)

Table 15.5 (continued)

Decimal equivalent	Binary	Teleprinter symbol
15	01111	I
16	10000	J
17	10001	K
18	10010	L
19	10011	M
20	10100	N
21	10101	P
22	10110	Q
23	10111	R
24	11000	S
25	11001	T
26	11010	U
27	11011	V
28	11100	W
29	11101	X
30	11110	Y
31	11111	Z

All but two of the symbols in the third column of Table 15.5 are printed by the teleprinter when in *Letter Shift*. The exceptions are `f` and `l`. These two cannot be printed because the bit-pattern 00000 is the teleprinter code for *Figure Shift* and the bit-pattern 00010 is the teleprinter code for *Letter Shift*. When in *Figure Shift*, a teleprinter can print the decimal digits 0–9, certain other symbols such as +, −, %, ?, etc., and can perform the actions *Carriage Return, Line Feed, Space*. Programmers wishing to print the symbols `f` and `l` have to choose alternative representations in *Figure Shift* such as ? and %.

Number representations for both the Mark I and the Mark I* used two's complement representation for negative numbers. In the Mark I*, however, the binary point is *two* places from the left-hand (more-significant) end, so that data is held in fractional form. Here are some simple five-bit illustrative examples.

Decimal number	Representation in the Ferranti Mark I*
+0	00.000
+0.5	00.100
+0.25	00.010
−2	10.000
−1	11.000
−0.5	11.100

Both the Mark I and the Mark I* had similar instruction formats, although the written form appears reversed since we now use normal binary rather than

Table 15.6 The instruction set for the Ferranti Mark I*

Code	Binary	Function	Comment	Is there a roughly equiv. Mark I instruction?
f	00000	Stop	Halts the machine; a steady hoot issues from the console's speaker	No (but see /L and /G)
£	00001	b′ = b − s	B subtract. (*This instruction is not itself modified*)	Yes: TG
1	00010	–	(Not to be used; effect undefined)	–
O	00011	b′ = b + s	B addition. (*This instruction is not itself modified*)	No
@	00100	s′ = c	The contents of C are placed in the ls ten digits of s; the ms ten digits of S are cleared	No
:	00101	s′ = b	B store. (*This instruction is not itself modified*)	Yes: TZ
$	00110	Output b	The ms five bits of b are sent to the output equipment	No (but see the use of a 'magnetic instruction')
A	00111	b′ = s	Load b. (*This instruction is not itself modified*)	Yes: TO
B	01000	Print a line on the Bull lineprinter	The 64 characters to be printed are assumed to have been pre-loaded into the output tube. The instruction then causes the contents of this tube to be sent to the Bull lineprinter	No. There is no provision for a lineprinter on the Mark I.
C	01001	If a ≥ 0, goto s	Conditional absolute jump	Yes, but see Note (a)
D	01010	If a < 0, goto s	Conditional absolute jump	No
E	01011	c′ = s	Unconditional absolute jump	Yes, but see Note (a)
F	01100	Input	The five-bit character from the paper tape	No (but see the use of a

(continued)

Table 15.6 (continued)

Code	Binary	Function	Comment	Is there a roughly equiv. Mark I instruction?
			reader are placed in the ls five positions of s. The rest of s is cleared and the tape reader moves on to the next position	'magnetic instruction')
G	01101	s′ = store switches	Copy the bit-pattern set up on the 20 'store' switches on the console into s	No
H	01110	–	(Not to be used; effect undefined)	–
I	01111	c′ = s if b ≥ 0, else c′ = c + 1	Conditional jump on the ms digit of the last-referenced B line	Similar: see /T and /O
J	10000	am′ = am + s		Yes: /J
K	10001	am′ = am − s		Similar: see TN
L	10010	am′ = am NEQ s	Logical 'not equivalent to' (exclusive OR)	Similar: TJ
M	10011	s′ = am		Yes: /A
N	10100	al′ = al + s; am cleared	Add	Similar: see TC
P	10101	am′ = s; al cleared	Load 40 bits into upper 40 bits of the double-length accumulator	Similar: see T/
Q	10110	am′ = am AND s	Logical AND	Similar to TR
R	10111	s′ = am; a cleared		Yes: /A
S	11000	am′ = am + (number of 1 s in s)	Sideways add into m ('population count'). I'm unclear whether this is an 'add' or just a 'load'	Yes: see /R
T	11001	a′ = a − d × s	Multiply & subtract	Yes: /D
U	11010	A shifted up or down by a function of n	Logical shift up or down, according to the value of n when Treated as a signed number	No (but see TK)
V	11011	A standardised	Shift until a '1' appears in the	No

(continued)

Table 15.6 (continued)

Code	Binary	Function	Comment	Is there a roughly equiv. Mark I instruction?
			appropriate digit-position	
W	11100	s is obeyed as a magnetic instruction	Used for reading/writing between the primary memory (CRT) and the secondary memory (magnetic drum)	Yes: /: See also note (c)
X	11101	a′ = a + d × s	Multiply and add	Yes: /F
Y	11110	d′ = s		Yes: /K
Z	11111	Dummy stop	Halt if console switch set	Yes: /L and /G

'backwards binary'. The Mark I* only had five function bits, as shown below and in Table 15.6.

Function		Spare		B		Address	
19	15	14	13	12	10	9	0

When a 40-bit operand is considered, the contents of the first-occurring line-address (the one given in the instruction) goes into the ls end of the 40-bit accumulator and the contents of the second line goes into the ms end of the accumulator.

In the previous list of Mark I* instructions (Table 15.6), the notation is much the same as given previously in Table 15.2.

The precise assignment of a particular instruction code in Table 15.6 may have been adapted for some Ferranti Mark I* installations. For example, code 01110 which is undefined in Table 15.6 was later used for an OVERFLOW instruction on the INAC Mark I* computer in Rome. The operation was: "Test the setting of the OW flip-flop; if the flip-flop is set, transfer control to the instruction whose address is specified and reset the OW flip-flop, otherwise use the instruction next in numerical order in the usual way". The necessary extra hardware for 01110 is listed in Chap. 9.

15.12 The Input Routine for the Ferranti Mark I*

The principal Mark I* system was known as the *Radix 32 Input and Organisation Scheme*. It was based upon "Schemes A and B of the present Manchester University computer, and on the FERUT Input Scheme, but incorporates other ideas as well".[21]

At the heart of the Radix 32 Scheme were the following code sequences:

1. The Input Routine, which includes subroutines *Write, Duowrite* and *Clear ½ track.*
2. The Routine Changing Sequence (RCS);
3. The Cueform routine (enables cues to be inserted at run-time);
4. The Rollcall routine (uses checksums to check that drum tracks that shouldn't have been changed have remained unaltered);
5. The Pagepunch routine (punches out code in a form acceptable to the Input routine).

These were held on drum tracks that were write-protected by a manual switch. In addition, a read/write area of the drum held data for the Directory, Checksums and Cueform.

The Input routine offered a repertoire of 15 warning characters, by which users could control the input of code and data and also cause certain actions to take place via the paper tape reader. An example is the frequently-used warning character K, whose effect is:

Kabn: read the characters which follow and assemble them into short lines which are deposited in locations ab to ab + n respectively, where $0 \leq n \leq 31$.

Other warning characters are, in brief:

- L return control to master routine;
- M Magnetic instruction;
- N input (a sequence of) numbers, either integers or fractions;
- P used for transferring cues;
- R used to form a magnetic read cue;
- S allows a standard subroutine to operate in 'any' page of primary store;
- T print (a sequence of) characters;
- U (two versions available). This calls subroutine Duowrite, which causes information to be written to, or read from, the magnetic drum.

---

[21] *A Radix 32 input and organisation scheme for Ferranti Mk I* digital computer.* Technical note (no number) from Ferranti Ltd. Moston. Foolscap typed manual, 37 pages. Undated but deduce about 1953 since "a number of modifications have been included at programmer's requests since the original version was produced a few months ago, and consequently it is believed that this final version has eliminated many of the disadvantages which have been hitherto inherent in the various Input Routines".

The Input routine performs a limited number of simple syntax checks and run-time checks (eg checksum error). Errors result in a series of *absolute*, *dummy* or *dynamic* stops. Each stop is at a particular address, with a particular pattern displayed. Users can therefore identify the problem and, if practical, recover from the error.

In summary, the Mark I*'s Radix 32 Scheme provided a primitive operating environment in which communication between the system and the user was carried out via the tape reader and the console's display tubes and handswitches. In this respect, it was a distinct improvement on the Mark I's operating environment.

15.13 Other Systems Software for the Ferranti Mark I*

The seven Ferranti Mark I* installations and their applications are described more fully in Chaps. 6–12. Below we highlight examples of some software systems above the level of machine code that were put to use at some of the seven sites. As far as is known, no such 'higher-level' programming took place at either Shell Amsterdam (Chap. 8) or GCHQ Cheltenham (Chap. 10). Of the other sites, the Mark I* installation at the Armaments Research Development Establishment (ARDE) at Fort Halstead in Kent was perhaps the most enthusiastic.

Intercode.

The people taking the initiative in developing system software at ARDE Fort Halstead included John Berry and James Gawlick. In 1956 John Berry introduced the Intercode scheme, described[22] as "an interpretive simplified coding scheme, designed for small computations only" … and "inspired chiefly by the Mark I Autocode". It is, however, quite modest in comparison with Brooker's work.

An Intercode program basically consists of a list of three-address instructions:

<function><operand 1><operand 2><operand 3>

The <function> is a single letter, of which the following are examples:

A = add; M = subtract; P = multiply; Q = divide;
J = jump; T = input from paper tape; E = print a number;
V_i = use the specified intrinsic function i (eg square root, exp, etc.)
W, U and I: these are used for manipulating B-lines.

Further details are given in Berry.[23] Only 100 named operands are allowed, each a floating-point number whose 'address' (name) is represented by a two-digit decimal number.

[22]Berry, F.J. 1959. Intercode, A Simplified Coding Scheme for AMOS. *Computer Journal* 2 (2): 55–58.

[23]Berry, F.J., Intercode, A Simplified Coding Scheme for AMOS.

Here is an example Intercode sequence for evaluating the sum of the squares of the numbers in locations 10–50 inclusive and placing the result in location 50. The program starts at store line 10 of page 01 of primary store.

Line	Fn	Op1	Op2	Op3	Interpretation/comment
10	A	00	00	51	[00] + [00] → [51]
11	W	B1	00	10	Set B-line 1 to the value ten
12	P	B1	B1	99	[B1] x [B1] → [99]
13	A	99	51	51	[99] + [51] → [51]
14	U	B1	00	01	Add 1 to B-line 1
15	I	B1	00	50	If [B1] > 50 skip the next order
16	J	01	12	00	Jump to page 1 line 12 if [00] ≥ 0

Notes

It is arranged that the contents of the variable whose name (address) is 00 is always zero. In other words, [00] = 0. This was a device for using a conditional jump as an unconditional jump.

A modified version of Intercode was used on the Ferranti Mark I* computer at Armstrong Siddeley Motors Ltd.

COSMOS and MIRFAC.

By 1959 the Mark I* at Fort Halstead was becoming unable to cope with ARDE's computing load. Because of the investment in programs coded for the Mark I*, Fort Halstead decided to design and build a new machine that was code-compatible with the Ferranti Mark I* but much more powerful. The new computer, called COSMOS, used transistors (not thermionic valves), was parallel (not serial) and was about 50 times faster as is described in Chap. 11. COSMOS was equipped with an operator's console that mimicked (simulated) the original Ferranti Mark I*'s console. There was a development period at Fort Halstead when COSMOS was running in parallel with the Mark I*. Then during January 1967 all the ARDE workload was transferred smoothly to COSMOS.

At ARDE James Gawlik created a new language for COSMOS called MIRFAC (Mathematics in Readable Form Automatically Compiled)[24] which became the main programming tool at Fort Halstead for 10 years. To quote[25]: "MIRFAC's central aim is to offer a language which makes it as easy as possible for the non-specialist to present problems to a computer. In support of this aim a special purpose typing machine–costing very little more than the standard [Friden Flexowriter] product–has been developed. The language is thus able to accept for compilation any mathematical equation written in standard text-book notation.

[24]Gawlik, H.J. 1963. MIRFAC: A Compiler Based on Standard Mathematical Notation and Plain English. *Communication of ACM* 6 (9): 545–547. See also: Gawlik, H.J., and F.J. Berry. 1967. *Programming in MIRFAC*. 2nd ed.

[25]Gawlik, H.J., MIRFAC: A Compiler Based on Standard Mathematical Notation and Plain English. See also: Gawlik, H.J., and F.J. Berry., *Programming in MIRFAC*.

Those statements in a problem which do not fall into this category–input, output, jumps and the like–are written in 'plain English' forms designed so as to avoid special conventions". It has to be said that mainstream programmers outside Fort Halstead such as Edsger Dijkstra did not think much of MIRFAC's philosophy!

INTINT.

At the Instituto Nazionale per le Applicazioni Calcolo (INAC) in Rome, it is believed that practically all the programming of their Ferranti Mark I* was done in machine code. An exception was a system called INTINT (INTerpretation-INTegration) for handling vectors designed by Corrado Böhm.[26] No further details have come to light. At INAC an attempt was made to prolong the life of their Ferranti Mark I* programs by providing a follow-on computer called CINAC with a special custom-built console and user interface similar to that of the Mark I*. By this means, existing Mark I* programs could be used directly via a simulator running on CINAC. For more details, see Chap. 9. FINAC was finally switched off sometime in 1966.

Tabular Interpretive Programme.

At the Atomic Weapons Research Establishment (AWRE) at Aldermaston in Berkshire their Ferranti Mark I* was mostly used in machine language. This was also initially true of the English Electric DEUCE computer that AWRE acquired in 1956 to run alongside the Mark I*. In 1957 AWRE acquired an IBM 704 which was dedicated to Aldermaston's larger programs whilst from 1958 to 1961 the DEUCE (and the Mark I*?) were used on short jobs for which easier programming became desirable. AWRE therefore began to use the General Interpretive Program written by NPL, the Tabular Interpretive Program written by Bristol-Siddeley Engines (see below) and DEUCE Alphacode written by English Electric. Meanwhile, the IBM 704 was being used with the Symbolic Assembly Program (SAP) written by the United Aircraft Corporation.

At A V Roe's Chadderton factory in north-east Manchester, their Ferranti Mark I* was used heavily for stress calculations. After a while, the aero designers at AVRO made use of what was described at the time as a 'tabular operating system'—most probably the Tabular Interpretive Programme (TIP) that had been written by the Mathematical Services Group of Bristol Siddeley Engines Ltd., Filton, Bristol in 1957.[27] TIP was made available for the English Electric DEUCE computers (as installed at Filton), the Ferranti Mark I* computers installed at Avro

---

[26]*INTINT programmazione indiretta per calcolatrici elettroniche.* Manuali per le Applicazioni Tecniche del Calcolo, Vol. 3, Cremonese, Roma (1958).

[27]*Tabular Interpretive Programme.* Bristol Siddeley Engines Ltd., Filton, Bristol. June 1961. 8-page printed manual, approx.. A4 size, bound in buff card. Tabular Interpretive Programme, TIP, was devised by the Mathematical Services group of Bristol Siddeley Engines Ltd. in 1957. TIP3, to which this manual refers, was completed in August 1960.

and at Armstrong Siddeley Engines Ltd. at Ansty, Coventry. All these aero companies, sooner or later, became members of the Hawker Siddeley Aircraft Group.

TIP formalised the tabular methods in common use in the 1940s and 1950s when setting out calculations for manual evaluation by electro-mechanical desk-top calculators. No specialised knowledge of a digital computer was needed. Thus TIP was a practical bridge between the old paper-and pencil traditions and the new world of automatic universal computers. In this respect, TIP had much in common with Spreadsheet systems developed ten years later for accounting purposes.

The TIP compiler reads in decimal *codewords*, checks these, and punches them out in a binary version for input to the TIP Interpreter. Each TIP codeword represented a sub-task in a larger calculation, where the calculations are in general performed on columns (vectors) of numerical data. Here are some example TIP codewords:

Codeword	Interpretation
a b c 2	(col. a) + (col. b) putting result in (col. c)
a b c 0	(col. a) x (col. b) putting result in (col. c)

TIP allows users to express data as floating-point decimal numbers to six significant figures, in the range $2^{-66} < x < 2^{62}$. The digit in the fourth column in the above examples denotes an operation. TIP has a repertoire of 32 operations, including: thc four arithmetic functions (+, −, x, /); standard functions such as square root, log and sine; looping constructs, subroutine entry/exit, input, output, and more.[28] TIP has 32 pre-set useful constants, N0 to N31, one of which is used to denote a 'dash' in a codeword (see example below). Here are two sample calculations to give the flavour of TIP:

Example 1 Form the expression $ax^2 + bx + c$ for a number of values of x. Assume the values of x are stored in column 0 and the coefficients a, b, c in N32, N33, N34.

Codeword					Description	
N32	0	1	0	form	ax	in column 1
1	N33	1	2	"	ax + b	in column 1
1	0	1	0	"	$ax^2 + bx$	in column 1
1	N34	1	2	"	$ax^2 + bx + c$	in column 1

[28]*Tabular Interpretive Programme*. Bristol Siddeley Engines Ltd., Filton, Bristol. June 1961. 8-page printed manual, approx. A4 size, bound in buff card. Tabular Interpretive Programme, TIP, was devised by the Mathematical Services group of Bristol Siddeley Engines Ltd. in 1957. TIP3, to which this manual refers, was completed in August 1960.

Example 2: Evaluate $y = \sinh(\cos^{-1}(x \log x))$ for a set of values of x, assuming the initial values of x are in column 1. The pre-set constants N1 = 1 and N2 = 2.

Codeword					Description	
0	–	1	8	form	log x	in col. 1
0	1	0	0	"	x log x	in col. 0
0	–	0	24	"	$\cos^{-1}(x \log x) = t$	in col. 0
0	–	0	9	"	e^t	in col. 0
N1	0	1	1	"	e^{-t}	in col. 1
0	1	0	3	"	$e^t - e^{-t}$	in col. 0
0	N2	1	1	"	$y = \sinh t = ½(e^t - e^{-t})$	in col. 1

References

Bennett, J.M., D.G. Prinz, and M.L. Woods. 1952. *Interpretative Sub-routines.* In *Proceedings of ACM National Conference*, 81–87, Toronto, Sept 1952.

Berry, F.J. 1959. Intercode, A Simplified Coding Scheme for AMOS. *Computer Journal* 2 (2): 55–58.

Brooker, R.A. 1955. An Attempt to Simplify Coding for the Manchester Electronic Computer. *British Journal of Applied Physics* 6: 307–311.

Campbell-Kelly, Martin. 1980. Programming the Mark I: Early Programming Activity at the University of Manchester. *Annals of the History of Computing* 2 (2): 130–168.

Gawlik, H.J. 1963. MIRFAC: A Compiler Based on Standard Mathematical Notation and Plain English. *Communication of ACM* 6 (9): 545–547.

Gawlik, H.J., and F.J. Berry. 1967. *Programming in MIRFAC.* 2nd ed.

Gotlieb, Calvin C. 1954. The Cost of Programming and Coding. *Computers and Automation* 25: 14 ff.

Ralston, Anthony. 1980. Random Number Generation on the Ferranti Mark I. *Annals of the History of Computing* 2: 270–271.

Sumner, F.H.(Frank). 1994. Memories of the Manchester Mark I. *Resurrection* 10: 9–13.

Tootill, G.C. 1949. *Informal Report on the Design of the Ferranti Mark I Computing Machine.*

Chapter 16
Appendix C. Mark I and Mark I* Hardware Details

16.1 Physical Layout of Sub-units in the Ferranti Mark I

The following explanation refers specifically to the first installation in the Computing Machine Laboratory at Manchester University in 1951. The general arrangements have been described in Chap. 2; these only differ in minor details from the Ferranti Mark I in Toronto (Chap. 3). There were also modest hardware improvements in the progression to the seven Ferranti Mark I* computers—improvements that we note later.

The first Ferranti Mark I is shown in Fig. 16.1, which should be compared with the view in Fig. 2.9 in Chap. 2. In addition to the equipment in the photograph there were motor-generators and the power supply control panel, placed in a separate room. These are described later.

The disposition of sub-units within the two main bays of the Ferranti Mark I are shown in Fig. 16.2.

In Fig. 16.2 the bolder horizontal bars either represent doors containing logic circuits or boxed CRT units. Photographs of these appear in Figs. 16.5, 16.6 and 16.7. There are 18 longer and 12 shorter bars in the upper bay and 20 longer and 8 shorter in the lower bay. Reading from left to right, the bars in the bays in Fig. 16.2 are labelled thus in the original diagram:

S. Lavington, *Early Computing in Britain*, History of Computing,
https://doi.org/10.1007/978-3-030-15103-4_16

Fig. 16.1 The console and the two main bays of the Ferranti Mark I. The lower sections of the right-hand bay in the photo house the CRT units of the primary store (RAM) and the control register. The magnetic drum secondary store is located at the far end of the left-hand bay, out of sight in the photo

Upper bay in *Fig.* 16.2:

Top row	main stores (4 shorter bars and two un-labelled (spare storage slots)); store X&Y time base; EHT; three un-labelled (spare slots); control store.
Second row	clock & basic waveform (100 kc/s); monitor X time base; basic waveforms 4 kc/s approx..; pulse separator; function & tube staticisor; B staticisor and B and C waveforms.
Third row	magnetic transfer check; inward and outward transfer gates; action waveforms (1 kc/s approx.); magnetic transfer control waveforms; line staticisors; control circuits.
Bottom row	main store gate circuits (3 + 1 longer bars); Y raster counters; C Y time base and gate circuit.

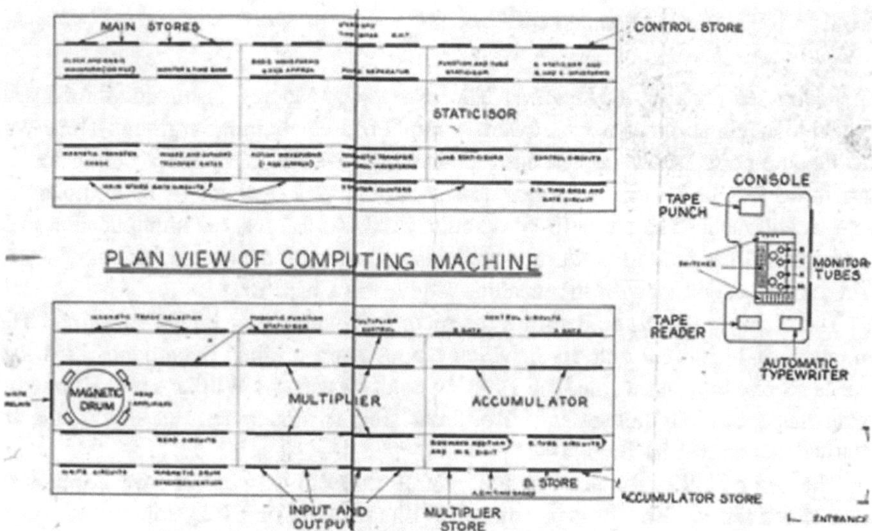

Fig. 16.2 Location within the two bays of the storage, logic and timing circuits. The magnetic drum may be seen at the left-hand end of the lower bay in the diagram. The units labelled *main stores, control store, B store, multiplier store* and *accumulator store* were all implemented as Williams/Kilburn electrostatic tubes (CRTs). *Staticisor* indicates flip-flop registers. Refer to the text for more explanation

Lower bay in *Fig.* 16.2:

Top row	magnetic track selection (2 longer bars); magnetic function staticisor; multiplier control; control circuits: B gate; control circuits: A gate.
Second row	one long unlabelled bar; magnetic function staticisor; multiplier; accumulator; accumulator.
Third row	one unlabelled bar; read circuits; multiplier; multiplier; sideways addition and m.s. digit; B. tube circuits.
Bottom row	write circuits; magnetic drum synchronisation; input and output (4 shorter bars); multiplier store; A.B.M. timebases; B. store; accumulator store.

Notice the relatively large proportion of equipment taken by the multiplication circuits in Fig. 16.2. The multiplier was reckoned to consume almost a quarter of the 4050 thermionic valves in the Ferranti Mark I.[1] Of the 4050 total valves, over half were EA90 (VR91) miniature signal diodes and most of the rest were EF50 (VR92) and EF55 RF pentodes and 12AT7 (CV455) which were RF double triodes. In all, the computer dissipated 25 kW of heat.

[1]Robinson, A.A. 1953. Multiplication in the Manchester University High-Speed Digital Computer. *Electronic Engineering* 6–10.

16.2 Register-Level Architecture

In modern terminology, the Ferranti Mark I had a single-address instruction format with a double-length accumulator, hardware multiplier and eight index registers. There was no floating-point hardware. The instruction-length was 20 bits. The accumulator, of length 40 bits or 80 bits according to instruction, was used for two's complement integer arithmetic. The basic 40-bit addition time was 1.2 ms, the multiplication time 2.16 ms and the time to produce a hardware-generated random number was 5.8 ms. The programmers' view of the machine is given in Chap. 15.

The diagram in Fig. 16.3 shows the main data paths between the sub-units. The notation of Figs. 16.2 and 16.3, which comes from original documents such as,[2] needs more explanation since the 1948/49 notation used by Williams and Kilburn at Manchester has long fallen out of use. Translating to modern terminology yields the equivalents shown in Table 16.1.

The use of CRTs for most of the central registers was adopted on grounds of cost-effectiveness. The cost-per-bit of Williams/Kilburn CRT storage was very much less than that of contemporary valve flip-flop registers, though flip-flops were faster. The whole timing and rhythm of the Mark I was linked to the requirements

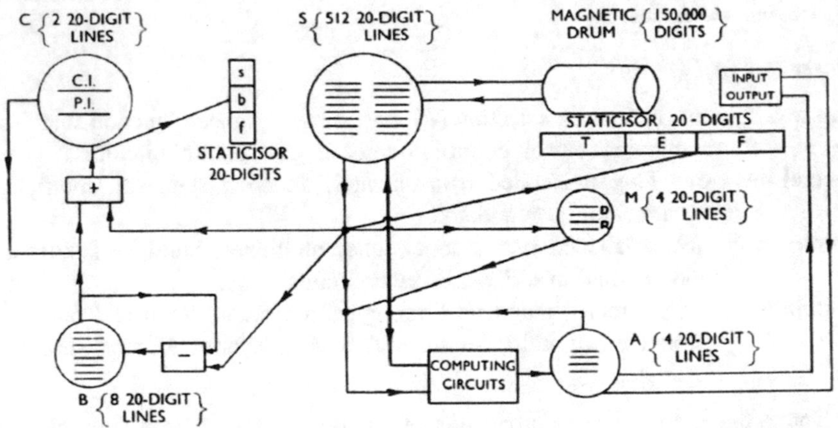

Fig. 16.3 The low-level architecture of the Ferranti Mark I, as presented in (Williams, F.C., and T. Kilburn., The University of Manchester Computing Machine.)

[2]Williams, F.C., and T. Kilburn. 1951. The University of Manchester Computing Machine. *Proceedings of the Manchester Computer Inaugural Conference*, 5–11, 9th–12th July 1951. This paper was also presented at the joint AIEE-IRE Computer Conference, Philadelphia, December 1951.

Table 16.1 An explanation of the terminology of Fig. 16.3

Notation in Fig. 16.3	Modern explanation
CI	20-bit Control line (Program Counter)
PI	20-bit Present Instruction
s/b/f staticisor	Flip-flop register holding the current 20-bit instruction: s = 12-bit address; b = 3-bit index register no.; f = 5-bit function (op code)
S	Main memory (RAM), initially 8 CRTs—see text
T/E/F staticisor	Control word for transfers to/from drum and for input/output to/from paper tape. For drum: T = track no.; E = CRT identity; F = direction of transfer
M	Multiplier logic
D	40-bit operand register for the multiplier
R	40-bit multiplier register for the multiplier
A	80-bit double-length Accumulator
B	8 index registers, each 20 bits

of the CRT storage. The digit frequency was a conservative 100 kHz. More on storage technologies is given below in Sect. 16.3.

16.3 Circuit Technology

Turning to the Ferranti Mark I's electronics, a basic flip-flop circuit using EF 50 pentodes is shown in Fig. 16.4. Negative-going logic pulses were employed, going from about +2.5 V to −15 V or more. The practical implementation of this and similar circuits for timing and control owed much to the experience gained in the development of pulse circuits for radar during the Second World War. A fuller account of the Mark I's electronic circuitry is given.[3] The principle thermionic valves used were EF50 and EF55 pentodes and EA90 diodes. EF50s had been employed in great numbers for radar equipment during the war and quantities were readily available during the immediate post-war years. The EF55 was used where greater current-drive or higher-frequency response was needed.

[3]Williams, F.C., A.A. Robinson, and T. Kilburn. 1952. Universal High Speed Digital Computers: Serial Computing Circuits. *Proceedings of IEE* 99 (Part 2, 68).

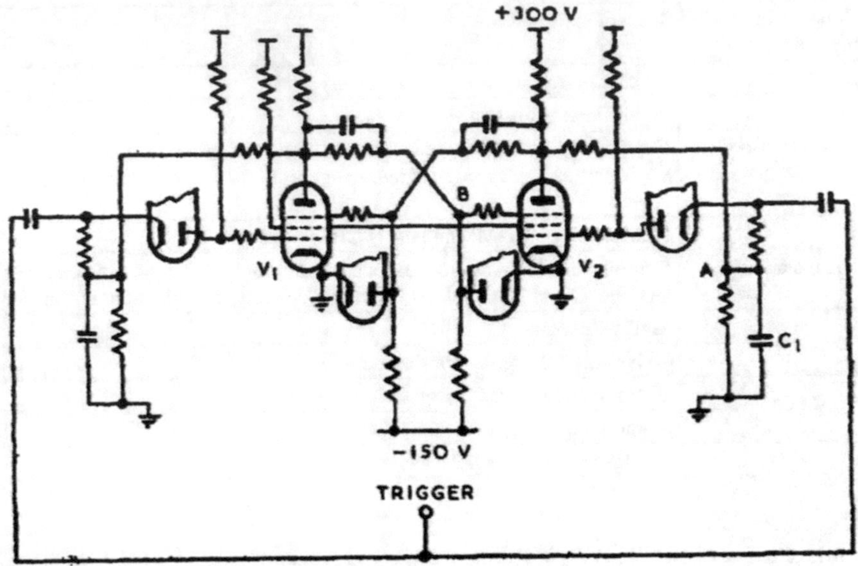

Fig. 16.4 A basic flip-flop circuit, using two EF50 pentodes and four EA90 diodes

16.4 Storage Technology

The choice of primary memory technology was the single most important problem facing the designers of early digital computers.

In the Williams/Kilburn system, information was stored as small areas of electrostatic charge on a phosphor-coated glass screen. The phosphor coat was on the inside of one end of a cathode ray tube (CRT), at the other end of which was a controlled source of electrons (the cathode) which produced a focussed beam. This beam of electrons was made to bombard selected areas of the phosphor screen, forming an array of small charged spots. The spots were made to be of two types, for example focussed/defocussed or dot/dash, each type corresponding either to a binary 0 or binary 1. Thus, an array representing binary digits could be 'written' as charged spots on the phosphor screen.

On the outside (front) of the CRT's screen was a metal pick-up plate. Each time a charged spot was bombarded, a small voltage was induced in the pick-up plate. Two types of voltage signal were detected, depending upon whether the selected spot was storing a 0 or a 1. So two types of stored binary digit could be 'read' and passed to the main computer.

The electrostatic charge in each stored spot unfortunately leaked away within a few milliseconds and so a mechanism had to be found for regenerating the charged spots. The crucial invention of F. C. Williams was to devise a neat way of

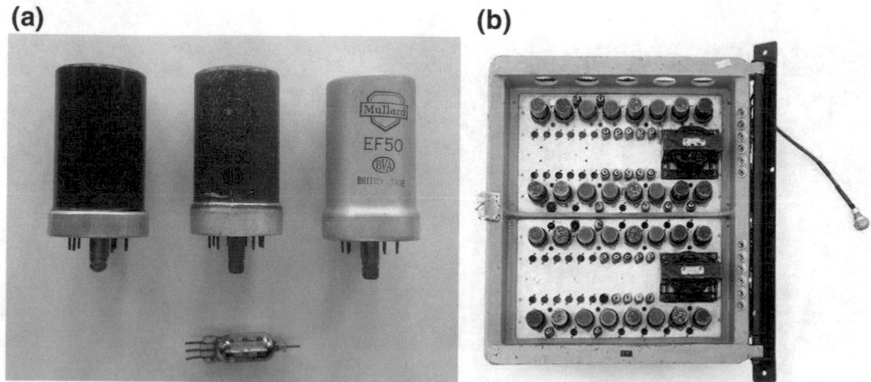

Fig. 16.5 **a** Three sample EF50 thermionic valves and an EA90 diode. The left-hand EF50 has the stamp of the Air Ministry, an indication that large quantities of these pentodes became available after the end of the Second World War. **b** A smaller logic door from a Ferranti Mark I*. The red valves are EF50 s and the black valves are EF55s

Fig. 16.6 **a** and **b** The front and back of a larger logic door from a Ferranti Mark I* computer. Units such as this could contain up to 12 flip-flop circuits. The three black boxes on the left-hand photo are 50 Hz heater transformers. The Mark I had smaller 1600 Hz transformers, which can be seen in the photo of Fig. 2.8 in Chap. 2

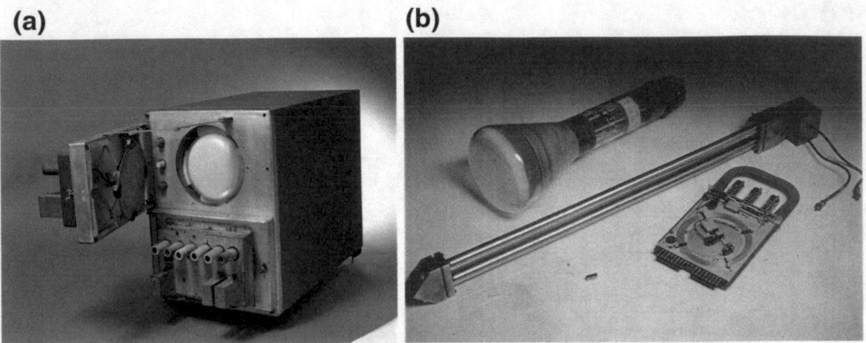

Fig. 16.7 **a** A twin Williams/Kilburn CRT storage unit from a Ferranti Mark I* computer. The pick-up plate and pre-amplifier of the upper tube has been opened back to reveal the CRT. **b** Memory technologies of the 1950s compared. A CRT is shown with (centre) a long mercury delay line from the **English Electric DEUCE** computer and (lower) a magnetostrictive delay line from a Ferranti Pegasus computer. A 1970s dual-in-line silicon integrated circuit chip is shown in the centre foreground. This could store more information than the other three devices combined

spot-regeneration, based on his observation of the precise shape of voltage signals induced in the CRT's pick-up plate. He realised that, because of the distribution of electrostatic charge, a scanning or 'interrogating' electron beam produced a voltage pulse on the pick-up plate that anticipated the state of each charged spot and steps could be taken to regenerate a 0 or 1 as appropriate. Although any off-the-shelf Cathode Ray Tube could be used, best results were achieved from tubes with very low internal phosphor contamination and such tubes were developed by GEC at their Wembley Laboratories in north London and supplied to Manchester. For more details, refer to the original papers.[4]

In summary, a Williams/Kilburn Tube and its surrounding electronics was able to write and read binary information at electronic speeds. Furthermore, since the electron beam could be aimed at any chosen spot on the screen, the time taken to select a particular spot was independent of that spot's position on the screen. In a word, the Williams Tube was what we now call a Random Access Memory (RAM). This contrasted with other contemporary storage systems such as mercury delay lines or magnetostrictive delay lines whose access was strictly sequential. Mercury delay lines were well-proven but were expensive per bit and required strict temperature control. Williams/Kilburn tubes used off-the shelf components and were

[4]There are many papers describing the Williams/Kilburn electrostatic storage scheme, for which the first patent was filed by Williams on 11th December 1946. Here are two of the early reports: (a) Kilburn, T. 1948. *A Storage System for Use with Binary Digital Computing Machines*. Ph.D. thesis, University of Manchester. A first version of this thesis with the same title was written as an internal report for TRE. This report was then circulated by the Dept. of Electrotechnics, University of Manchester, dated 1st December 1947. Several copies are known to have reached the USA.; (b) Williams, F.C., and T. Kilburn. 1949. A Storage System for Use with Binary Digital Computing Machines. *Proceedings of IEE* 98 (Part 2, 30): 183 ff.

cheapest per bit but were sensitive to electromagnetic interference, for example from nearby electrical rotating machinery. Magnetostrictive lines were the most robust. Some early computers used magnetic drums as their primary memory. Drums were essentially sequential and much slower than the above technologies.

In a practical computer such as the Ferranti Mark I, a monitoring CRT (without the pick-up plate) could be driven in parallel with the Williams/Kilburn Tube, enabling the stored charge pattern to be observed on the monitoring screen. Furthermore, the spots on a monitoring tube could be displayed in 5-bit group, making it easy to associate bit-patterns with teleprinter characters—(see also Chap. 15 and Fig. 15.1). A Ferranti Mark I programmer therefore had an immediate visual representation of the instructions and data as held in memory.

The Ferranti Mark I as initially installed at Manchester University has eight CRTs in its primary store, holding a total of 10,240 digits, equivalent to 32 40-bit words or 64 20-bit half-words per tube. Digits were displayed as lines of 5-bit characters on the operator's monitoring screens, as shown in Chap. 15. The Mark I at Manchester and later Ferranti Mark I* computers each had more CRTs, up to a total of 16 per computer. An additional CRT buffer store was provided later, when a Bull lineprinter was installed—(see for example the Rome Mark I* of Chap. 9.

A survey of nine American pioneering computers to have come into operation in the period 1950–1952 reveals that five used mercury delay lines, two used Williams/Kilburn electrostatic storage, one (Whirlwind at MIT) used another type of electrostatic storage and one (ERA 1101) used a magnetic drum. Of six British pioneering computers to have come into operation in the period 1948–1953, three used mercury delay lines, two used Williams/Kilburn electrostatic storage and one (the Elliott Nicholas) used magnetostrictive delay lines. Further details are given.[5]

The secondary, or backing, storage for the Ferranti Mark I consisted of the magnetic drum shown in Fig. 16.8. Magnetic drums had been used for digital storage from at least 1947 in America (for example at Engineering Research Associates (ERA)). The Manchester University drum used a read/write head structure similar to one developed by A D Booth at Birkbeck College, London, for his ARC2, SEC and APE(X)C computers. However, the rest of the Mark I's drum electro-mechanics and the unique recording system (phase modulation) were developed at Manchester. Unusually for drum systems of the 1950s, the Manchester drum motor was synchronised to the computer's clock rather than the other way round. This meant that, in principle, several drums could be added to one computer —though in practice all the Ferranti Mark I and Mark I* machines only had one drum each. More technical details are presented.[6]

[5]Lavington, Simon. 1980. *Early British Computers*. Manchester: Manchester University Press. This is out of print but has helpfully been made available at: http://ed-thelen.org/comp-hist/EarlyBritish.html.

[6](a) Thomas, G.E. 1950. Magnetic storage. *Report of a Conference on High Speed Automatic Calculating Machines*, 75–80. Cambridge, 22nd–25th June 1949. Published by the Cambridge: University Mathematical Laboratory.; (b) West, J.C., and F.C. Williams. 1951. The Position Synchronisation of a Rotating Drum. Proceedings of IEE 98 (Part 2, 61).

The initial plan was for the Ferranti Mark I at Manchester to have a six-inch diameter nickel plated drum of capacity 150,000 bits.[7] The final installation had a ten-inch diameter drum with a capacity of up to 655,360 bits on 256 tracks at a packing density of 165 bits/inch. One of the frequent comments in engineers' log books in the early days was that the number of tracks available to users was considerably less that the maximum, probably due to imperfections in the drum's magnetised surface. The data-transfer rate was such that it took 36 ms to transfer a track.

16.5 Input/Output Equipment

Initially, input was via a 200 character/s 5-track paper tape reader which used photo-electric cells to sense the punched holes. The transfer rate was later increased to 300 characters/s. Output was to a paper tape punch and/or to a teleprinter, in both

Fig. 16.8 The Ferranti Mark I's 10" diameter original magnetic drum store

[7]Williams, F.C., and T. Kilburn., The University of Manchester Computing Machine.

cases at the slow rate of 10 characters/s. The Ferranti Mark I at Manchester and many of the Mark I* installations were later equipped with Bull lineprinters, printing at a rate of 250 lines/min.

All the Ferranti Mark I and Mark I* sites naturally had off-line paper tape editing equipment. The standard initial Ferranti package consisted of two keyboard perforators each with a tape reader, a keyboard transmitter and a re-perforator, one teleprinter and one control unit. This was added to by most sites.

16.6 Power Supplies

The main HT power rails and with their current drains were: +300 V at 10 A, +200 V at 12 A, −150 V at 12 A. In the Ferranti Mark I these voltages were supplied by separate motor-generators, suitably stabilised electronically. Interlocked control circuits were provided to ensure correct power-up and power-down, everything being brought together to a central PSU control and monitoring cabinet. A photo of a similar PSU control cabinet may be seen in Chap. 3, Fig. 3.6

The thermionic valve heaters took 1300 A at 6.3 VAC. Since it was at first thought that the CRT stores would be affected by electromagnetic interference from heavy currents at 50 Hz, a higher heater frequency was chosen at which EM screening would be simpler. Suitable motor-generators giving 115 V at 1600 Hz were available, together with miniature sealed transformers originally developed for the aircraft industry. 1600 Hz was therefore chosen as the heater supply frequency. A miniature transformer was mounted on each logic chassis.

By the end of 1952, experience had shown (i) that CRT stores could be readily screened from 50 Hz EM interference and (ii) that commutator ripple on the generator sets was becoming a maintenance annoyance. The power supply arrangements for the Mark I* computer were therefore rationalised so that *all* power was derived from a single motor-alternator set of high inertia providing stabilised 50 Hz AC. Metal rectifiers and electronic smoothing were used for the HT supplies and 50 Hz transformers, such as those shown in Fig. 16.3, were used for the valve heaters.[8]

For the Ferranti Mark I* machines, a simple arrangement was designed that depended upon a single motor/alternator as the power source for the whole computer installation. It may be instructive for modern readers to have an idea of the procedures involved in switching on a Mark I* computer at the start of the day and then switching it off – (though most installations ended up by keeping the machine running continuously). Here is a summary of the instructions, as defined by a

[8]Pollard, B.W., and K. Lonsdale. 1953. The Construction and Operation of the Manchester University Computer. *Proceedings of IEE* 100 (Part 2): 501–512.

GCHQ engineer for the Mark I* installation at Cheltenham.[9] The instructions refer to a 'control and monitoring cubicle'; a photo of one of these cubicles or cabinets is given in Chap. 8, Fig. 8.7.

Start the alternator. Switch on the power to the PSU control and monitoring cubicle. After hearing the main contactors operate for the star-delta starter, switch on the valve heater controller—which will bring the heaters up to their full power in about 80 s. The refrigeration equipment automatically comes on. After the required 80 s, switch on the HT voltages. Check that the correct voltages and currents are registering on the control panel. The magnetic drum should achieve synchronism with the CPU clock within 15 min of switching on the HT. Two neon indicators on the console show whether the drum is running slow, fast, or in synch. When it is clear that the drum is in synchronism, the computer is ready to use.

When switching off the computer, first switch off the HT voltages. When the indicator meters have fallen to zero, switch off the heaters. Wait 80 s. Then switch off power to the control and monitoring cubicle. The refrigeration equipment will power down. (Note that the refrigerator works on the basis of constant heat extraction rather than constant temperature. Thus if the HT is left off but power is left on to the control and monitoring cubicle, the machine will freeze after a short time). Finally, switch off the alternator.

16.7 Reliability

The initial reliability of the Ferranti Mark I at Manchester University was measured for the 61-week period starting 1st September 1951.[10] 343 faults were logged during a scheduled operating period of 3791 h, giving a failure on average every 11.1 h. Overall, the computer operated at about 85% efficiency. The formula used was:

$$\text{Operating efficiency} = (\text{TO} - \text{TF})/\text{TF} \times 100\%,$$

where TO = total scheduled operating time and TF = fault time. A detailed analysis of the various types of fault is given.[11] A set of detailed test and diagnostic programs had been written by an M.Sc. student, R. L (Dick) Grimsdale.

The efficiency of Manchester's Ferranti Mark I went up as the machine settled down and preventive maintenance procedures were improved. By 1958, an examination of the log books reveals that users were usually signing off with the

[9] *Notes on the operation of CLEOPATRA*. J. H. Ellis. Internal GCHQ memo X/4832/2002, 13th July 1954. Three typed pages.

[10] Pollard, B.W., and K. Lonsdale., The Construction and Operation of the Manchester University Computer.

[11] Pollard, B.W., and K. Lonsdale., The Construction and Operation of the Manchester University Computer.

words "machine perfect". This remark should be interpreted according to the norms of computing in the 1950s, where all installations had scheduled daily maintenance periods and the maintenance engineers were normally on stand-by 24 h a day. All virtue is by comparison.

Reliability figures for other Mark I and Mark I* installations have mostly not come to light. An exception is the Mark I* installed at the Instituto Nazionale per le Applicazioni Calcolo (INAC) in Rome. It was recorded that: "During the first three years of operation (30 June 1955 - 30 June1958) the efficiencies were 89.1, 91.2, and 92.2%"—see Chap. 9.

In Chap. 8 there is an indirect comment on the reliability of the Ferranti Mark I* at Shell's Amsterdam Laboratory. An American employee at the US Embassy in London prepared a detailed survey of West European computer installations in 1957.[12] Of Shell's Ferranti Mark I* at Amsterdam he commented: "their utilisation record is one of the best the writer has ever encountered".

16.8 Later Hardware Modifications

It is known that from time to time, minor in-service hardware improvements were made to certain Ferranti Mark I and Mark I* installations. These were in addition to the installation of conventional input/output equipment such as lineprinters and magnetic tape decks. The surviving records describing the hardware improvements are sparse.

For the Mark I* that went to the Instituto Nazionale per le Applicazioni Calcolo (INAC) in Rome, Chap. 9 mentions three hardware improvements, implemented respectively during 1956, 1957 and 1958. In brief:

In 1956: a new overflow instruction was introduced that used the unassigned function code 01110. "The components involved in the construction of the new [overflow] circuitry and in the modification of the existing circuits are the following: 36 valves (10 pentodes, 14 double triodes and 12 double diodes), 10 capacitors, 10 crystal diodes and 140 resistors. A whole new chassis is needed".

In 1957, new circuits were added to aid double precision arithmetic. "For double precision two orders are used to set or reset a flip-flop. When the flip-flop is in the reset state the computer works as usual, that is, the instructions regarding accumulator and multiplier refer to 40 bits. When the flip-flop is in the set state the same instructions refer to 80 bits. A "jump if overflow occurred" instruction controls both simple and double precision arithmetic. This allows fixed point operation also in

[12]Hoffman, A.J. 1957. *New Computers in France and the Netherlands*, 1. This document is archived as Box 216, Backman papers, National Museum of American History, Mathematical branch, Smithsonian Institution, Washington DC. The quotation is cited as given on page 164 of: Cordata, J.W. 2012. *The Digital Flood: The Diffusion of Information Technology Across the US, Europe and Asia*. Oxford. ISBN 978-0-19-992155-3.

case of ill conditioned systems of linear equations, speeding up computing time. For the new circuits 106 valves have been installed and 37 removed".

In 1958, a transistorized flip-flop register was installed to monitor the address of the last Jump-instruction performed by the machine. "This facility is particularly useful in debugging and trouble shooting. A set of transistor counters driving electromechanical counters is being installed for the purpose of analyzing the frequencies of occurrence for the various types of instructions. The connection between valve circuits and transistor circuits is done through more or less conventional crystal diode circuits".

References

Cordata, J.W. 2012. *The Digital Flood: The Diffusion of Information Technology Across the US, Europe and Asia*. Oxford. ISBN 978-0-19-992155-3.

Hoffman, A.J. 1957. *New Computers in France and the Netherlands*, 1.

Kilburn, T. 1948. *A Storage System for Use with Binary Digital Computing Machines*. Ph.D. thesis, University of Manchester.

Lavington, Simon. 1980. *Early British Computers*. Manchester: Manchester University Press.

Pollard, B.W., and K. Lonsdale. 1953. The Construction and Operation of the Manchester University Computer. *Proceedings of IEE* 100 (Part 2): 501–512.

Robinson, A.A. 1953. Multiplication in the Manchester University High-Speed Digital Computer. *Electronic Engineering* 6–10.

Thomas, G.E. 1950. Magnetic storage. *Report of a Conference on High Speed Automatic Calculating Machines*, 75–80. Cambridge, 22nd–25th June 1949. Published by the Cambridge: University Mathematical Laboratory.

West, J.C., and F.C. Williams. 1951. The Position Synchronisation of a Rotating Drum. Proceedings of IEE 98 (Part 2, 61).

Williams, F.C., and T. Kilburn. 1949. A Storage System for Use with Binary Digital Computing Machines. *Proceedings of IEE* 98 (Part 2, 30): 183 ff.

Williams, F.C., and T. Kilburn. 1951. The University of Manchester Computing Machine. *Proceedings of the Manchester Computer Inaugural Conference*, 5–11, 9th–12th July 1951. This paper was also presented at the joint AIEE-IRE Computer Conference, Philadelphia, December 1951.

Williams, F.C., A.A. Robinson, and T. Kilburn. 1952. Universal High Speed Digital Computers: Serial Computing Circuits. *Proceedings of IEE* 99 (Part 2, 68).

Chapter 17
Appendix D. Naming Names

The ten-year period 1948–1958 was of some significance to the people who found themselves involved with computers at Manchester University and at Ferranti's Moston factory. All were to look back in later life and remember the unexpected challenges thrown their way. In retirement they reminisced, sometimes ruefully but more often with secret satisfaction, on the small contribution they made to the promotion of the modern computer. And some were aware, even at the time, that they were breaking new ground.

The University's staff records have survived. And the people who worked in and around the University's Computing Machine Laboratory have their memorials: mentions by name in scientific papers, government reports, letters and history books. All that remains is to add anecdotes and photos—which is what we do in this Appendix.

With Ferranti Ltd. the case is much less clear. Few fragments of the Ferranti company's personnel records survive. Internal and external Ferranti reports are now thin on the ground, with the names of contributors frequently omitted. This Appendix is therefore firstly about Ferranti staff—a compilation of anecdotes that will go some way to remembering those whose collective efforts pulled academic research out of the Lab and into products and then helped to apply these products in the wider world.

In presenting the material, an arbitrary division has been made between hardware and software, between the 'engineers' and the 'programmers' (who were usually mathematicians). This crudely represents the initial backgrounds of the individuals mentioned. In truth, as the years went by there came to be a considerable overlap in activities.

S. Lavington, *Early Computing in Britain*, History of Computing,
https://doi.org/10.1007/978-3-030-15103-4_17

17.1 Ferranti Engineers Working on the Mark I and Mark I*

Naming all those who contributed to the Ferranti Mark I and Mark I*'s hardware cannot be done with certainty but Table 17.1 goes some way in recognising individual contributions.

The annotations *D* or *M* in column 3 of Table 17.1 stand for *Development* or *Manufacturing*. Adopting the convention in use at West Gorton in 1958, *Development* signified staff in the Labs who designed the computer, made the prototype work and released it to *Manufacturing* where the customers' machines were manufactured, commissioned, dispatched and brought into full operation on customer's premises. Referring back to Moston and the early Mark I/Mark I* efforts, D or M possibly indicate those with and without a university degree qualification—though precise evidence is sparse. In any case, as with all pioneering projects, there was some overlap in de facto contributions and many who were initially involved in Mark I maintenance joined the design teams of later Ferranti computers.

The following *Notes on* Table 17.1 start with a general comment. Then a few personal details, where known, are given on named individuals.

1. The name appears in the Log Books of the Ferranti Mark I at Manchester University as a maintenance engineer.[1] In several cases, this probably signified part of an initial training period. People named in the Log Books such as Roy Duffy, Allan Ellson and Ted Hodgkinson certainly went on to contribute more significantly to Ferranti's computing endeavours.
2. **Vivian Bowden**, later Lord Bowden, obtained a Ph.D. in Physics from Cambridge and worked at TRE during the war. He joined Ferranti after being a partner for a couple of years in Sir Robert Watson-Watt's firm of consulting engineers. At Ferranti he both promoted the sale of computers and influenced the company's computer marketing strategy in the 1950s. Olaf Chedzoy remembers[2] that: "He could get a good overall view and expound clearly and persuasively. He was the perfect man to deal with politicians. The idea to write a book [*Faster than Thought*, published in 1953] was extremely timely, and no doubt raised the interest level in computers". Of Bowden, Mary-Lee Berners-Lee said[3] "He was an amazingly able man, with a very good sense of humour". Bowden left Ferranti in 1953 to become Principle of the University

[1]Some surviving log books from the Manchester University's Ferranti Mark I computer, 1951–58, are preserved in the National Archive for the History of Computing, catalogued as NAHC/MUC/2/C6.

[2]Olaf Chedzoy, e-mail dated 6th November 2015 to Simon Lavington.

[3]Mary Lee Berners-Lee: An Interview Conducted by Janet Abbate for the IEEE History Center, 12 September 2001. Interview #578 for the IEEE History Center, The Institute of Electrical and Electronic Engineers, Inc. See: https://ethw.org/Oral-History:Mary_Lee_Berners-Lee#About_Mary_Lee_Berners-Lee.

Table 17.1 Tentative list of all Ferranti engineers who contributed to the hardware design, maintenance and operation of the Mark I and Mark I* computers

Name	Dates at Ferranti	Initial role	Comments
? Bainbridge	(1952)	M	Note 1
K. P. (Ken) Balme	(1953)	M	Note 1
B. V. (Vivian) Bowden	1950–53	Marketing; R&D	Note 2
Bruce Brown	?	M	Note 3
J. D. Carter	?	General Manager	Note 4
Adrian Clarke	Joined 1952?	M	Note 1
Frank Cooper	?	D	Notes 4 and 5
Roy Duffy	Nov. 1951 on	M	Note 1 and 6
Allan Ellson	May 1951 on	M	Notes 1 and 7
Owen Ephraim	1953 on	M	Note 8
G. Fox	?	D	Note 4
John Freer	From 1950?	D	
Eric Grundy	1931–1971	D; Manager	Note 9
Peter Hall	1950 on	D	Note 10
Len Hewitt	1955–57	M	Note 11
R. E. (Ted) Hodgkinson	1951 on	M	Notes 12
John Leech	1951 ?	D	Note 13
Keith Lonsdale	Sep. 1948	D	Note 14
H. Malbon	?	D	Note 4
Mike Moore	?	M	Note 15
Brian Pollard	1948 ?	D; Manager	Notes 4 and 16
Charlie Portman	1954 on	D	Note 17
Alec Robinson	Oct. 1949 on	D	Note 18
Dave Robinson	?	M	Note 19
Ron St. John	?	M	Note 20
Thompson, John E	?	D	-
Geoff Tootill	1949	D	Note 21
Bill Wallace	1950?	D	Note 22
Ken Wallis	1950?	M	Note 23
E. T. (Ianto) Warburton	1948 on	D	Notes 4 and 24
Brian Welby	1951?	D	Note 4
David Wilde	?	M	Note 25
Martin Wingstedt	Nov. 1951 on	M, D	Notes 1 and 26
Dick Vogel	1951	M	Note 27

Fig. 17.1 B. V. (Vivian) Bowden, later Lord Bowden, promoted Ferranti's computers vigorously in the UK and America from 1950 to 1953. He was, in effect, Britain's first computer salesman

of Manchester's Institute of Science and Technology (UMIST). In 1964/65 he served as Minister for Education and Science in Harold Wilson's Labour government, after which he returned to UMIST.

3. **Bruce Brown** was present during the commissioning of the Shell Mark I* at Amsterdam and for the commissioning of FINAC in Rome.
4. In the Ferranti paper on the engineering production of the Mark I[4] there is an acknowledgement of thanks to "the following members of the Ferranti computer section for their many contributions to the design of this machine: G. I. Thomas, B. G. Welby, H. Malbon, E. T. Warburton, F. Cooper, G. Fox and Dr. A. A. Robinson". More information on most of these individuals is given below. The inclusion of G. I. Thomas is perhaps a political gesture. Thomas, Manager of Ferranti's Radio Department at Moston, was the senior relevant person when Manchester University began to liaise with Ferranti. However he quickly faded, to be replaced by Brian Pollard who was Manager of the Moston Instrument Department and later of the Computer Department.
5. **Frank Cooper** was present at Manchester University late in 1951 when the BBC recorded the Mark I playing melodies.
6. **Roy Duffy** joined Ferranti in November 1951, thereafter being involved for most of his life in University/Ferranti projects. In the 1970s he joined the

[4] Pollard, B.W., and K. Lonsdale. 1953. The Construction and Operation of the Manchester University Computer. *Proceedings of IEE* 100 (Part 2): 501–512.

University's Regional Computing Centre, working with the ICL 1906A and other Regional Centre computers.

7. **Allan Ellson** was involved in the initial commissioning of the Mark I in the summer of 1951 and in commissioning all subsequent Mark I/Mark I*machines at Moston and later at Gem Mill where he was Head of Commissioning.
8. **Owen Ephraim** Joined Ferranti in 1953 after leaving the RAF. His first job was being trained to maintain the Mark I at the University. He then became a commissioning engineer on Mark I*s at Moston and Gem Mill. He took a Mark I* to Rome, returning in July 1966 to work on Pegasus etc.[5]
9. **Eric Grundy** had worked with Arthur Porter during the war. According to[6] Arthur Porter, by then at the Military College of Science, recommended to Grundy in about 1947/8, that D. G. Prinz should investigate computers—see Chap. 1.
10. **Peter Hall** became Manager of Ferranti's Computer Department in December 1958, following the resignation of Brian Pollard. See Chap. 13.
11. **Len Hewitt** spent two years with Marconi after his National Service in the RAF where he obtained radio experience. He joined Ferranti in mid-1955 and was based at Gem Mill. Len helped to install the Mark I* at Avro Chadderton. He left Ferranti in mid-1957 to maintain a Ferranti Pegasus at ICI. When Len joined Ferranti in 1955, he says that units of the Mark I* were "made at Moston and then brought over to the commissioning floor at Gem Mill for the final assembly and testing".[7]
12. **R. E. (Ted) Hodgkinson**. There is scope for a little confusion here, because Ted had a brother Neil. However, concentrating on R. E. (Ted), he was the first active maintenance engineer on the University's Ferranti Mark I and was, according to Tony Brooker, 'quite a bright person'. He was amongst several Ferranti people to contribute to an initial course on the Mark I*, held at Fort Halstead from 1st to 12th November 1954. The subject of Ted's lecture was "The functional operation of the Mark 1* computer".
13. **John Leech** graduated from Cambridge in the summer of 1951 and thereupon joined Ferranti. John left Moston in 1954 to return to Cambridge as an EDSAC research student, then left in 1959 to go to Glasgow University, and from thence in 1968 to the University of Stirling to found its Computer Science Department.
14. **Keith Lonsdale** graduated from the College of Technology, University of Manchester, in 1945. He worked at Salford Electrical Instruments, then for AC. Cossor and then joined Ferranti in September 1948 as a development engineer. He was appointed Senior Engineer in 1950 and Chief Engineer of the newly formed Computer Department in 1953. In 1961 he was appointed Manager of

[5]Owen Ephraim, letter dated October 1999 to Simon Lavington.

[6]Wilson, J.F. 2001. *Ferranti: A History. Building a Family Business, 1882–1975.* Carnegie Publishing Ltd. ISBN 1-85936-080-7.

[7]Len Hewitt, e-mail of 22nd September 2915 to Simon Lavington.

the Technical Services Department and then Manager of West Gorton Factories in 1963. When Ferranti's Computer Department was sold to ICT in September 1963 Keith transferred to ICT Letchworth where he remained as Manager of the Computer Division of the Engineering Services Organisation until he moved to Scotland in 1971. Keith Lonsdale was described as 'a good engineer' by a colleague, Martin Wingstedt. In traditional engineering culture, to be described by a mate as 'good' is a rare accolade.

15. **Mike Moore** was present during the commissioning of the Shell Mark I* at Amsterdam.
16. **Brian Pollard** was one of the first group of about four Ferranti engineers who were assigned to computer development in November 1948.[8] Pollard served as the Computer Department Manager at Moston from 1949 to 1958. Upon graduating from Cambridge, Pollard had joined Ferranti in 1948 after applying unsuccessfully to join Elliott's Borehamwood Computing Division where he was interviewed by the Head of Section W. S. (Bill) Elliott. An unfortunate animosity developed between Pollard and Elliott that had grave consequences for Ferranti in the mid-1950s, by which time Bill Elliott was striving to set up a Pegasus development group in London and Pollard, his nominal boss in Moston, was being uncooperative. The technical, financial and legal consequences of the clash are described in Hendry[9] Pollard resigned in 1958 and moved to America where he worked first for Burroughs and then for RCA.
17. **Charlie Portman** joined Ferranti's Computer Department in 1954 as a circuit engineer. He went on to become a logic designer, test programmer and team leader, working in due course on the hardware design of Sirius and Orion, the 1900 Series and the 2900 Series on computers at West Gorton. In 1972 Charlie was appointed Manager of ICL's newly formed Software Division with staff at Manchester, Kidsgrove, Stoke and Stevenage. He served on various committees for the Department of Trade & Industry and for the Science and Engineering Research Council.
18. **Alec Robinson** was a Ph.D. student, working under F. C. Williams from 1947 to 49 on the design of a fast multiplier for the Mark I. In October 1949 he transferred from Manchester University's computer design team to Ferranti's design team at Moston. In 1962 he moved to London to be Director of the University of London Computing Centre, which installed a Ferranti Atlas computer. Alec became Director of the National Computing Centre in Manchester in 1969. In 1974 he became Director of the Computing Centre at the University of Wales, College of Cardiff, where he stayed until his retirement in 1991. Alec married Sylvia J Wagstaff, who was Alan Turing's secretary at Manchester University from about 1951 until Turing's death in June 1954.
19. **Dave Robinson**, amongst other things, maintained FINAC in Rome.

[8]Pollard, B.W. 1957. The Rise of the Computer Department. *Ferranti Journal* 15 (3): 20–23.

[9]Hendry, John. 1989. *Innovating for Failure: Government Policy and the Early British Computer Industry.* Published by the MIT Press.

20. **Ron St John** helped to install the Avro Mark I*.
21. **Geoff Tootill** graduated in mathematics from Cambridge, worked at TRE during the war and, from 1947, worked with Tom Kilburn on CRT storage and the design of the SSEM (Baby computer) at Manchester. He transferred to Moston from Manchester University's design team in about November 1949. He left Ferranti after a few months to be a senior lecturer at the Military College of Science, Scrivenham, where he stayed until 1954. He subsequently worked at RAE Farnborough, the European Space Agency, the DTI, the Executive Office of European Informatics Network Project and the National Maritime Institute.
22. **Bill Wallace** joined Ferranti in late November 1951. He remained with Ferranti/ICT/ICL and retired in 1981. In Chap. 3 there is a photo of Bill Wallace working on the Toronto Mark I (FERUT) whilst it was being commissioned at Moston.
23. **Ken Wallis** went with the Ferranti Mark I (FERUT) to Canada, where he married the Ferranti programmer Audrey Bates.
24. **Ianto Warburton** graduated in about 1946 from the Electro-Technics Department at Manchester University and joined Ferranti soon after. He was one of a few Ferranti employees to be attached to Manchester University's computer design team in 1949. Ianto remained with Ferranti/ICT/ICL all his working life, being associated throughout with University projects from Meg (aka Mercury) through to MUSE/Atlas and MU5. He was described by Dai Edwards as 'a very good engineer' and by Allan Ellson as "the Welsh wizard who was endlessly knowledgeable about all aspects of the digital circuitry of the Mark I*".[10]
25. **David Wilde** worked on the Toronto Mark I (FERUT).
26. **Martin Wingstedt** recalled[11] that his first assignment (autumn 1951) was as a maintenance engineer on the University Ferranti Mark I. He then went to work at Moston. In June 1954 he was sent to commission the Shell Amsterdam machine and stayed until the summer of 1955 when he was called back to join the Pegasus team. He went on to work on the design of other Ferranti computers. Martin once wrote a program to simulate a dog (e.g. greyhound) race, with 20 dogs whose progress was determined with the aid of the Mark I's hardware Random Number Generator. As output, he arranged a screen display on the console that simulated the movement of the dogs along the track.
27. **Dick Vogel** left Ferranti Ltd. in the late 1960s to join the engineering team in the Department of Computer Science at Manchester University.

Many of the above engineers remained with Ferranti Ltd. after the Mark I/Mark I* developments. When computer design at Moston was discontinued, they were

[10]Ellson, Allan. 2015. *Ferranti MkI* Commissioning and Installation: Some Recollections*. Two typed pages, prepared at Simon Lavington's request.

[11]Martin Wingstedt, 'phone conversations with Simon Lavington on 18th and 19th September 2015 and 22nd March 2017.

re-located to West Gorton, Bracknell, etc., where they worked variously on the Ferranti Pegasus, Perseus, Sirius, Orion (s) and Atlas computers.

17.2 Ferranti Programmers

At the start of 1951 Ferranti's Moston programmers probably numbered three or four, led by John Bennett and including Dietrich Prinz, Cyril Gradwell and Audrey Bates. Thereafter, the group expanded rapidly. A tentative list is given in Table 17.2 of all Ferranti programmers who were based, at least initially, in the Tin Hut at Moston. After 1954 some of these people began to migrate to temporary accommodation in London until Ferranti's new London Computer Centre was opened in January 1955 at 21 Portland Place. Those early programmers who had stayed in the Tin Hut were moved to Ferranti's new West Gorton premises in 1956.

Ferranti never had a Mark I* machine of its own so Moston programmers were probably not that much involved in producing programs for it, apart obviously for Test and Acceptance test programs. The applications written at Moston in the early days were run on the Mark I at Manchester University.

The following *Notes on* Table 17.2 include such personal information as has come to light.

Fruitful sources of Tin Hut anecdotes.[12,13,14,15] Other sources are as quoted in the text below.

Audrey Bates (later Clayton) graduated in 1949 with a first-class Mathematics degree from Manchester University. She then did research for an M.Sc. using the prototype Mark I computer. Her Supervisor was Alan Turing—she was his first research student. Audrey joined Ferranti Moston in November 1950. She went with

---

[12]Mary Lee Berners-Lee: An Interview Conducted by Janet Abbate for the IEEE History Center, 12 September 2001. Interview #578 for the IEEE History Center, The Institute of Electrical and Electronic Engineers, Inc. See: https://ethw.org/Oral-History:Mary_Lee_Berners-Lee#About_Mary_Lee_Berners-Lee.

[13]Olaf Chedzoy organised a reunion of Ferranti's Tin Hut programmers (and two or three engineers) at Curdon Mill, Williton, Somerset, on 21st April 1993. 20 ex-Ferranti people attended with their partners and a further six people sent informative letters. Each participant provided some biographical notes of their life after Ferranti. This material, together with some Mark I technical information, was put together by Olaf in a 40-page A5 typed booklet, specially produced for the reunion.

[14]Chedzoy, Olaf. 2015. *Ferranti Memories*. Five typed pages. This document contains additional information on Ferranti programmers, not included in the above *Reunion* booklet. It was produced by Olaf in response to questions asked by Simon Lavington in 2015.

[15]Joan Travis (neé Kaye): two long e-mails to Simon Lavington, dated 25th October and 11th November 2015 and spanning about five A4 pages in total (including a long extract from a Meteorological Office report—see Footnote 23). Joan's memories were detailed and vivid. She was, she said, "a mere minnow. However because it was my first job that might have made it a bit easier for me to remember some things".

Table 17.2 List of Ferranti programmers who were initially based in the Tin Hut at Moston

Name	Date joined Ferranti
Audrey Bates (later Clayton)	Nov. 1950
John Bennett	Dec. 1950
Ted Braunholtz	Summer 1953
Betty Broadbent	Mid-1950s?
Olaf Chedzoy	Oct. 1952
Harry Cotton	Sept. 1952
Betty Dyke (later Hall)	Sept. 1951
Sheila Fletcher (later Hawton)	Before 1953?
Cyril Gradwell	1950?
Vera Hewison (later Brooker)	Sept. 1952
Joan Kaye (later Travis)	April 1953
Margaret Lewin (later Marrs)	Sept. 1951
Dietrich Prinz	1947?
E. K. (Erik) Robertson	March 1953
Dorothy Steele	Autumn 1953
Mary Tunnell (later Shenton)	Before April 1953?
Joyce Ward (later Clarke)	?
Wendy Walton (later MacDonald Smith, then Breeze)	Autumn 1953
Mary-Lee Woods (later Berners-Lee)	Sept. 1951

the second production Ferranti Mark I (FERUT) to Toronto—perhaps initially for a spell in the autumn of 1952 but finally settling in Toronto in September 1953.[16] She married the Ferranti engineer Ken Wallis and they remained in Canada. Later, she married a computer programmer called Leigh Clayton. Audrey spent her working life in the North American computing industry.

John Bennett, an Australian, joined the Cambridge EDSAC team in 1947. He was the first EDSAC research student to obtain a Ph.D., in 1950. He then joined Ferranti Moston. He took a leading role in proposing the Mark I modifications that led to the Ferranti Mark I* computer. John married Mary who worked in the Moston Estimating Department and they moved to Ferranti's London Computer Laboratory late in 1953. In London, amongst other things, John was in due course responsible for the specification with Tony Baker of the Ferranti Perseus. John and Mary returned to Australia in 1956, where John spent the rest of his working life in Sydney University. John retired in 1986 and died in 2010.

Ted Braunholtz joined the Moston Tin Hut in the summer of 1953 after graduating from Cambridge. He did not have much to do with programming either the Mark I or the Mark I* but soon moved to London to help with the design of a magnetic tape system for Pegasus. He then went to Ferranti's Bracknell Lab to help

[16]Audrey Clayton (neé Bates), e-mail exchanges with Simon Lavington in the spring of 2010.

Fig. 17.2 Audrey Bates, one of the first people to write a program for the prototype Manchester University Mark I computer in 1949, joined Ferranti Moston in 1950. She went with the FERUT computer to Toronto University, where she remained for some years

with Orion. He eventually moved to California in about 1963. He returned to England towards the end of the 1960s.

Betty Broadbent joined Ferranti in Manchester in approximately mid-1955, "after she had programmed the Mark 1* in Rome".[17]

Olaf Chedzoy joined Ferranti Moston in October 1952 after two years' National Service in the RAF. "After the summer of 1953, I transferred to working on the Mark I*. Some of the minor features tended to change as the machine developed, and I was given the task of writing two documents. One of these was the description of the machine code for programmers, and the other was designing and explaining the Input Program—in essence, I suppose that one would now call it the Operating System. I left Moston to work in London with Ferranti, where I was assigned to work with Gordon Welchman. I left Ferranti in December 1954".[18] Sometime later

[17]Joan Travis (neé Kaye): two long e-mails to Simon Lavington, dated 25th October and 11th November 2015 and spanning about five A4 pages in total (including a long extract from a Meteorological Office report). Joan's memories were detailed and vivid. She was, she said, "a mere minnow. However because it was my first job that might have made it a bit easier for me to remember some things".

[18]Chedzoy, Olaf., *Ferranti Memories*.

Olaf became a Statistics Lecturer at Bath University. In 1993 he organised a reunion in Somerset for Ferranti's Tin Hut programmers.

Harry Cotton joined Ferranti in September 1952. Amongst other things, Harry investigated the application of computers to the work of the Ordnance Survey. He stayed on in Manchester whilst other Tin Hut programmers moved to Ferranti's London Computer Centre. Harry remained with Ferranti/ICT/ICL in the Manchester area all his working life.

Betty Dyke joined in September 1951. Olaf Chedzoy remembers that "She took all newcomers under her wing for teaching the basic Mark I machine operation and code. She was a brilliant teacher: I think she taught mathematics in schools before she joined Ferranti, and after she left".[19] Betty did indeed continue as a maths teacher after leaving Moston. She retired in 1985.

Sheila Fletcher joined some time before 1953. Olaf recalls that "I knew her for two years, but in a way, I didn't feel I knew her at all. She married a Baptist minister"[20] Mary-Lee Woods said[21] "I remember Sheila being asked to do a wages program, and my thinking "Oh, poor girl!" And that was the very first business program".

Cyril Gradwell was born in about 1923, so was a little older than most programmers. He remembers working in 'the crummy office' before moving into the Tin Hut. Cyril started on the Mark I by writing system software routines (for example *Input G* and *Reciprocal G*) that had some advantages over the original versions written by Alan Turing.[22] He then wrote Mark I programs for Ferranti's guided missile work for RAE Farnborough and on cotton spinning applications for the British Cotton Industry Research Association's Shirley Institute, Didsbury, Manchester.[23] He retired in 1987, having worked continuously for Ferranti/ICT/ICL at Moston, West Gorton, Putney, Reading and finally at Bracknell.

Vera Hewison joined Moston in September 1952, having been a student at Manchester University. Vera then left Ferranti 18 months later and returned to the University to complete her Maths and Physics degree, followed by an M.Sc. in

---

[19]Chedzoy, Olaf., *Ferranti Memories.*

[20]Chedzoy, Olaf., *Ferranti Memories.*

[21]Mary Lee Berners-Lee: An Interview Conducted by Janet Abbate for the IEEE History Center, 12 September 2001. Interview #578 for the IEEE History Center, The Institute of Electrical and Electronic Engineers, Inc. See: https://ethw.org/Oral-History:Mary_Lee_Berners-Lee#About_Mary_Lee_Berners-Lee.

[22]Joan Travis (neé Kaye): two long e-mails to Simon Lavington, dated 25th October and 11th November 2015 and spanning about five A4 pages in total (including a long extract from a Meteorological Office report). Joan's memories were detailed and vivid. She was, she said, "a mere minnow. However because it was my first job that might have made it a bit easier for me to remember some things".

[23](a) *XIV: Theoretical study of ring and cap spinning balloon curves (with and without air drag).* C. Mack. A publication of the British Cotton Industry Research Association Shirley Institute, Didsbury, Manchester.; (b) Gradwell, Cyril and Joan Kaye. 1955. *Electronic Calculation of Critical Whirling Speeds*. The Engineer, 4th March 1955. The British Thomson-Houston Company Ltd. provided data for this paper.

astronomy which entailed computational work on the 3-body problem. She married Tony Brooker. When their children had reached secondary school, Vera became a mathematics teacher. She retired in 1984 and died in 2018.

Joan Kaye joined Moston in April 1953 and in 1960 married John Travis, a Ferranti engineer. Joan and John eventually moved south, where Joan worked for a couple of years with Ferranti Bracknell. But back in 1953 in Moston, Joan was a test programmer, working on all of the Mark I* machines from Shell's Amsterdam installation onwards. She recalls that "Generally, I acted as a sort of engineers-mate (analogous to a plumbers-mate) for the design engineers when various 'exotic' peripherals (line printers or mag. tape) were being attached to the computers …. It would appear I had four separate visits to Holland (Shell) and one to Italy (Rome) …. We were working one weekend [in Amsterdam] and the Dutch engineers decided to give us a treat. The normal tea, which was served from the tea trolley, was herbal—very weak and without milk. The engineers tried to make us what they thought was real 'English' tea; this involved them boiling up tea leaves in a container for a long time! When we had to work overnight we would keep our packed food in the computer's drum compartment to keep it cool …. I was involved with the last Mark 1* at Gem Mill. It was situated in a big, otherwise empty, space on the ground floor there …. I think we were trying to attach 35 mm. mag. tape to the Mark 1*—unsuccessfully …. During the period of Mark 1* deliveries, other Ferranti computer ranges were being designed and manufactured. I seem to remember doing trial programming for the Pegasus (for Chris Strachey I think) … Later in West Gorton I did acceptance test programming for peripherals for Pegasus. I also did the engineers test programs for Mercury and Perseus".[24]

Margaret Lewin, a Cambridge mathematics graduate, joined Moston in September 1951 straight from university. Margaret left Moston at the end of August 1952 for EDSAC and the Mathematics Laboratory at Cambridge where, in due course, she married Eric Mutch. During her year in the Tin Hut, Margaret's task was to work with Cyril Gradwell on guided missiles, as part of an RAE Farnborough contract. This was one of three 'contract jobs' at the time.[25]

Dietrich Prinz and his role in encouraging Ferranti to get involved with digital computers in 1948 has been described in Chap. 1. Prinz used the prototype Mark I at Manchester University before using the production Ferranti Mark I in connection with some of Ferranti's defence contracts.

E. K. (Erik) Robertson joined the Tin Hut in March 1953. He had a background of working as an actuary, although he joined Ferranti as a programmer. He left Ferranti in 1960.

[24]Joan Travis (neé Kaye): two long e-mails to Simon Lavington, dated 25th October and 11th November 2015 and spanning about five A4 pages in total (including a long extract from a Meteorological Office report). Joan's memories were detailed and vivid. She was, she said, "a mere minnow. However because it was my first job that might have made it a bit easier for me to remember some things".

[25]Margaret Lewin, 'phone conversation with Simon Lavington, 2nd November 2015.

Dorothy Steele was at Moston from the autumn of 1953 until December 1955. At that point, she and her husband Bernard moved to Lytham St. Annes, where he had a job with the Atomic Energy Authority. By the mid-1960s, having started a family, Dorothy became a maths teacher.

Mary Tunnell joined Moston sometime before 1953.

Joyce Ward (later Clarke) married the Ferranti Mark I maintenance engineer Adrian Clarke. In the Tin Hut Joyce "tended to be an organiser of small group social events (apart from the play-reading)".[26]

Wendy Walton. Wendy joined Ferranti in the summer of 1953 after graduating from Bristol. She left Moston in the summer of 1954 and moved to Ferranti's London Computer Centre, finally leaving Ferranti in May 1955. She married John Macdonald Smith. They eventually divorced and Wendy then married Richard Breeze.

Mary-Lee Woods graduated from Birmingham with a two-year 'war degree', was drafted to TRE for two years, returned to take her final 'honours' year and then went to Australia to work on stellar classification at the Canberra Observatory. She had thus seen life by the time she joined Moston in September 1951. Olaf Chedzoy remembers that Mary-Lee "was the instigator of a play-reading group. We borrowed plays from Manchester library, and met—with spouses—in each other's flat. We were quite a group". By 1954 Mary-Lee and a number of other Ferranti programmers were moving to the company's London Computer centre at Portland Place. She married Conway Berners-Lee, who was on the sales side seeking out applications.

The large open-plan office in the Tin Hut occasionally hosted other visiting programmers. Four of these are shown in Table 17.3.

Also sharing the open-plan programmers' office in the Tin Hut for a period was Doug Chatt, a Ferranti technical writer. Finally, there was John Lech. "He was employed by the Ferranti Instrument Division to write test programs and service other engineer's requirements. He did not have a seat in the Tin Hut, which perhaps emphasized that there was a gulf between the computer engineering section and the programming group. Ironically, John was probably the best mathematician amongst us all".[27]

[26]Chedzoy, Olaf., *Ferranti Memories*.

[27]Chedzoy, Olaf., *Ferranti Memories*.

Table 17.3 List of programmers visiting Ferranti on temporary assignment and based in the Tin Hut at Moston

Name	Dates	Comments
Tony Baker	Approx. 1953	On secondment from the Royal Insurance Company, Liverpool. Tony qualified as an Actuary in 1948. He retired in 1983
Rosabelle Edge (later Bolton)	1953, 1954	A Cambridge vacation student at Moston in 1953 and 1954. She became a mathematics teacher in the Manchester area
Don Leapman	1953?	On secondment from Legal & General Insurance Co. After the time he spent at Moston, Don wrote that he "has since had no business relationship with computers"[a]
Tony Ralston	Summer 1953	A post-graduate student from MIT, working on testing the Mark I's hardware random Number Generator.[b] In 1965 Tony became Professor of Computer Science and Mathematics at the State University of New York

[a]Olaf Chedzoy organised a reunion of Ferranti's Tin Hut programmers (and two or three engineers) at Curdon Mill, Williton, Somerset, on 21st April 1993. 20 ex-Ferranti people attended with their partners and a further six people sent informative letters. Each participant provided some biographical notes of their life after Ferranti. This material, together with some Mark I technical information, was put together by Olaf in a 40-page A5 typed booklet, specially produced for the reunion

[b]Ralston, Anthony. 1980. Random Number Generation on the Ferranti Mark I. *Annals of the History of Computing* 2: 270–271. This short article is part of the *Comments, Queries and Debates* section occupying pages 268–271 and is therefore not mentioned explicitly in the journal's normal index of papers

17.3 Comparison with Other UK Industrial Programming Groups

The Tin Hut has entered Ferranti folk law as a magical place. "Young people were recruited …. It was a very good team—very exciting! And we were a good group too, socially".[28] More generally, Sheila Hall remembers that "Ferranti was a fantastic firm to work for. The firm was very attentive to its workers, and everyone was always very kind, polite, and with good manners. It was a privilege to work for Ferranti's, a happy family atmosphere, and the happiest working days of my life!" [29]

[28]Mary Lee Berners-Lee: An Interview Conducted by Janet Abbate for the IEEE History Center, 12 September 2001. Interview #578 for the IEEE History Center, The Institute of Electrical and Electronic Engineers, Inc. See: https://ethw.org/Oral-History:Mary_Lee_Berners-Lee#About_Mary_Lee_Berners-Lee.

[29]Letter written on 27th August 2018 by Sheila Hall to the team of volunteers running the SSEM (Baby computer) exhibit at the Science & Industry Museum, Manchester. Sheila Hall (neé Bamblin) joined Ferranti at Hollinwood in about 1944 and moved to the Moston factory in about 1947. She worked in the Drawing Office and remembers being involved in producing engineering drawings for Ferranti's Computer Department.

Compared with other UK computer manufacturers at the time, Ferranti's Moston programming group was out in front in respect of size and gender equality (they enjoyed equal pay!). Here are the statistics and the historical context.

In April 1953 Ferranti's Tin Hut team numbered about fourteen programmers, namely:

Audrey Bates, John Bennett, Olaf Chedzy, Harry Cotton, Betty Dyke, Sheila Fletcher, Cyril Gradwell, Vera Hewison, Joan Kaye, Dietrich Prinz, Erik Robertson, Mary Tunnell, Joyce Ward and Mary-Lee Woods. Over half the programmers were women. Ferranti's recruitment strategy, and indeed the career-choice of *programmer,* may have been influenced by the relatively large percentage of female *computers* (i.e. operators of electro-mechanical desk calculators) employed in scientific and engineering organisations in the 1930s and 1940s. To avoid confusion with the data-preparation and computer operator grades of the 1960s and 1970s, it is worth emphasising that the Moston women were highly qualified. Typically, they had maths degrees. When Sheila Cooper (neé Gray) was interviewed by Ferranti in 1954 during her final year at the University of London, she was offered a job conditional upon her obtaining a first-class Maths degree.[30] Sheila did get her *first* but, perhaps over-awed by Moston, she chose to work as a Mark I* programmer at Avro's nearby Chadderton factory—(see Chap. 6).

Vivian Bowden has set the context for the Tin Hut team. He remembered[31] that, in 1950, "there were only a few dozen programmers in the world. They were all very competent professional mathematicians and I think I knew most of them myself". Bowden went to consult Professor Douglas Hartree, who advised that the new digital computers "were exceedingly difficult to use, and could not be trusted to anyone who was not a professional mathematician"…. "We [Ferranti] decided to recruit some good chess players who could do crossword puzzles in their heads… We actually recruited some of the prettiest, most delightful young women I have ever had the good fortune to know …. They tackled our programs with skill, determination and enthusiasm. We began to get results more and more often. We were in business at last".

How does Ferranti's total of 14 programmers in April 1953 compare with the size of similar groups in other British computer manufacturers in the same year? Only two companies stand comparison: J. Lyons & Co. Ltd. at Cadby Hall, London and the Borehamwood Laboratories of Elliott Brothers Ltd. The Lyons' LEO computer ran a simple program in February 1951 and the first useful application, a Bakeries Valuation job, was being run routinely by the end of November 1951. At

[30]*Sheila Cooper's memories of working with a Ferranti Mark I* computer at Avro Chadderton.* Transcript produced by Simon Lavington in April 2017, based on an audio interview of Sheila by Buxton's local historian Vivienne Doyle recorded on 14th April 2016 and supplemented by telephone conversations and e-mails with Sheila in the spring of 2017 and with her daughter Judie Adnett.

[31]Bowden, B.V. 1970. The Language of Computers. *American Scientist* 58 (1): 43–53. Paper originally delivered at the Brighton College of Technology as the first Richard Goodman lecture, 2nd May 1969.

Borehamwood, after earlier work on the Elliott 152 real-time computer, the Elliott Nicholas computer came into operation in December 1952 followed by the Elliott 401 in the spring of 1953. The analyses below indicate that Ferranti's early programming group was twice the size of either the Lyons group or the Elliott group.

Lyons.

From information provided by Frank Land,[32] the first crop of programmers were all recruited from Lyons own staff, mainly trainees with no university degrees. David Caminer, already head of the Lyons Systems Analysis office, was appointed Manager responsible for applications (systems and programing). In 1950 he recruited two of Lyons Management Trainees as programmers, namely John Grover and Tony Barnes. By the end of 1951 two more were added: Derek Hemy and Leo Fantl. In 1952 two more people joined: Mary Blood (with a French degree) and Frank Land (with an economics degree from LSE). Next year came John Gosden, Jim Smith and Eric Cavannah, all with mathematics degrees, bringing the total programmers to nine. During 1954–1955 about five more programmers were added, most of whom had degrees. In conclusion, by mid-1953 Lyons was only employing about half the number of programmers as Ferranti. Only one of the early Lyons programmers was female.

Elliott.

In 1949 the first tentative steps to program the Elliott 152 computer were taken by the small Borehamwood *Theory Group* of just three people which, as remembered by S. E. (Ed) Hersom,[33] "was headed by N. D. Hill with myself and a *computer*, i.e. a girl who operated a Marchant electro-mechanical calculator ....". By the end of 1952 when the Borehamwood in-house Nicholas computer was becoming operational, there were four active programmers: Ed Hersom, Brighid Rose (later Simpkin), Bobby (Roberta) Lewis (later Hersom) and Ruth Holt (later Felton). By the end of 1954 George Felton, Hugh Devonald, Bruce Bambrough, Dina Vaughan, Albert Wakefield and Roger Cook had joined but Brighid Rose, Ruth Holt and Bobby Lewis had left, bringing the active programmers to about seven. In 1955 a change of emphasis occurred with the specific recruitment of the first batch of six *programmer-investigators*—in effect, technical salesmen.[34] Borehamwood programmers were typically mathematics graduates, with the females coming straight from university but the males often via National Service or industry. In conclusion, by mid-1953 Elliotts was only employing about half the number of programmers as Ferranti Ltd.

[32]Frank Land, e-mail dated 19th September 2016 to Simon Lavington.

[33]Hersom, S.E. 2002. Nicholas, The Forgotten Elliott Project. *Resurrection* the *Bulletin of the Computer Conservation Society* (27): 10–14.

[34]Lavington, Simon. 2011. *Moving Targets: Elliott-Automation and the Dawn of the Computer Age in Britain, 1947–67*. Springer. ISBN 978-1-84882-932-9.

17.4 Relevant Manchester University Staff

Historically speaking, the University's interest in digital computers started off in a small way in 1945 in the Mathematics Department—(see Chap. 14). By the time Ferranti became involved in 1948, the centre of activity was firmly in the Electrical Engineering Department, where it remained until a spin-off Department of Computer Science was formed in 1964. By 1975 the full-time staff (faculty) in Computer Science numbered 32, including four Professors who had been involved in one way or another with the Ferranti Mark I (Tom Kilburn, Dai Edwards, Frank Sumner and Derrick Morris).

17.4.1 The Early Days of Ferranti Collaboration

In the creative period 1948–1950, when prototype hardware was being developed and then the hardware/software know-how was being transferred to Ferranti, the principle academic players were as shown in Table 17.4. The individual contributions noted in Table 17.4 mostly concerned hardware innovations—which had resulted in a total of 34 patents being taken out between 1946 and 1949. The right-hand column in the Table indicates the number of patents of which an individual was an inventor or co-inventor during this period. Newman's name featured on the famous B-line (index register) patent, along with Williams, Kilburn and Tootill. Cliff West contributed to the magnetic drum synchro servomechanism.

Apart from Max Newman, Alan Turing and F C Williams, most of those in Table 17.4 were comparatively young and inexperienced. Examples are Dai Edwards and Tommy Thomas. Tommy later recalled the scene.[35] "The arrival of two young Welshmen straight from the first post-war degree course in the Physics Department at Manchester was in itself a curiosity. Dai and I had chosen the electronics option in our third year and had come under the spell of Freddie Williams. Following graduation in the summer of 1948 we were given the choice of National Service, Coal Mining, Nuclear Engineering or Control/Electronic engineering. We had no difficulty in deciding to move along the yellow tiled corridor and enter the wonderful world of bottoming pentodes and *pugnatrons* [a term coined by FC for describing a type of feedback circuit used in the control of motors] that F.C. had fascinated us with in our final year as undergraduates. Therefore in September 1948, resplendent in our new white lab coats, we arrived in the period immediately following the birth of the *Baby* and were soon involved in its

[35] *Recollections of Manchester University (1945–1955)*. G. E. (Tommy) Thomas. Illustrated web page dated 18th June 1998, and written for the University's Anniversary celebrations for the Baby computer. See: http://curation.cs.manchester.ac.uk/computer50/www.computer50.org/mark1/gethomas/index.html.

Table 17.4 Main academic contributors to computer design (hardware and software) at Manchester University in the period 1948–1950, the period when know-how was being transferred to Ferranti Ltd.

Name	Dates at Manchester University	Number of Manchester computer patents
D. B. G. (Dai) Edwards	Sept. 1948–Aug. 88	2
Tom Kilburn	Dec. 1946–Aug. 81	27
M. H. A. (Max) Newman	Oct. 1945–Sept. 64	1
A. A. (Alec) Robinson	Apr. 1947–Apr. 49	7
G. E (Tommy) Thomas	Sept. 1948–Aug. 55	3
G. C. (Geoff) Tootill	Sept. 1947–Nov. 49	12
A. M. (Alan) Turing	Oct. 1948–June 54 (died in post)	–
J. C. (Cliff) West	Sept. 1946–Aug. 57	1
F. C. (Freddie) Williams	1936–39; then war work at TRE; Dec. 46–Aug. 77 (died in post)	32

development into the working prototypes used by Ferranti to design and produce their first commercial machine, the Ferranti Mark I".

At several levels above Thomas and Edwards, older and wiser heads were also busy. Behind the scenes at Manchester during the period of Table 17.4 was the influential figure of Patrick Blackett, later Lord Blackett, who had won a Nobel Prize for Physics in 1948. He had served on many important government committees during the war and, despite his socialist leanings, was definitely one of the country's Great and Good. Blackett was Professor of Physics at Manchester from 1937 to 1953. He was convinced of the need to promote the design and implementation of the new kind of high-speed, general-purpose, digital computing machines. There is evidence to suggest that Blackett was influential in bringing Newman to Manchester in 1945 and Williams to Manchester in 1946. He also arranged the visit of Sir Ben Lockspeiser, Chief Scientist at the Ministry of Supply, to see the SSEM (Baby) computer in 1948. To cap his behind-the-scenes influence, Blackett was *Best Man* at Douglas Hartree's wedding in 1923 and at Max Newman's wedding in 1934. Douglas Hartree had been Professor of Applied Mathematics at Manchester from 1933 to 1946 and, by the end of the war, was acknowledged on both sides of the Atlantic as an expert in computational methods. The social and intellectual world of Blackett, Hartree and Newman at Manchester is amply described.[36]

[36]*Alan Turing's Manchester.* Jonathan Swinton to be published in the spring of 2019 by Deodands. ISBN 9780993178924.

Fig. 17.3 The Manchester University Mark I computer in the summer of 1949 with *(left-to-right)* Dai Edwards, F. C. Williams, Tom Kilburn, Alec Robinson and Tommy Thomas. *(Inset)* Geoff Tootill, who was writing up his M.Sc. thesis at the time the photo was taken

Table 17.5 Staff at the Computing Machine Laboratory, Manchester University, who influenced the development of software and applications for the Ferranti Mark I computer

Name	Dates at Manchester
R. A. (Tony) Brooker	Oct. 1951–Aug. 1967
N. E. (Nick) Hoskin	Aug. 1952–approx. Aug. 1954
R. K. (Richard) Livesley	Aug. 1952–approx. Aug. 1954
C. M. (Cicely) Popplewell	Oct. 1949–approx. Aug 1968
A. M. (Alan) Turing	Oct. 1948–June 1954 (died in post)

Once the Computing Machine Laboratory had achieved a physical presence at the start of 1951, the influence of University staff on the Ferranti Mark I and Mark I* developments, and the wider UK programming community depended upon the efforts of the five people mentioned in Table 17.5. Their efforts were directed at systems software and applications. Of course, Williams and increasingly Kilburn continued to provide overall direction of the Computing Machine Laboratory.

Fig. 17.4 Cicely Williams *(neé Popplewell)* and her husband in the early 1970s, after her retirement from Manchester University. Cicely was probably the only graduate female programmer on the staff of the University's Computing Machine Laboratory from 1950 to 1960. Cicely died in 1995

Compared with the number of Ferranti women programmers mentioned in Table 17.2, the female gender was significantly under-represented at the University. There is actually one more woman, omitted from most accounts, whose contribution to early hardware developments earned her the sobriquet *Fabulous:* Ida Fitzgerald. Ida was a so-called 'wireman' on secondment from TRE for a few months in about 1947/8 who skilfully wired up chassis of electronic components to the design of Williams and Kilburn. In some sense, Ida Fitzgerald was the forerunner of the many unsung but essential female data-preparation and operating staff who were to support the Computing Service at Manchester University from the late 1950s until on-line user terminals began to appear in the mid-1970s.

17.4.2 Later Collaborations with Ferranti Ltd

On the hardware side, F. C. Williams soon moved into other areas of electrical engineering and Tom Kilburn took over the de facto lead. The collaborations blossomed into the Ferranti Mercury and Atlas computers, as described in Chap. 13. In 1960 Tom was appointed Professor of Computer Engineering, and in 1964 the

Fig. 17.5 The original SSEM (Baby) engineers photographed in June 1998 at a celebration of the computer's 50th Anniversary. *(top row, left-to-right)* Tommy Thomas, Dai Edwards, Laurie Allard. *(Bottom row, l-to-r)* Geoff Tootill, Tom Kilburn, Alex Robinson. Missing is F. C. Williams, who led the SSEM (Baby) project; he died in 1976. Allard joined the General Electric Company's Research Laboratories at Wembley, Middlesex, in 1940. He was promoted to Group Leader of the Cathode-Ray Tube Group in 1965. He was instrumental in enabling GEC to produce CRTs with low internal impurities, thus more suitable for use in the Williams/Kilburn storage system

computer group split from Electrical Engineering to form the UK's first Department of Computer Science with Tom, now Professor of Computer Science, at its head. He was elected a Fellow of the Royal Society in 1965, was awarded a CBE in 1973 and retired in 1981.

Of the other original engineers in Table 17.4, Dai Edwards worked alongside Tom Kilburn throughout the Mercury and Atlas and later projects. Dai became a Professor in 1966 and was later Head of the Computer Science Department. The other engineers mentioned in Table 17.4 eventually left the University. Thomas left in 1955 to found the Central Digital Computing facility at Imperial Chemical Industries Ltd. (ICI). In 1966, he moved to the University of Edinburgh to set up and direct the first of the UK's three Regional Computing Centres. Alec Robinson and Geoff Tootill both left in 1949 to join Ferranti Ltd. Their subsequent careers have been described earlier. Cliff West continued to work in control theory until leaving the University in 1957 to take up a Chair of Electrical Engineering at Queen's University, Belfast.

In due course, new engineers joined Kilburn's team to work on collaborations with Ferranti Ltd. An early example is R. L (Dick) Grimsdale. He graduated in Electrical Engineering at Manchester in 1950, whereupon he was given the task of writing test and diagnostic programs for the Ferranti Mark I. He went on to develop the Experimental Transistor Computer, which first ran a program in November 1953. Amongst his other projects were a Graphical Output Unit for the Mercury computer and the fast Read-Only memory for Atlas. Grimsdale left the University in 1962 to join AEI Ltd. (Metropolitan Vickers Division).

Of course, on the software side many Manchester research students used the Ferranti Mark I at the University. Some of them stayed on to contribute to the Computing Machine Laboratory's service to users. An example is Frank Sumner. After getting a Ph.D. in Chemistry in 1954 Frank became a Research Assistant in the Computing Machine Laboratory until going to Aldermaston in 1956. He returned to the University of Manchester in 1957 and used the Ferranti Mark I and then the Mercury. Amongst other projects, he simulated the Atlas one-level store (later to be called *Virtual Memory*) and in this connection developed the *Drum Transfer Learning Program* for Atlas. He also worked on the logical design of the Atlas Central Processor.

Another early user was Bernard Richards, who started in about 1953 as Alan Turing's research student to work on morphogenesis. In about 1955 Bernard began to assist ICI's newly-established Central Digital Computing facility to program the University's computer in Mark I Autocode (see Chap. 13). Bernard soon joined ICI and helped the company install a Ferranti Mercury computer at its Wilton works in North Yorkshire.

References

Bowden, B.V. 1970. The Language of Computers. *American Scientist* 58 (1): 43–53.

Chedzoy, Olaf. 2015. *Ferranti Memories*.

Ellson, Allan. 2015. *Ferranti MkI* Commissioning and Installation: Some Recollections*.

Gradwell, Cyril and Joan Kaye. 1955. *Electronic Calculation of Critical Whirling Speeds*.

Hendry, John. 1989. *Innovating for Failure: Government Policy and the Early British Computer Industry*. Published by the MIT Press.

Hersom, S.E. 2002. Nicholas, The Forgotten Elliott Project. *Resurrection the Bulletin of the Computer Conservation Society* 27: 10–14.

Lavington, Simon. 2011. *Moving Targets: Elliott-Automation and the Dawn of the Computer Age in Britain, 1947–67*. Springer. ISBN 978-1-84882-932-9.

Pollard, B.W. 1957. The Rise of the Computer Department. *Ferranti Journal* 15 (3): 20–23.

Pollard, B.W., and K. Lonsdale. 1953. The Construction and Operation of the Manchester University Computer. *Proceedings of IEE* 100 (Part 2): 501–512.

Ralston, Anthony. 1980. Random Number Generation on the Ferranti Mark I. *Annals of the History of Computing* 2: 270–271.

Swinton, Jonathan. 2019. *Alan Turing's Manchester*. Deodands. ISBN 978-0-9931789-2-4.

Wilson, J.F. 2001. *Ferranti: A History. Building a Family Business, 1882–1975*. Carnegie Publishing Ltd. ISBN 1-85936-080-7.

Chapter 18
Appendix E. Performance, Cost and Delivery Details of Other Computers

18.1 Analysis of UK Computer Deliveries, 1955–1957

The Tables that follow in this section give the background to the market discussions in Chap. 5.

The three Tables on the next pages give the destinations and, where known, the applications of all computers to have been delivered by UK computer manufacturers in the years 1955, 1956 and 1957. The evidence comes from the Computer Conservation Society's *Our Computer Heritage* project, data for which was gathered in the period 2004–2012. See: http://www.ourcomputerheritage.org/.

18.2 American and British High Performance Computers, 1954–1964

The material that follows gives the technical background to the Ferranti Atlas project and the challenges that it faced in the market place. In the period 1954–1964 at least five American computers were produced whose performance was aimed at the top end of the scientific market—which usually meant that government research establishments were the target users. Atomic energy and nuclear weapons were typical target applications. Table 18.5 compares the instruction times of these five American computers with Ferranti's two offerings in this market area, Mercury and Atlas.

In terms of raw instruction speed, it is seen that Atlas was a little slower than its close rival, the IBM STRETCH (also known as the IBM 7030). Raw speed was not the only criterion. The rate of processing jobs in a multi-programming environment was also important.

F. R. A. (Bob) Hopgood spent his working life in the employment of the UK's Atomic Energy Research Authority. He wrote compilers for both STRETCH and

S. Lavington, *Early Computing in Britain*, History of Computing,
https://doi.org/10.1007/978-3-030-15103-4_18

Table 18.1 The hardware capabilities and the cost of the nine British production computers that were on the market in the period 1955–1957

Computer	Primary memory	Drum secondary memory	Fxpt Add time, min./max. in ms	Approx. price
BTM HEC2M	–	1K words	1.25/21.25	£20K
BTM 1201	–	4K words	1.25/21.25	£30K
Elliott 402	15 words	3K words	0.204/13.1	£25K–£35K
Elliott 405	512 words	32K words	0.306/13.1	£50K–£125K
EE DEUCE	402 words	8K words	0.064/1.064	£42K–£50K
Ferranti Mk I*	512 words	16K words	1.2	£85K–£90K
Ferranti Pegasus	56 words	5K words	0.3	£32K–£45K
Ferranti Mercury	1024 words	32K words	0.06	£100K
LEO II	1038 words	64K words	0.34	£90K

This is a repeat of the Table given in Chap. 5

Table 18.2 An analysis of all computers delivered by UK manufacturers in 1955

		Applications, where known
HEC2M	GEC Research Laboratories, Wembley ESSO Refinery, Fawley, Hampshire	
1201	–	
402	Institut Blaise Pascal, Paris, France Army Operational Research Group, West Byfleet	Research and numerical analysis Operational research
405	–	
DEUCE	Eng. Elec. Nelson Research Lab., Stafford National Physical Lab., Teddington Royal Aircraft Establishment, Farnborough Eng. Elec., Marconi House, Strand, London Eng. Elec., Main Works, Stafford	Internal computing service for EE NPL Computing Service Aircraft research Computing Bureau Transformer design
Mark I*	Instituto Nazionale per le Applic. Calc. (INAC), Rome Atomic Weapons Research Est., Aldermaston	(see Chap. 9) (see Chap. 12)
Pegasus	–	
Mercury	–	
LEO II	–	

Table 18.3 An analysis of all computers delivered by UK manufacturers in 1956

		Applications, where known
HEC2M	Royal Aeronautical Establishment, Bedford Aeronautical Research Association, Bedford Ministry of Aviation, London Royal Aeronautical Establishment, Boscombe Down	Aircraft research
1201	ICT Ltd., Technical Division, Bradenham, Bucks The Morgan Crucible Co. Ltd., London	Programme testing, training Payroll, stores control, etc.
402	Imperial Chemical Industries Ltd., Co Durham	Op. research; molecular res.
405	Elliott's Computing Services Division NCR Co. Ltd., Computing Services, London	Computing service bureau Demo. & computing service
DEUCE	Bristol Aeroplane Co., Filton New South Wales University of Technology, Australia Atomic Weapons Research Est., Aldermaston Royal Aircraft Establishment, Farnborough	Aircraft design Aircraft research
Mark I*	A. V. Roe Ltd. Chadderton, Manchester	(see Chap. 6)
Pegasus	Ferranti Ltd. Computing Service, Portland Place Hawker Aircraft Ltd., Kingston-upon-Thames Sir W. G. Armstrong Whitworth Aircraft Ltd., Coventry	General computing service Aviation design calculations Aviation design; payroll
Mercury	–	
LEO II	–	

Table 18.4 An analysis of all computers delivered by UK manufacturers in 1957

		Applications, where known
HEC2M	Indian Statistical Institute, Calcutta	
1201	Monsanto Chemical Ltd., London; British Railways, Western Region, Paddington; Imperial Chemical Industries Ltd., Northwich; Ministry of Aviation. Accountant Div., Chessington; Provincial Admin., Cape of Good Hope, South Africa; Irish Sugar Co. Ltd., Thurles, Eire; Shell Refining Co. Ltd., Shellhaven (no. 1); Shell Refining Co. Ltd., Shellhaven (no. 2); Guest Keen & Nettlefolds Ltd., Birmingham; West Riding of Yorkshire County Council, Wakefield; British Railways, Western Region, Bristol	Payroll, accounting Payroll Job costs; sales research Payroll Payroll Accounting Oil stock control; payroll Oil stock control; payroll Payroll Payroll; labour costing payroll
402	Rank, Taylor and Hobson Ltd., Leicester Bomber Command, RAF High Wycombe	Optical lens systems design Operational research
405	Norwich City Treasurer's Department National Gas Turbine Establishment, Farnborough Unilever Ltd., London Board of Trade, Census Office, Pinner	Accounting; eventually payroll On-line engine testing Linear programming, etc. Census analysis
DEUCE	Bristol Siddeley Aero-Engine Co., Patchway, Bristol. BP, Aldgate, London Short Bros & Harland, Belfast Eng. Elec., Warton, near Preston	Aero engine design Seismic work. Aircraft design Aircraft design
Mark I*	Armstrong Siddeley Aero Engines, Ansty, Coventry	(see Chap. 7)
Pegasus	Admiralty Research Laboratory, Teddington	Research work

(continued)

Table 18.4 (continued)

		Applications, where known
	Royal Aircraft Establishment, Farnborough Vickers-Armstrongs (Aircraft) Ltd., Weybridge NRDC, Northampton Polytechnic, London De Havilland Aircraft Co Ltd., Hatfield British Thomson-Houston Co. Ltd. (AEI), Rugby Leeds University Durham University British Iron & Steel Research Association, London ICI Dyestuffs Division, Manchester	Aircraft research Aviation design calculations Research and training Research; payroll; budgeting Turbine design; costing Research & service; admin Research & service; admin Operational research work Research; sales; stock control
Mercury	Norwegian Defence Research Est., Kjeller Manchester University French Atomic Energy Authority, Saclay	Atomic energy work; classified Research and service Atomic energy work
LEO II	J. Lyons & Co. Ltd., London	Internal EDP; service bureau

Atlas and implemented a large Quantum Chemistry package (MIDIAT) on both computers. In December 2012, on the 50th Anniversary of the first Atlas, he compared STRETCH and Atlas as follows:

"STRETCH could run extremely fast if you had the code set up just right and it remained in core memory. It had some terrible deficiencies as well. It made guesses as to which way a conditional jump would go and if you got it wrong it had to backup all the computation it had done. So the same conditional jump could be as much as a factor of 16 different in time between guessing right and wrong. The STRETCH nuclear weapon codes at AWRE Aldermaston probably outperformed Atlas by quite a bit. On the other hand Atlas ran some large number theory and matrix manipulation calculations much faster than STRETCH. My codes were pretty similar in performance but on large calculations where intermediate results had to be stored on magnetic tape, Atlas was significantly faster due to the Ampex tape decks. I think on an untuned general purpose workload Atlas was faster and if the code was tuned to STRETCH it would be faster. In conclusion, I would say that

Table 18.5 Instruction times in microseconds for five American and two British computers that were first installed between 1954 and 1964

Instruction	IBM 704 (1954)	Ferranti Mercury (1957)	IBM 7090 (1959)	IBM STRETCH (1961)	Univac LARC (1960)	Ferranti ATLAS (1962)	CDC 6600 (1964)
FXPT ADD	24	60	4.8	1.5	(4?)	1.59	?
FLPT ADD	84	180	16.8	1.38–1.5	4	1.61–2.61	0.3
FLPT MPY	204	300 (360)	16.8–40.8	2.48–2.7	8	4.97	1
FLPT DIV	216	?	43.2	9.0–9.9	28	10.66–29.8	3.4

in 1962 'Atlas was reckoned to be the world's most powerful general-purpose computer'".

This last comment echoes what was said in 1962 by Hugh Devonald, who for a time led Ferranti's Atlas program development team: "*Atlas is in fact claimed to be the world's most powerful computing system. By such a claim it is meant that, if Atlas and any of its rivals were presented simultaneously with similar large sets of representative computing jobs, Atlas should complete its set ahead of all other computers.*"

It is on this basis that, in 1962, Atlas was claimed to be the world's most powerful computer.

Picture Credits

Permission-giver	Figure numbers
BAE Systems	1.1, 4.9, 5.4, 6.3, 6.4a, b, 6.5, 6.6a, b
Bill Blakemore via John McNamara (formerly at Bristol Siddeley). Original photographer unknown	7.7, 7.8, 7.9
Images taken from: *Setting up a computer department*. D. R. Evans. British Communications and Electronics, vol. 5, no. 12, December 1958, pp. 918–922. Publisher: Heywood & Co., London, now defunct	7.2, 7.3, 7.4, 7.6
Key Publishing Ltd., with thanks to Ben Dunnell. Photos originally appeared in *The Aeroplane*, 27th June 1958, pp. 893–896	7.1, 7.5
Andrea Celli, L'Istituto per le applicazioni del calcolo 'Mauro Picone' (IAC)	9.1
Elisabetta Mori. Photos taken at the Museum of Computing Machinery, Pisa, in 2010	9.8, 9.9
Images are held in the Archive of L'Istituto per le applicazioni del calcolo 'Mauro Picone' (IAC) and in the Archivio Centrale dello Stato. With thanks to Andrea Celli (IAC)	9.4, 9.5, 9.6, 9.7
The British Library, with thanks to Dr. Thomas Lean	4.3
Olaf Chedzoy	4.4, 4.5
Copyright Keith Williamson	4.8
The Computer Conservation Society, with thanks to Rachel Burnett	4.10
Crown Copyright, GCHQ Cheltenham, with thanks to Tony Comer, GCHQ Historian	10.3a, b, 10.4a, b

(continued)

S. Lavington, *Early Computing in Britain*, History of Computing,
https://doi.org/10.1007/978-3-030-15103-4

(continued)

Permission-giver	Figure numbers
Crown Copyright, Ministry of Defence. With thanks to Glynn Libberton, AWE Archivist	12.1, 12.2a, b, 12.3
Crown Copyright, Ministry of Defence. With thanks to Gordon Pattison, DSTL Fellow	11.1, 11.2, 11.3, 11.4, 11.5, 11.6
Historical Photograph Collection, Special Collections, University Libraries, Virginia Polytechnic Institute and State University, Blacksburg, Va	10.1
Reproduced with permission from Raytheon Company. Original image produced by Engineering Research Associates	10.2
STFC Rutherford Appleton Laboratory	10.5
Prof. Brian Randell	12.4
Crown Copyright 1981. Information provided by the National Meteorological Library and Archive—Met Office, UK. With thanks to Duncan Ball	13.3b
Special Collections and Western Manuscripts Section, Bodleian Library, University of Oxford	13.4
Prof. R. A. (Tony) Brooker	13.5
Manchester University Engineering Society. With thanks to James Peters, University of Manchester	13.7
Iain MacCallum	13.9
Copyright Hulton-Deutsch Collection/CORBIS/Corbis via Getty Images	1.7
Fujitsu Services Limited, with additional thanks to the Ferranti Archive at the Science & Industry Museum, Manchester and the School of Computer Science, University of Manchester	1.5, 1.6a, b, 1.9, 1.12, 2.3, 2.4b, 2.5, 4.1a, b, 4.2, 4.6, 4.7, 5.3, 5.5, 5.6, 5.7a, b, 8.3, 9.2, 9.3, 13.8, 16.8
NRC Media (Algemeen Handelsblad; Kim van Steeg). With acknowledgements to Fujitsu Services Ltd. and Croon Elektrotechniek BV	1.8
Leonard Hewitt	6.1, 6.2
The Institution of Engineering and Technology, With thanks to James Sutherland	2.4a, 16.4
Copyright Historic England Archive (Aerofilms Collection). Annotations by Simon Lavington	2.1

(continued)

(continued)

Permission-giver	Figure numbers
Image obtained from the University of Manchester via Prof. D. B. G. Edwards. Photographer unknown. Annotations by Simon Lavington	2.2
Prof. D. B. G. Edwards	2.7
Simon Lavington	5.1, 5.2, 13.1, 13.12, 16.5a
Simon Lavington, with thanks to Fujitsu Services Limited and Peter Freeman	16.2
Adapted by Simon Lavington, with the permission of Prof. Martin Campbell-Kelly	13.10
Simon Lavington, with thanks to the National Computing Museum	13.11a, b, 16.6a, b
Simon Lavington, with acknowledgement to the Royal Army Pay Corps' historical collection	1.3
Copyright, National Museums Scotland, with thanks to Tacye Phillipson	6.8
Copyright Mark Richards. Courtesy of the Computer History Museum, Mountain View, California. With thanks to Massimo Petrozzi	16.5b
Robin Bowden and Ginny Murray (neé Bowden)	17.1
Peter Readman	17.4
By permission of the Master and Fellows of St. John's College, Cambridge	14.1, 14.3a, b
Courtesy of The Turing Archive for the History of Computing, with thanks to Prof Jack Copeland	14.2
Courtesy of The National Museum of Computing, with thanks to Prof. Martin Campbell-Kelly	2.11
Courtesy of the University of Manchester, with additional thanks to Prof. Jim Miles, School of Computer Science, James Peters, National Archive for the History of Computing and Fujitsu Services Limited	2.6, 2.8, 2.9, 2.10, 6.7, 13.2, 13.3a, 13.6, 14.4, 14.5a, b, 15.1, 15.2, 15.3, 16.1, 16.3, 16.7a, b, 17.5
Courtesy of the University of Manchester, School of Computer Science, and with acknowledgement to the Illustrated London News	1.10, 1.11
Courtesy of the University of Manchester, School of Computer Science, and Chris Burton and G. C. Tootill.	17.3
National Research Council Canada Archives	3.6, 3.7, 3.9
	3.2, 3.4, 3.5

(continued)

(continued)

Permission-giver	Figure numbers
University of Toronto Archive, with thanks to Prof. Michael Williams and courtesy of Alva Worsley	
C. Calvin (Kelly) Gotlieb Fonds, University of Toronto Archive, with thanks to Prof. Michael Williams and to the Computer Science Department. Original sources unknown	1.4, 3.3, 17.2
Brian Jeffrey	3.8
Allan Ellson	3.1
Bart Hofmeester (AeroCamera)	8.1a
Rosalie van Egmond, Curator of Shell's historical heritage and archive, with thanks to Onno Zweers	8.1b
CWI, Amsterdam, with thanks to Rob van Rooijen	8.2
Onno Zweers, with acknowledgement to Alida (Lidy) Zweers—de Ronde	1.2, 8.4, 8.5, 8.6, 8.7, 8.8, 8.9, 8.10a, b
Nationaal Archief, Collectie Spaarnestad, Nationaal Foto	8.11

References

Ahmed, F.R., and D.W.J. Cruickshank. 1952. Crystallographic Calculations on the Manchester University Electronic Digital Computer (Mark II). *Acta Crystallographica* 5: 765–769.

Alberts, Gerard, and Bas, van Vlijmen. 2017. *Computerpioniers: het begin van het computertijdperk in Nederland*. Amsterdam: Amsterdam University Press. ISBN 978-94-6298-378-6.

Alberts, Gerard. 2000. *Collectie Nederland oude computers. Werkgroep Verzamelbeleid Computerhistorie*. (Dutch collections of old computers. Workgroup for Collection Policy, Computer History). Produced by Schiedam, Holland: Dynamiek Productions.

Ashdown, G.L., and K.L. Selig. 1951. A General Purpose Differential Analyser: Part I, Description of Machine. *The Elliott Journal* 1 (2): 44–48.

Bell, C.G., and A. Newell. 1971. *Computer Structures: Readings and Examples*. McGraw-Hill.

Bennett, J.M., and J.C. Kendrew. 1953. The Computation of Fourier Syntheses with a Digital Electronic Calculating Machine. *Acta Crystallographica* 6 (10): 109–116.

Bennett, J.M., D.G. Prinz, and M.L. Woods. 1952. *Interpretative Sub-routines*. In *Proceedings of ACM National Conference*, 81–87, Toronto, Sept 1952.

Berry, F.J. 1959. Intercode, a Simplified Coding Scheme for AMOS. *Computer Journal* 2 (2): 55–58.

Bigelow, J. 1976. Computer Development at the Institute for Advanced Study. In *International Research Conference on the History of Computing*, Los Alamos, 10th–15th June 1976.

Boslaugh, David, Peter Marland, and John Vardalas. 2017. The Information Age & Naval Command and Control. In: *Presented at the McMullen Naval History Symposium, at USNA Annapolis*, 14–15 Sept 2017. Possibly to be re-published by USNA.

Bowden, B.V. (eds). 1953. *Faster than Thought*. Published by Pitman.

Bowden, B.V. 1970. The Language of Computers. *American Scientist* 58 (1): 43–53.

Broadbent, T.E. 1998. *Electrical Engineering at Manchester University: The Story of 125 Years of Achievement*. Published by The Manchester School of Engineering, University of Manchester. ISBN 0-9531203-0-9.

Brooker, R.A. 1955. An Attempt to Simplify Coding for the Manchester Electronic Computer. *British Journal of Applied Physics* 6: 307–311.

Budiansky, Stephen. 2001. Codebreaking with IBM Machines in World War II. *Cryptologia* 25 (4): 241–255.

Campbell, Scott M. 2003. *Beatrice Helen Worsley: Canada's female computer pioneer*. (University of Toronto). *IEEE Annals of the History of Computing* 25(4), 51–62.

Campbell-Kelly, M. 1989. *ICL: A Business and Technical History*. Oxford: Oxford University Press. ISBN 0-19-853918-5.

S. Lavington, *Early Computing in Britain*, History of Computing,
https://doi.org/10.1007/978-3-030-15103-4

Campbell-Kelly, Martin. 1980. Programming the Mark I: Early Programming Activity at the University of Manchester. *Annals of the History of Computing* 2 (2): 130–168.

Carpenter, B.E., and R.W. Doran. 1977. The Other Turing Machine. *The Computer Journal* 20 (3): 269–279.

Carter, R.H.A., and A.M. Uttley. 1953. The Telecommunications Research Establishment Parallel Electronic Digital Computer (Chap. 10). In *Faster than Thought,* ed. B.V. Bowden. Published by Pitman.

Celli, A., M. Mattaliano, and R. Spitaleri. 2012. Istituto per le Applicazioni del Calcolo, 1927–2012: 85 years in Computational and Applied Mathematics. In *Proceeding of MASCOT 12 & ISGG 12, The Joint 12th Meeting on Applied Scientific Computing and Tools and 12th ISGG Meeting on Numerical Grid Generation, Las Palmas de Gran Canaria*, 61–70, Oct 2012. Published by IMACS, the International Association for Mathematics and Computers in Simulation.

Coombs, A.W.M. (July & Oct 1955, Jan 1956, March 1953). An Electronic Digital Computer, Parts 1–4. *Post Office E E Journal* 48, 114, 137 & 212: 38–42; and 49, 18, 126 (April & July 1956).

Coombs, A.W.M. 1954. The Ministry of Supply Automatic Computer. In *Automatic Digital Computation, the Proceedings of a Symposium* held at NPL, Published by HMSO.

Cordata, J.W. 2012. *The Digital Flood: The Diffusion of Information Technology Across the US, Europe and Asia*. Oxford. ISBN 978-0-19-992155-3.

De Barr, A.E. 1953. Digital Storage Using Ferromagnetic Materials. *The Elliott Journal* 1 (4): 116–120.

Marco, De, Giovanni Mainetto Giuseppe, Serena Pisani, and Pasquale Savino. 1999. The Early Computers of Italy. *IEEE Annals of the History of Computing* 21 (4): 28–36.

Doornbusch, Paul. 2004. Computer Sound Synthesis in 1951: The Music of CSIRAC. *Computer Music Journal* 28 (1): 10–25.

Doornbusch, Paul. 2017. Early Computer Music Experiments in Australia and England. *Organised Sound* 22 (2): 297–307.

Evans, D.R. 1958. Setting up a Computer Department. *British Communications and Electronics* 5 (12): 918–922.

Franchi, Stefano, and Francesco Bianchini. 2011. *The Search for a Theory of Cognition: Early Mechanisms and New Ideas.* Published by Rodopi. ISBN 978-90-420-3427.

Gawlik, H.J. 1963. MIRFAC: A Compiler Based on Standard Mathematical Notation and Plain English. *Communication of ACM* 6 (9): 545–547.

Gotlieb, Calvin C. 1954. The Cost of Programming and Coding. *Computers and Automation* 25: 14 ff.

Gradwell, Cyril and Joan Kaye. 1955. *Electronic Calculation of Critical Whirling Speeds the Engineer*, 4th March 1955.

Hendry, John. 1989. *Innovating for Failure: Government Policy and the Early British Computer industry*. Published by the MIT Press.

Hersom, S.E. 2002. Nicholas, The Forgotten Elliott Project. *Resurrection the Bulletin of the Computer Conservation Society* 27: 10–14.

Hinds, Mavis K. 1981. Computer Story. *The Meteorological Office Magazine* 110 (1304): 69–81.

Hodges, Andrew. 1983. *Alan Turing: The Enigma.* Published by Burnett Books. ISBN 0-09-152130-0.

Hume, J.N.P. 1994. Development of systems software for the FERUT computer at the University of Toronto, 1952–1955. *IEEE Annals of the History of Computing* 16 (2): 13–19.

Hume, J.N.P., and Beatrice H. Worsley. 1955. T4R13a: transcode: a system of automatic coding for Ferut. *Journal of the Association for Computing Machinery* 2 (4): 243–252.

Kilburn, T. 1948. *A Storage System for use with Binary Digital Computing Machines.* Ph.D. thesis, University of Manchester.

Kilburn, T., D. Morris, J.S. Rohl, and F.H. Sumner. 1968. A System Design Proposal. In *Proceedings of IFIP Congress*, 76–80, Edinburgh, Section D.

Kilburn, T., D.B.G. Edwards, and G.E. Thomas. 1956. The Manchester University Mark II Digital Computing Machine. *Proceedings of IEE 103* (Part B, Supp. 1–3): 247–268.

Kilburn, T., D.B.G. Edwards, M.J. Lanigan, and F.H. Sumner. 1962. One Level Storage System. *IRE Transactions on Electronic Computers* EC-11 (2): 223–235.

Kilburn, Tom. 1990. From Cathode Ray Tube to the Ferranti Mark I. *Resurrection the Journal of the Computer Conservation Society* (2).

Krol, Gerrit. 1967. *Het gemillimeterde hoofd (The Close-Cropped Head)*. Em. Querido: Published in Amsterdam.

Lavington, Simon. 1975. *A History of Manchester Computers*. 1st ed. National Computing Centre. ISBN 0-85012-155-8.

Lavington, Simon. 1980. *Early British Computers*. Manchester University Press. ISBN 0-7190-0803-4.

Lavington, Simon. 1998. *A History of Manchester Computers*. 2nd ed. Published by the British Computer Society. ISBN 0-902505-01-8.

Lavington, Simon. 2000. *The Pegasus Story; A History of a Vintage British Computer*. Published by the Science Museum, London. ISBN 1-900747-40-5.

Lavington, Simon. 2006. In the Footsteps of Colossus: A Description of Oedipus. *IEEE Annals of the History of Computing* 28 (2): 44–55.

Lavington, Simon. 2011. *Moving Targets: Elliott-Automation and the Dawn of the Computer Age in Britain, 1947–67*. Berlin: Springer. ISBN 978-1-84882-932-9.

Lavington, Simon. 2017. Reflections on the Hoot. *Resurrection* 77: 13–20.

Link, David. 2012/13. Programming ENTER: Christopher Strachey's Draughts Program. *Resurrection The Journal of the Computer Conservation Society* (60).

Link, David. 2016/17. God Save the King—An Early Musical Program on the Ferranti Mark I. *Resurrection The Journal of the Computer Conservation Society* 76: 11–16.

Lonsdale, K., and E.T. Warberton. 1956. Mercury: A High-Speed Digital Computer. *Proceedings of IEE* 103 (Part B, Supp. 1–3): 483–490.

Maller, V.A.J. 1980. Information Retrieval Using the Content Addressable File Store. In *Information Processing 80, The Proceedings of IFIP Congress*, Tokyo/Melbourne, 187–192. Published by North-Holland.

McDonald, Jean K. 1954. FERUT, Canada's electronic computing machine. *Journal of the Royal Astronomical Society of Canada* 48 (5): 176–184.

McGregor Ross, Hugh. 1993. Obituary: Bernard Swann. *Resurrection (The Journal of the Computer Conservation Society*) (8).

McGregor Ross, Hugh. 2001. Ferranti's London Computer Centre. *Resurrection* (*The Bulletin of the Computer Conservation Society*) (25).

Metropolis, N., J. Howlett, and G.-C. Rota (eds.). 1980. *A History of Computing in the Twentieth Century*. Academic Press.

Millership, R., R.C. Robbins, and A.E. De Barr. 1951. Magnetostriction Storage Systems for High-Speed Digital Computers. *British Journal of Applied Physics* 2 (10): 304.

Morton, Peter. 1989. *Fire Across the Desert: Woomera and the Anglo-Australian Joint Project 1946–1980*. Canberra: Australian Government Publishing Service.

Nouwen, Pieter. 2003. Het mirakel van Amsterdam (The Miracle of Amsterdam). *Shell's Venster Magazine* 19–23.

Nouwen, Pieter. 2003. Het mirakel van Amsterdam (The Miracle of Amsterdam). *Shell's Venster Magazine* 19–23. Text (in Dutch) by Pieter Nouwen.

Pattison, Gordon. 2000. *Early Computers at the Fort.* DERA News, Oct 2000.

Pollard, B.W. 1950. Application of Digital Computing Machines. *Ferranti Journal* 8 (4): 94.

Pollard, B.W. 1955. Ferranti and Powers Samas. *The Ferranti Journal* 13 (1): 8.

Pollard, B.W. 1956. Rome, December 14th 1955: The Inauguration of a Ferranti Computer. *Ferranti Journal* 9–11.

Pollard, B.W. 1957. The Rise of the Computer Department. *Ferranti Journal* 15 (3): 20–23.

Pollard, B.W., and K. Lonsdale. 1953. The Construction and Operation of the Manchester University Computer. *Proceedings of IEE* 100 (Part 2): 501–512.

Pritchard, H.O., and J. Mol. 2001. Computational Chemistry in the 1950s. *Graphics Modelings* 19 (6): 623–627.

Ralston, Anthony. 1980. Random Number Generation on the Ferranti Mark I. *Annals of the History of Computing* 2: 270–271.

Reiners, C.A. 1953. *Survey of Computing Facilities in the UK*. Published by the Ministry of Supply.

Robinson, A. 1952. *The testing of cathode ray tubes for use in the Williams type storage system,* 42–45. Proceedings of the 1952 ACM National meeting, Toronto. Published by ACM Press.

Robinson, A.A. 1953. Multiplication in the Manchester University High-Speed Digital Computer. *Electronic Engineering* 6–10.

Robinson, Tim. 2008. Computer history museum, oral history of Arthur Porter. Interviewed by Tim Robinson, 8th Mar 2008. See http://archive.computerhistory.org/resources/access/text/2015/06/102658245-05-01-acc.pdf.

Rooijendijk, Cordula. 2010. *Alles moest nog worden uitgevonden (Everything had yet to be Invented):* de geschiedenis van de computer in Nederland. 2nd ed. Published by Olympus Pockets. (First edition published by Atlas Contact, Uitgeverij in 2007). ISBN 9789046744048.

Snyder, S.S. 1980. *ABNER:* The ASA Computer. Part II: Fabrication, Operation and Impact. Declassified 2017 with redactions. *NSA Technical Journal* 63–85.

Snyder, Samuel S. 1964. *History of NSA General-Purpose Electronic Digital Computers.* Washington DC: Dept of Defense.Strachey, Christopher. 1952. *Logical or non-mathematical programmes*, 46–49.

Sumner, F.H.(Frank). 1994. Memories of the Manchester Mark I. *Resurrection* 10: 9–13.

Swann, B.B. 1975. *The Ferranti Computer Department: A History*.

Taylor, JIM. 2001. History of Scientific Computing at AWE—Part 1. *Discovery, the Science & Technology Journal of AWE* (2): 52–55.

Thomas, G.E. 1950. Magnetic storage. *Report of a Conference on High Speed Automatic Calculating Machines*, 75–80. Cambridge, 22nd–25th June 1949. Published by the Cambridge: University Mathematical Laboratory.

Turing, A.M. 1937. On Computable Numbers, with an Application to the Entscheidungsproblem. *Proceedings of the London Mathematical Society,* Series 2 42 (1): 230–265.

Turing, A.M. 1950. Computing Machinery and Intelligence. *Mind* 59: 433–460.

Turing, A.M. 1952. The Chemical Basis of Morphogenesis. *Philosophical Transactions of the Royal Society* B237 (641): 37 ff.

Vardalas, John. 1994. From DATAR To The FP-6000 computer: Technological change in a Canadian industrial context. *IEEE Annals of the History of Computing* 16(2).

West, J.C., and F.C. Williams. 1951. The Position Synchronisation of a Rotating Drum. *Proceedings of IEE 98* (Part 2, 61).

Wheatley, L.H. 1953. *Cryptanalytic Machines in NSA,* Paper A60928. NSA34.

Wilkes, M.V., D.J. Wheeler, and S. Gill. 1951. *The Preparation of Programs for an Electronic Digital Computer.* Published by Addison-Wesley.

Williams, David. 2003. Information Technology at Ansty. *The Sphinx, The Coventry Branch Magazine* 53: 13–18.

Williams, F.C. 1956. Introductory Lecture: IEE Convention on digital computers. *Proceedings of. IEE* 103 (Part B, Supp. 1–3): 3–9.

Williams, F.C. 1975. Early Computers at Manchester University. *The Radio & Electronic Engineer* 45 (7): 327–331.

Williams, F.C., A.A. Robinson, and T. Kilburn. 1952. Universal High Speed Digital Computers: Serial Computing Circuits. *Proceedings of IEE* 99 (Part 2, 68).

Williams, F.C., and T. Kilburn. 1949. A Storage System for Use with Binary Digital Computing Machines. *Proceedings of IEE* 98 (Part 2, 30): 183 ff.

Williams, F.C., and T. Kilburn. 1951. The University of Manchester Computing Machine. *Proceedings of the Manchester Computer Inaugural Conference*, 5–11, 9th–12th July 1951. This paper was also presented at the joint AIEE-IRE Computer Conference, Philadelphia, Dec 1951.

Williams, F.C., E.R. Laithwaite, and L.S. Piggott. 1957. Brushless Variable-Speed Induction Motors. *Proceedings of IEE* 104 (14): 102–118.

Williams, Michael R. 1994. UTEC and FERUT: The University of Toronto's computation centre. *IEEE Annals of the History of Computing* 16 (2): 4–12.

Wilson, J.F. 2001. *Ferranti: A History. Building a Family Business, 1882–1975.* Carnegie Publishing Ltd. ISBN 1-85936-080-7.

Zielinsky, Siegfried, and David, Link. 2006. There Must Be an Angel: On the Beginnings of the Arithmetics of Rays. In *Variantology 2, On deep time relations of Arts, Sciences and Technologies*, 15–42. Published Cologne, Konig. See: http://www.alpha60.de/research/there_must_be_an_angel/DavidLink_MustBeAnAngel_2006.pdf.

Name Index

S. Lavington, *Early Computing in Britain*, History of Computing,
https://doi.org/10.1007/978-3-030-15103-4

L

M

N

O

P

R

S

Subject Index

S. Lavington, *Early Computing in Britain*, History of Computing,
https://doi.org/10.1007/978-3-030-15103-4

Printed in the United States
By Bookmasters